Data Preparation for Analytics

Using SAS®

Gerhard Svolba, Ph.D.

The correct bibliographic citation for this manual is as follows: Svolba, Gerhard. 2006. *Data Preparation for Analytics Using SAS®*. Cary, NC: SAS Institute Inc.

Data Preparation for Analytics Using SAS®

To my family

Martina, Matthias, Jakob, and Clemens,
for being the most valuable support I can imagine.

Contents

Part 3 Data Mart Coding and Content

Part 5   Case Studies

Preface

The Structure and Use of This Book

This book has been written for SAS users and all people that are involved in the data preparation process for analytics. The goal of the book is to give practical advice in the form of SAS coding tips and tricks as well as conceptual background information on data structures and considerations from the business point of view.

This book will allow the reader to

- View analytic data preparation in light of its business environment and to consider the underlying business questions in data preparation
- Learn about data models and relevant analytic data structures such as the one-row-per-subject data mart, the multiple-rows-per-subject data mart, and the longitudinal data mart
- Use powerful SAS macros to change between the various data mart structures
- Consider the specifics of predictive modeling for data mart creation
- Learn how to create meaningful derived variables for all data mart types
- Illustrate how to create a one-row-per-subject data mart from various data sources and data structures
- Learn the concepts and considerations for data preparation for time series analysis
- Illustrate how scoring can be done with various SAS procedures and with SAS Enterprise Miner
- Learn about the power of SAS for analytic data preparation and the data mart requirements of various SAS procedures
- Benefit from many do's and don'ts and from practical examples for data mart building

Overview of the Main Parts of the Book

This book has been structured into five parts plus an appendix.

Part 1, Data Preparation: Business Point of View

The intended audience for Part 1 is anyone who is involved in the data preparation process for analytics. This part addresses the three persona groups that are usually integrated in the data preparation process: business people, IT people, and analysts. This part of the book is free from technical considerations or SAS code.

In this part, we deal with the definition of the business questions and with various considerations and features of analytic business questions. We investigate different points of view from IT people, business people, and statisticians on analytic data preparation and also take a look at business-relevant properties of data sources.

Business people who are the consumers of analytic results rather than being involved in the data preparation and analysis process benefit from these chapters by getting a deeper understanding of the reasons for certain requirements and difficulties. Also, IT people benefit from this part of the book by exploring the reasons made by the analyst concerning data requirements.

Part 2, Data Structures and Data Modeling

The intended audience for Part 2 is IT people and analysts. This part deals with data models and structures of both the available source data and the resulting analysis table. In this part, we deal with data sources and their structures in the source systems. By exploring the terms *analysis subject* and *multiple observation*, we define the most common data mart structures for analytics, namely the *one-row-per-subject data mart,* the *multiple-rows-per-subject data mart,* and the *longitudinal data mart.*

Additionally we look at general features of data marts and special considerations for predictive modeling. In this part, we start considering data preparation from a technical point of view. We do this by showing a number of examples from a conceptual point of view, i.e., without SAS coding.

Whereas business people can benefit from the concepts in this part of the book, the chapters mainly address IT people and analysts by showing the data preparation concepts for analytics. Analysts will probably find in these chapters the explicit formulation and rationale of data structures, which they use intuitively in their daily work.

Part 3, Data Mart Coding and Content

The intended audience for Part 3 is analysts and SAS programmers. In this part of the book, we start to get practical. This part explains how to fill a data mart with content. By using a similar structure as Part 2, we show with SAS code examples and macros how to create an analytic data mart from the source data. Here we run through data extraction from the source data, restructuring of the data (transposing), aggregating data, and creating derived variables.

The aim of this part is to help the analyst and SAS programmer to speed up his or her work by providing SAS code examples, macros, and tips and tricks. It will also help the reader to learn about methods to create powerful derived variables that are essential for a meaningful analysis.

Part 4, Sampling, Scoring, and Automation

The intended audience for Part 4 is analysts and SAS programmers. This part contains chapters that deal with important aspects of data preparation, which did not fit into the previous part. Here we deal with topics such as sampling of observations, data preparation and processing considerations of scoring, and additional tips and tricks in data preparation.

Part 5, Case Studies

The intended audience for Part 5 is analysts and SAS programmers. However, business people might be interested to see practical examples of steps that are needed to get from business questions to an analytic data mart.

In this part we put together what we learned in the previous parts by preparing data for special business questions. These case studies present the source data and their data model and show all of the SAS code needed to create an analytic data mart with relevant derived variables.

Appendixes

The appendixes include two appendixes that are specific to SAS, and a third appendix that includes a programming alternative to PROC TRANSPOSE for large data sets. These appendixes illustrate the advantage of SAS for data preparation and explore the data structure requirements of the most common SAS procedures.

Graphical Overview of the Book

The following diagram gives an overview of the parts of the book.

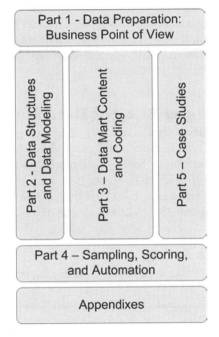

Part 1 is a chapter of general interest. Parts 2, 3, and 5 have a strong relationship to each other by dealing with data preparation from a conceptual point of view in Part 2, a SAS programming point of view in Part 3, and a case study point of view in Part 5.

The relationships among chapters are shown in more detail in the next section and diagram.

Relationships among the Chapters

General

In the next diagram, we use the same structure as shown in the first diagram. Additionally, the chapters of each part are shown. The relationships among Parts 2, 3, and 5 are illustrated. The following relationships among chapters exist.

Accessing Source Data

Accessing source data is the topic of Chapters 5, 6, and 13. Chapter 5 gives an overview of the source systems from a technical point of view. Chapter 6 explains the most common data models in the source systems and introduces the concepts of data modeling and entity relationship diagrams.

One Row per Subject versus Multiple Rows per Subject

The considerations of analysis subjects, multiple observations per subject, and their representation in a rectangular table are topics discussed in Chapters 7, 8, 9, 10, 14, 15, and Appendix C. Chapter 7 gives a general introduction to this topic. Chapters 8, 9, and 10 each deal with one-row-per-subject data mart, multiple-rows-per-subject data mart, and longitudinal data structures. Chapters 14, 15, and Appendix C use SAS code to show how to change the data mart structure, namely how to transpose data.

The One-Row-per-Subject Data Mart

The one-row-per-subject data mart is very common in data mining and statistical analyses. In this book this data mart type is specifically covered in Chapters 8, 14, 18, 19, 25, and 26. Chapter 8 shows general considerations for this data mart type and Chapter 14 shows how to restructure data. Chapters 18 and 19 show how data from multiple-rows-per-subject structures can be aggregated to a one-row-per-subject structure and how relevant and powerful derived variables can be created. Chapters 25 and 26 show case studies with concrete business questions.

Predictive Modeling

Predictive modeling is covered in Chapters 12, 20, and 25. Chapter 12 shows general considerations for predictive modeling. Chapter 20 gives code examples for some types of derived variables for predictive modeling. The case study in Chapter 25 also deals with aspects of predictive modeling, such as time windows and snapshot dates.

Data Preparation for Time Series Analysis

Data preparation for time series analysis is the topic of Chapters 9, 10, 15, 21, and 27. Chapters 9 and 10 show general considerations for multiple-rows-per-subject and longitudinal data marts. Chapter 15 shows how to convert from among different longitudinal data structures. Chapter 21 deals with aggregations and the creation of derived variables. Finally, Chapter 27 shows a case study for data preparation for time series analysis.

The following diagram shows the relationships among the chapters.

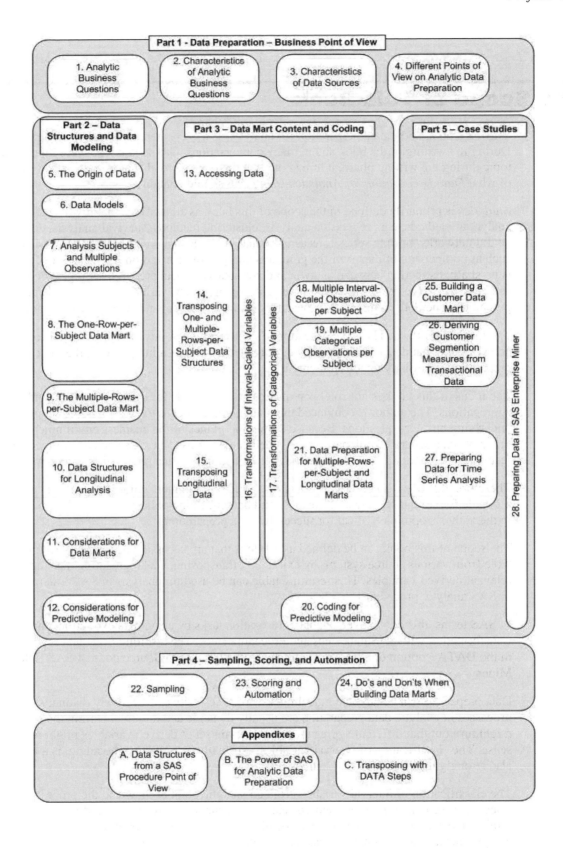

The Scope of This Book

Writing a book about data preparation for analytics is a very interesting and challenging task. During the writing of the book and in many conversations with people who were interested in this topic during the writing phase, it turned out that there are many different aspects and expectations of what *Data Preparation for Analytics Using SAS* should contain.

Analytics is primarily defined in the scope of this book as all analyses that involve advanced analytical methods, e.g., regression analysis, clustering methods, survival analysis, decision trees, neural networks, or time series forecasting. Analytics can also involve less sophisticated methods such as comparison of means or the graphical analysis of distributions or courses over time. There is no sharp distinction between analytic and non-analytic. Data prepared for regression analysis can also be used to perform a descriptive profiling of various variables. The focus of the data preparation methods in this book is primarily in the area of advanced analytical methods.

The motivation for this book was the author's conviction that this topic is essential for successful analytics and that despite a large number of analytic and data mining books, the topic of data preparation is strongly underrepresented.

The focus of this book is not solely on macros and SAS code for derived variables and aggregations. The author is convinced that data preparation has to be seen in the scope of its underlying business questions. Because of the importance of the business environment for data preparation, a separate book part has been dedicated to this topic. Also considerations about the resulting data mart structure are essential before coding for data preparation can start.

Therefore, this book is not only a coding book and programmers guide, but it also explains the rationale for certain steps from a business and data modeling point of view. This understanding is, in the author's opinion, critical for successful data preparation.

The scope of this book can be defined as all steps that are necessary to derive a rectangular data table from various source systems by extracting, transposing, and aggregating data, and creating relevant derived variables. This resulting table can be used for analysis in SAS Enterprise Miner or SAS analytic procedures.

In SAS terms, this book covers all data preparation tasks by using data extraction, and data transformation with SAS data steps and procedure steps. The resulting SAS data set can be used in the DATA= option of a SAS procedure or in the Input Data Source node of SAS Enterprise Miner.

Data preparation and analytical modeling are not necessarily consecutively distinct steps. The processes in analytic data preparation are usually dynamic and iterative. During the analysis it might turn out that different aggregations or a certain set of derived variables might make more sense. The classification of interval variables, especially in predictive modeling, is a topic that also strongly involves analytic methods.

The classification of interval variables from an analytic point of view and the process of variable selection based on correlations and association measures are already covered in a lot of data mining and analytics books. For the scope of this book, these two topics are considered part of the analysis itself and not as data preparation topics.

The topic of data quality checks and data cleansing would fill a book itself. For this topic, see other SAS Press titles on the topic.

The author tried to keep this book as practical as possible by including a lot of macros, SAS code examples, and tips and tricks. Obviously a large number of additional programming tricks that are not covered in this book exist in SAS and in SAS Press titles and SAS documentation. The objective of this book, however, is data preparation for analytics; it would go beyond the scope of this book to include voluminous programming tips and tricks.

This book is focused on data preparation for analyses that require a rectangular data table, with observations in the rows and attributes in the columns. These data mart structures are needed by most procedures in SAS/STAT and SAS/ETS as well as SAS Enterprise Miner. Data structures such as triangular data, which are used in PROC MDS in SAS/STAT, are not featured in this book.

Note that all SAS macros and code examples in this book can be downloaded from the companion Web site at http://support.sas.com/publishing/bbu/companion_site/60502.html.

The Data Preparation Process

Success Factors for Analytics

There are two main factors that influence the quality of an analytical model:

- The selection and tuning of the appropriate model
- The availability and preparation of data that are suitable to answer the business questions

In statistical theory and literature, there is much focus on the selection and tuning of an analytical model. A lot of books and courses exist for predictive modeling, e.g., to train neural networks, to create decisions trees, or to build regression models. For all these tasks, a strong knowledge of statistics is needed and has proven to be a critical success factor for good analytical results.

Besides the importance of analytical skills, however, the need for good and relevant data for the business questions is also essential.

Success Factors for Data

The success factor *data* itself can be divided into two subcategories, namely:

- The availability of data
- The adequate preparation of the data

One of the topics that this book addresses is *data availability* in Part 1. However, whether historic snapshots of the data that are necessary for an analysis are available or not is typically something the analyst has little, if any, control over. For statistical analysis, especially in the data mining area, much of the data are observational rather than data gathered through a designed data collection approach.

Another topic that this book addresses is *adequate preparation of the data*. This is an important success factor for analysis. Even if the right data sources are available and can be accessed, the appropriate preparation is an essential prerequisite for a good model.

The author's experience in numerous projects leads him to conclude that a missing or improperly created derived variable or aggregation cannot be substituted for by a clever parameterization of the model. It is the goal of the following chapters to help the reader optimally prepare the data for analytics.

Example

In predictive modeling at the customer level, a data mart with one-row-per customer is needed. Transactional data such as account transactions have to be aggregated to a one-row-per-customer structure. This aggregation step is very crucial in data preparation as we want to retrieve as much information about the client's transaction behavior as possible from the data, e.g., by creating trend indicators or concentration measures.

This Book

This book addresses the data preparation process for analytics in the light of the underlying business questions. In Part 1 we will start to consider the relation of the business questions to data preparation and will move in Part 2 to data structures and data models that form the underlying basis of an analysis data mart. In Part 3 we will fill the data mart with content, restructure data, and create derived variables. Sampling, scoring, and automation are the topics of Part 4. Finally in Part 5 we consider a number of data preparation case studies.

Data Mart, Analysis Table, Rectangular Table, or Data Set?

It is hard to name the *final product of data preparation*. In the data mining community, it is very common to use the term *data mart* for the resulting table that is used in data mining analysis. In statistical analysis, names such as *data matrix* and *analysis table* are very common and the term data mart is hardly known. Data mart is also a term that is used in a data warehousing context to refer to a set of database tables that are logically related and hold data for a certain business area such as sales reporting or controlling.

In the following chapters, all of these terms may be used. However they mean the same thing in the context of the book. When we use data mart, analysis table, or data set, we mean a single rectangular table to be one that holds the data for the analysis, where the observations are represented by rows and the attributes are represented by columns.

About the Author

Gerhard Svolba, PhD, has been programming in SAS since 1991. He worked at the University of Vienna, Department for Medical Statistics, where he performed biometrical analyses and built numerous analysis tables for clinical trials. In his PhD, he focused on statistical quality control in clinical trials, where he showed the application and data preparation for the control of quality characteristics in clinical trials.

Since 1999, he has worked as a consultant for SAS Austria, where he is involved in numerous analytic projects in the CRM, Basel II, and demand forecasting areas across various industries. He has product manager responsibility for the SAS analytic products and solutions in Austria and has constructed a number of analytic data mart creations as well as creating concepts for analytic projects and data preparation.

Gerhard lives in Vienna, Austria. He is married and is the father of three sons. Besides working for SAS and with SAS software, he likes to spend time with his family and to be out in nature, especially sailing on a lake.

Acknowledgments

Many people have contributed to the completion and to the content of this book in different ways.

Gerhard, Christian, Wolfgang, and Michael were the reason to start and to keep working for SAS. Peter supported my skills by allowing me to work in both disciplines, statistics and informatics.

Andreas, Bertram, Christine, Franz, Josef, Hendrik, Martin, Nico, Rami, and a lot of unnamed people took the time to discuss ideas for my book and to talk about questions in data preparation in general. Many SAS customers and SAS users have requested solutions from me for their data preparations tasks, which resulted in some of the ideas shown in this book.

Julie Platt, Donna Faircloth, and the team from SAS Press supported me throughout the process. The reviewers gave valuable feedback on the content of the book.

,

Part 1

Data Preparation: Business Point of View

Introduction

In Part 1 of this book we will deal with the business point of view of data preparation. This part is free from code and data structures. It contains only the nontechnical components of data preparation. Without reducing the importance of data structures, table merges, and the creation of derived variables, it is fundamental to understand that behind data preparation there always stands a business question.

This part of the book also helps to categorize business questions and to show their impact on data sources and data models.

In *Chapter 1 – Analytic Business Questions,* we will look at various business questions and will see what they all have in common—that data are needed for their analysis. We will also go through the analysis process and will learn that data preparation is influenced by the business question and the corresponding analysis method.

We will see that a business question itself poses many other questions that need to be discussed in order to prepare and perform the analysis.

In *Chapter 2 – Characteristics of Analytic Business Questions*, we will show the properties of a business question. We will see, for example, that data preparation is affected by the analysis paradigm, whether it is classical statistics or data mining. We will look at the data requirements, and we will examine whether the analysis is a repeated analysis (e.g., model recalibration over time), or a scoring analysis.

In *Chapter 3 – Characteristics of Data Sources*, we will go into detail about the properties of data sources from a business standpoint. We will focus on properties such as the periodic availability of data and the advantages of a data warehouse as an input data source.

In *Chapter 4 – Different Points of View on Analytic Data Preparation*, we will define three roles that are frequently encountered in the analysis process. We will look at the business, analytical, and technical points of view. The different backgrounds and objectives of these roles can lead to conflicts in a project, but they can also provide many benefits for a project if different points of view are exchanged and knowledge is shared.

The intention of Part 1 is to illustrate the business rationale in the data preparation process for analytic data preparation. Part 1 introduces data preparation in general (i.e., from a data modeling and coding point of view). We should keep in mind that all data models, features, derived variables, and tips and tricks we will encounter in this book are provided in order to solve a business problem and are to be seen in the context of the points we discuss in the first four chapters.

Chapter 1

Analytic Business Questions

1.1 Introduction

Before we go into the details of analytic data preparation, we will consider analytics from a business point of view. In this chapter we will examine three business questions that deal with analytics in the background. This chapter deals with situations in which analytics is applied and consequently data are prepared. We will list business questions from various industries and will provide a definition of the business point of view.

Answering business questions is done through a process we call the *analysis process*. The steps and features of this analysis process will be described. The fact that the analyst must not step back from acquiring domain-specific knowledge will be discussed as well as the fact that even a simple business question needs to be evaluated for more details.

1.2 The Term *Business Question*

Business questions for analytics can be derived from various industries and research disciplines. In this book, we will call them all *business questions*, even if their organizational background is not a business area such as banking, insurance, or telecommunications but instead a research organization.

For our purposes, business question also describes the content and rationale of the question for which an analysis is done.

Business does not primarily reflect the fact that a monetary benefit needs to be linked to a business question. In our context, business knowledge can also be referred to as *domain expertise*.

With business questions we also want to separate the rationale for the analysis from methodological questions such as the selection of the statistical method.

1.3 Examples of Analytic Business Questions

A complete list of all possible analytic business questions could fill infinite volumes. The following list gives an overview of representative business questions that incorporate analytical methods:

- Which risk factors have a strong influence on the occurrence of lung cancer?
- What is the average time until the occurrence of a disease for breast cancer patients?
- Is the success rate for the verum treatment significantly higher than that of the placebo treatment (given e.g., alpha = 0.05)?
- How many overnight stays of tourists in Vienna hotels on a daily basis can we expect in the next six months?
- How will economic parameters like GDP or interest rates behave over time?
- How do factors such as time, branch, promotion, and price influence the sale of a soft drink?
- What is the daily cash demand for automatic teller machines for banks in urban areas?
- How many pieces of a new movie will be produced by sales region?
- Which products will be offered to which customers in order to maximize customer satisfaction and profitability?
- Which customers have a high cancellation risk in the next month?
- Which customer properties are correlated with premature contract cancellation?
- Does a production process for hard disk drives fit the quality requirements over time?
- What is the expected customer value for the next 60 months?
- What is the payback probability of loan customers?
- How can customers be segmented based on their purchase behavior?
- Which products are being bought together frequently?
- What are the most frequent paths that Web site visitors take on the way through a Web site (clickstream analysis)?

- How can documents be clustered based on the frequency of certain words and word combinations?
- Which profiles can be derived from psychological tests and surveys?
- How long is the remaining customer lifetime in order to segment the customers?
- What is the efficiency and influence of certain production parameters on the resultant product quality?
- Which properties of the whole population can be inferred from sample data?
- Is the variability of the fill amount for liter bottles in a bottling plant in control?
- How long is the average durability of electronic devices?
- Which subset of properties of deciduous trees will be used for clustering?
- How does the course of laboratory parameters evolve over time?
- Is there a difference in the response behavior of customers in the test or control group?
- Is there a visual relationship between age and cholesterol (scatter plot)?

This list shows that business questions can come from various industries and knowledge domains. In *Chapter 2 – Characteristics of Analytic Business Questions*, we will show characteristics that allow for a classification of these business questions.

1.4 The Analysis Process

General

The preceding examples of business questions are different because they come from different industries or research disciplines, they involve different statistical methods, and they have different data requirements, just to mention a few.

However, there are two similarities among them:

- Data are needed to answer them.
- They follow an analysis process.

The analysis process usually involves the following steps:

1. Select an analysis method.
2. Identify data sources.
3. Prepare the data.
4. Execute the analysis.
5. Interpret and evaluate the results.
6. Automate (to the extent possible) data preparation and execution of analysis, if the business question has to be answered more than once (e.g., at a later time).

The preceding steps are not necessarily performed in sequential order. They can also be iterated, or the process can skip certain steps.

- The identification of data sources can result in the fact that the data needed for the analysis are not available or can only be achieved with substantial effort. This might lead to a redefinition of the business questions—not that we are dealing mostly with observational data but rather with data that are specifically gathered for an analysis.
- Currently unavailable but essential data sources might lead to a separate data-gathering step outside the analysis process.
- In many cases the process iterates between data preparation and analysis itself.
- The interpretations of the results can lead to a refinement of the business questions. From here the analysis process starts again. This can possibly lead to another data preparation step if the detailed analysis cannot be performed on existing data.

In this book we will only deal with the step 4 "Prepare the data." We will not cover the selection of the appropriate analysis method for a business question, the analysis itself, or its interpretation.

Success Factors for Good Analysis Results

In many cases only the availability of data and the analytic modeling itself are considered key success factors for good analysis results. Certainly the general availability of data is an important success factor. Without data no analysis can be performed. Also, analytic modeling where parameters are estimated, models are tuned, and predictions, classifications, and cluster allocations are created, has high importance.

However, the adequate preparation of data also makes available data accessible and usable to analytic methods. For example, the fact that the data representation of customer behavior is available in a transactional table does not necessarily mean that analytic methods can automatically derive indicators of customer behavior. In this case data preparation is essential to the quality of the analysis results as relevant derived variables such as indicators, transition flags, or ratios are created and made available to the analysis methods.

1.5 Challenging an Analytic Business Question

There are of course simple questions that can hardly be misinterpreted. For example, "Has the market research study shown a significant difference in age between product users and non-users?" If in the data we have an age column and a product usage YES/NO column, the analysis can easily be performed.

The analysis world is, however, not always so easy that simply a question is posed and it is obvious which data are needed or which method should be applied. There are many analytic business questions that have to be asked again or challenged.

A statistician or analyst, when confronted with an analytic business question, does not need to be an expert in the respective business area, such as marketing, medicine, or life science. However, he has to be able to pose the questions that help to solve potential misunderstandings or ambiguities. In most cases the statistician or analyst works with experts in the business area. One of the goals is to eliminate alternative and competing explanations for the observed phenomena.

This process of questioning leads to a deeper understanding of the business question, which is mandatory for a correct analysis and interpretation of the results. Even though analysis and interpretation are not in the scope of this book, the process is very important for data preparation as data history or data structures are influenced.

This shows that the analyst who is in charge of the data preparation has to become involved with the business process. The analyst cannot ignore these questions as they are crucial for the data basis and therefore to the success of the analysis.

The preceding points are illustrated by three example questions.

Which customers are likely to buy a product NEWPROD?

Assume that a product manager approaches the statistician and requires an analysis of the question, "Which customers are likely to buy product NEWPROD?" Even if this question is very clear to the product manager, the statistician will ask a number of questions in order to decide which data are needed and how the analysis will be performed.

- What do you mean by "are likely to buy"?
  - Do you want to have a prioritized list of customers who have a high purchase probability, and are you going to select the top 10,000 customers from this list?
  - Do you want a list of customers whose purchase probability is higher than the average, or do you want some threshold?

- What do you mean by "to buy"?
  - Is the event you are interested in just the buying event, or are you also interested in people who simply respond to the campaign?

- What is the property of NEWPROD?
  - Is it a new product, or is it an existing one?
  - If it is an existing product, does the campaign run under the same conditions as in the past?
  - Can we assume that responders to the campaign in the past are representative of buyers in this campaign?
  - In the case of a new product, which customer behavior in the past can we take to derive information about expected behavior in this campaign?

- Customers, households, or contracts
  - What is the analysis entity that you want to address in the marketing campaign? Do we want to send multiple offers to households if more than one person is eligible for the campaign?

Can you calculate the probabilities for me that a customer will cancel product x?

Similar to the preceding example, there are some questions that need to be answered in order to plan data preparation and analysis.

- What do you mean by "cancel"?

  - Do you mean the full cancellation of the product or a decline in usage?

  - Do you want to include customers that have canceled the product but have started to use a more (or less) advanced product?

  - Do you also want to consider customers who did not themselves cancel but were canceled by our company?

- What is the business process for contacting customers?

- What do you want to do with customers with a high cancellation probability?

- How quickly can you contact customers with a high probability (e.g., to execute an intervention)?

  - Which latency period should we consider between differential availability of customer data and cancellation data?

Can you calculate a survival curve for the time until relapse per treatment group for melanoma patients?

In clinical research statisticians cooperate with health professionals in order to plan and evaluate clinical trials. The following questions can be discussed.

- What is the start date for the time interval?

  - What is the date of admission to the hospital?

  - What is the date of treatment?

  - What is the date of primary surgery?

- How will patients without relapse be treated?

  - If they are "lost to follow-up" will the last date they have been seen be assumed to be an event date (pessimistic assumption)?

  - Will they be censored with their last-seen date?

We see that our three example questions seem to be clear at first. However, we then see there are many points that must be queried in order to have relevant details for data preparation and project planning.

1.6 Business Point of View Needed

Some of the preceding questions result from the fact that business people do not mention certain information as it is implicitly clear to them. These questions also result from the fact that the statistician sees things from a different point of view and knows that certain information has to be clarified in advance.

It is important for business people to understand that these kinds of questions are not posed in order to waste their time, but to allow for better data preparation and analysis.

In Part 3 and Part 4 of this book we will emphasize that coding plays an important role in data preparation for analytics.

- Data have to be extracted from data sources, and quality has to be checked, cleaned, joined, and transformed to different data structures.
- Derived variables have to be calculated.

Coding itself is only part of the task. We want to emphasize that data preparation is more than a coding task; there is also a conceptual part, where business considerations come into play. These business considerations include the following:

- business requirements, e.g., definition of populations and possible actions
- selection of the data sources, their quality status, and their refreshment cycles
- availability of historical data in the sense of a historic snapshot of data
- identification of the analysis subject
- definition of the data structure of the analysis data mart
- analytical methods that are derived from the business questions
- necessary length of the time history

For the statistician and analyst it is important to understand that they have the duty to pose these questions and get into a conversation about them with the business people. The statistician and the analyst have to acquire that amount of domain expertise so that they are in a position to pose the right questions.

In *Chapter 4 – Different Points of View on Analytic Data Preparation*, we will take a more detailed look at different roles in the data preparation process and their objectives.

Chapter 2

Characteristics of Analytic Business Questions

2.1 Introduction

The business questions that are considered in this book need data in order to answer them. The need for data is therefore a common theme. There are also certain characteristics that allow differentiating among these business questions.

In this chapter we will look at the following characteristics and will also see that some of them have an impact on how the data will be prepared in certain cases. We will refer to these characteristics from time to time in other chapters.

- analysis complexity: real analytic or reporting
- analysis paradigm: statistics or data mining
- data preparation paradigm: as much data as possible or business knowledge first
- analysis method: supervised or unsupervised analysis
- scoring needed, yes/no
- periodicity of analysis: one-shot analysis or re-run analysis
- historic data needed, yes/no
- data structure: one row or multiple rows per subject
- complexity of the analysis team

2.2 Analysis Complexity: Real Analytic or Reporting?

When talking about analytic data preparation we might think primarily of data preparation for *advanced* analytics, where we use various analytical methods such as regressions, multivariate statistics, time series, or econometrics. In these cases we build a data table, which is used by various procedures residing in the SAS/STAT, SAS/ETS, or SAS Enterprise Miner package.

There are, however, many cases where data are prepared for analyses that are not complex analytical methods. These methods are often referred to as *reporting* or *descriptive analyses*. Examples of these methods include the calculations of simple measures such as means, sums, differences, or frequencies as well as result tables with univariate and multivariate frequencies, and descriptive statistics and graphical representations in the form of bar charts, histograms, and line plots.

The following are examples of descriptive or reporting analyses:

- line plot per patient; the course of laboratory parameters over time
- crosstabulation of 4,000 customers for the categories test/control group and yes/no response
- scatter plots to visualize the correlation between age and cholesterol

We want to emphasize here that analytic data preparation is not exclusively for complex analytic methods, but is also relevant for reporting analyses. In practice, analytics and reporting are not an "either/or" proposition but practically an "and" proposition.

- Descriptive or exploratory analysis is performed in order to get knowledge of the data for further analysis.
- Analytic results are often presented in combination with descriptive statistics such as a 2x2 cross table with the statistics for the chi-squared test for independence.

Therefore, the considerations and the data preparation examples that are mentioned in the following chapters are not only the basis for elaborate analytic methods, but also for reporting and descriptive methods.

2.3 Analysis Paradigm: Statistics or Data Mining?

Analytic business questions can generally be categorized by whether the analysis paradigm is the classic analytic approach or the data mining approach. A short example illustrates the difference. Consider the following two business questions:

- In a controlled clinical trial, which risk factors have a strong influence on the occurrence of lung cancer?
- In marketing, which customers have a high cancellation risk in the next month? This analysis is sometimes called *churn analysis*.

Both business questions can be solved using the same analytic method, such as logistic regression. In the data table, one column will hold the variable that has to be predicted. The difference between these two questions is the analysis paradigm.

- The analysis of the risk factors for lung cancer is usually performed using classic statistical methods and follows a predefined analysis scheme.

  - There are usually between 10 and 20 risk factors or patient attributes that have been collected in the clinical trial.

  - The influence of these risk factors is evaluated by looking at the p-values of their regression coefficients.

  - Some of the input variables are not necessarily variables that serve as predictors, but can be covariates that absorb the effect of certain patient properties.

  - In this case we analyze in the statistical paradigm: a list of potential explanatory variables is defined a priori, their influence is statistically measured in the form of p-values, and the evaluation of the results is performed on predefined guidelines such as the definition of an alpha error.

- The analysis of the premature cancellation of customers is usually performed by the data mining approach.

  □ Data for the analysis table is gathered from many different data sources. These different data sources can have a one-to-one relationship with the customer but can also have repeated observations per customer over time.

  □ From these data, an analysis table with one row per customer is built, where various forms of aggregations of repeated observations per customer are applied.

  □ The emphasis is on building an analysis table with many derived customer attributes that might influence whether a customer cancels the contract or not. Very often these analysis tables have more than 100 attributes.

  □ In the analysis, emphasis is not on *p*-values but more on lift values of the predicted results. It is evaluated by how many hits—successful predictions of churn—are achieved if 2% of the customer base is contacted that have the highest predicted probability from the logistic regression.

We see that both examples incorporate the same statistical method, but the analysis paradigm is different. This also leads to different scopes of the analysis tables. The analysis table for the lung cancer example has a predefined number of variables. In the marketing example, the list of input variables includes cleverly derived variables that consolidate a certain product usage behavior over time from a few separate variables.

Note that the difference between statistics and data mining is not primarily based on the analytic method that we use but on the analysis paradigm. We have either a clear analysis question with a defined set of analysis data, or we have a more exploratory approach to discover relationships.

2.4 Data Preparation Paradigm: As Much Data As Possible or Business Knowledge First?

In the previous section we explained that analysis tables for data mining usually have a large number of variables. If we have a lot of input data sources and time histories per customer, there is theoretically no limit to the number of variables for an analysis table that can be built. In practice a lot of analysis tables have hundreds of input variables. The rationale for this large number of input variables is that any possible property of a subject that can be potentially in relationship with the target will be considered. In the analysis itself, only a small subset of these variables is used in the final model.

Concerning the number of variables and the philosophy of how analytic tables are built, two paradigms exist:

- In the *as much data as possible* paradigm, the classic data mining idea is followed. We try to get as much data as possible, combine them, and try to find relationships in the data. All data that seem to have some relationship to the business question are included.

- A contrast is in the *business knowledge first* paradigm. The problem is approached from a business point of view. It is evaluated which data can best be used to explain certain subject behavior. For example, "What causes a subject to perform a certain action?" The aim is to determine with business knowledge which attributes might have the most influence on the business questions.

In the case of the modeling of the contract cancellation of a telecommunications customer (also called churn), the following predominant variables can be explained from a business point of view:

- The number of days until the obligation (contract binding) period ends.
  Interpretation: The sooner the client is free and does not have to pay penalties for contract canceling, the more he will be at risk to cancel the contract.

- The number of loyalty points the customer has collected and not cashed.
  Interpretation: The more points a customer has collected, the more he will lose when he cancels his contract, as he cannot be rewarded any more. Additionally, the more loyalty points a customer has collected indicates that he has a long relationship with the company or he has a very high usage, which lets him collect a number of points.

The importance and explanatory power of these two variables usually supersedes sets of highly complex derived variables describing usage patterns over time.

Competing Paradigms?

Note that these two paradigms do not necessarily exclude each other. On the one hand, it makes sense to try to get as much data as possible for the analysis in order to answer a business question. The creative and exploratory part of a data mining project should not be ended prematurely because we might leave out potentially useful data and miss important findings.

The fact that we select input characteristics from a business point of view does not mean that there is no place for clever data preparation and meaningful derived variables. On the other hand, not every technically possible derived variable needs to be built into the analysis paradigm.

We also need to bear in mind the necessary resources, such as data allocation and extraction in the sources systems, data loading times, disk space for data storage, analysis time, business coordination, and selection time to separate useful information from non-useful information. Let us not forget the general pressure for success when a large number of resources have been invested in a project.

2.5 Analysis Method: Supervised or Unsupervised?

Analytic business questions that require advanced analytical methods can be separated into supervised and unsupervised analyses.

In *supervised analyses* a target variable is present that defines the levels of possible outcomes. This target variable can indicate an event such as the purchase of a certain product, or it can represent a value or the size of an insurance claim. In the analysis, the relationship between the input variables and the target variable is analyzed and described in a certain set of rules. Supervised analysis is often referred to as *predictive modeling*. Examples include the following:

- daily prediction for the next six months of the overnight stays of tourists in Vienna hotels
- prediction of the payback probability of loan customers

In *unsupervised analyses,* no target variable is present. In this type of analysis, subjects that appear similar to each other, that are associated with each other, or that show the same pattern are grouped together. There is no target variable that can be used for model training. One instance of unsupervised modeling is segmentation based on clustering. Examples include the following:

- segmentation of customers based on their purchase behavior
- segmentation based on market basket analysis, or which products are frequently bought together

2.6 Scoring Needed: Yes/No?

General

Analyses can be categorized by whether they create scoring rules or not. In any case, each analysis creates analysis results. These results can include whether a hypothesis will be rejected or not, the influence of certain input parameters in predictive modeling, or a cluster identifier and the cluster properties in clustering. These results are the basis for further decision making and increased business knowledge.

It is also possible that the result of an analysis is additionally represented by a set of explicit rules. This set of rules is also called the *scoring rules* or *scorecard.* In the case of a decision tree, the scoring rules are built of IF THEN/ELSE clauses, such as "If the customer has more than two accounts AND his age is > 40 THEN his purchase probability is 0.28." In the case of a regression, the scoring rules are a mathematical formula that is based on the regression coefficients and that provides an estimated or predicted value for the target variable.

The types and usage of scoring can be divided into two main groups:

- Scoring of new analysis subjects. *New* means that those analysis subjects were not available during training of the model in the analysis, such as new customers or new loan applicants that are being scored with the scorecard. These subjects are often called *prospects.*
- Scoring of analysis subjects at a later time. This includes new and old analysis subjects whose score is actualized periodically (e.g., weekly) in order to consider the most recent data. The subjects are said to have their scores recalibrated with the freshest data available.

Example

Assume we perform predictive modeling in order to predict the purchase event for a certain product. This analysis is performed based on historic purchase data, and the purchase event is explained by various customer attributes. The result of this analysis is a probability for the purchase event for each customer in the analysis table.

Additionally, the calculation rule for the purchase probability can be output as a *scoring rule.* In logistic regression, this scoring rule is based on the regression coefficients. The representation of the scoring rules in code is called the *score code.* The score code can be used to apply the rules to another data table, to other analysis subjects, or to the same or other analysis subjects at a later time. Applying the score code results in an additional column containing the purchase probability.

Usually a scorecard for buying behavior is applied to the customer base in certain time intervals (e.g., monthly) in order to allow the ranking of customers based on their purchase probability.

Scorecards and score code are frequently used in predictive modeling but also in clustering, where the assignment of subjects to clusters is performed using the score code.

Other examples of scorecards are churn scorecards that calculate a probability that the customer relationship is terminated during a certain period of time, credit scoring scorecards that calculate the probability that a loan or credit can be paid back, scorecards that assign customers to behavior segments, propensity-to-buy scorecards, or scorecards that predict the future customer value.

Data

Considering scoring, we can talk of an analysis table and a scoring table.

- The *analysis table* holds data that are used to perform the analysis and to create the results. The result of the analysis is the basis for the score code.

- The *scoring table* is the table on which the scoring rules are applied. In predictive modeling, the scoring table usually has no target variable, but a predicted value for the target variable is created on the basis of the score code. During scoring, the score is created as an additional column in the table.

We will deal with scoring in *Chapter 14 – Transposing between One and Multiple Rows per Subject Data Structures*, and also come back to the requirements for periodic data later in this chapter.

2.7 Periodicity of Analysis: One-Shot Analysis or Re-run Analysis?

Analyses can be distinguished by the frequency in which they are performed. Some analyses are performed only once, for example, as an ad-hoc analysis. Other analyses are performed repeatedly. In this section, we will distinguish between one-shot analysis and re-run analysis.

2.7.1 The One-Shot Analysis

One-shot analysis means that the analysis is performed only once. For this analysis, data are retrieved from an existing system or are newly collected. The analysis is performed on these data in order to answer a certain business question. The analysis is usually not repeated at a later time, as there are either no new data available or data are not newly collected due to time or resource constraints. For the business questions behind this analysis, a periodic update of results is not foreseen or not feasible. Periodic actualization of the results or scoring new analysis subjects is not the focus of these types of analyses.

The following are examples of one-shot analysis:

- In biometrical analyses such as medical statistics, patient data are collected in a trial or study. Data are entered into a table and passed to the statistician who performs the appropriate statistical tests for certain hypotheses.

- In market research, survey data are entered into a table. Derived variables are calculated and the appropriate analysis, such as cluster analysis, factor analysis, or conjoint analysis, is performed on that data.

One shot does not mean that the final analysis results are derived in only one analysis step. These analyses might be complex in the sense that pre-analysis or exploratory analysis is needed where properties of the data are evaluated. Analysis can also get more complex in the analysis process because more detailed business questions on subgroups arise. However, *one sho*t does mean that the analysis and data preparation will most likely not be repeated in the future.

2.7.2 Re-run Analysis

In contrast to a one-shot analysis there are analyses that are planned to be re-run over time, when additional or more recent data on the analysis subjects are available. The most frequent cases for *re-run analyses* are the following:

- The analysis is re-run at a later time when data for more analysis subjects are available. This case is frequently seen when analysis subjects enter the study or survey at different times. Time after time, more data are available on the analysis subjects and the analysis results are updated.

- The analysis is re-run at a later time when more recent data for the analysis subjects are available. This also includes the case where analyses are re-run at a later time to compare the results over time, such as for new groups of subjects.

Because re-run analyses will be performed more than once or on additional data not only on the original data, all other data sources they will be applied to need to have the same data structure as the original data.

In the previous section we looked at scoring. With periodic scoring the analysis is not re-run, but the results of an analysis are re-applied. However, to be able to apply the score code to other data, they need to be prepared and structured as the original data. We are, therefore, in almost the same situation as with re-run analyses. We have the same data requirements except we do not need to prepare a target variable as in the case of predictive modeling.

2.8 Need for Historic Data: Yes/No?

Analyses can have different requirements for historic data—mainly, whether they need historic data from data sources or not. Some business questions can be answered with actual data, such as the evaluation of results from a marketing survey.

Other analyses, such as time series forecasting, require historic data because the course over time needs to be learned for historic periods in order to forecast it for the future. This difference is relevant because the extraction of historic data from data sources can be both resource- and time-intensive.

2.9 Data Structure: One-Row-per-Subject or Multiple-Rows-per Subject?

Analyses can also be distinguished by the data structure that is needed by the respective analysis. For example, data per analysis subject can be structured into a single row or can be organized in multiple rows in the table. The following data structures are frequently found in analytics:

- one-row-per-subject data structures
- multiple-row-per-subject data structures
- longitudinal data structures

We will go into more detail in Part 2 of this book.

2.10 Complexity of the Analysis Team

The number of persons that are involved in the analysis of a business question is also an important factor for consideration. If only one person is in charge of data preparation, analysis, and interpretation of results, then no interactions exist and there is almost no dispersion of know-how (although eventually a few other people might be contacted for specific details).

In the analysis of a complex business question usually a team of people are working together on a project. The involvement of several people also has an effect on the data preparation of a project. The effect is not on the data content or data structure itself but on the necessary communication and resulting project documentation.

If data preparation and analysis are done by different people, the transfer of details that are gathered during data preparation to the analysis people is important. This transfer might include the documentation of the meaning of missing values, the source systems, and their reliability and data quality.

Note that the complexity of the analysis team is not necessarily directly linked to the business question. It is also a resource decision in organizations as to how many people are involved in analysis.

2.11 Conclusion

Some of the preceding characteristics are predefined concerning certain business questions. Other characteristics are defined by the analysis approach. The selection of the data preparation paradigm—for example, whether all possible derived variables are built or if only a few variables that are derived from business knowledge are created—is decided by the analyst. The need for historic data is inherent to the business problem and is in most cases mandatory.

This chapter also aimed to give more insight into various characteristics as they allow for differentiating among business questions. The preceding characteristics also allow business people to gain more understanding of characteristics and details the analyst has to consider in formulating the best methodological approach to address the business objective under consideration.

Chapter 3

Characteristics of Data Sources

3.1 Introduction

After having discussed the characteristics of business questions in the previous chapter, we will now go into more detail about the characteristics of data sources. In this chapter we will continue with the business point of view of data and will discuss properties of data sources such as the rationale of the source system and periodic data availability. We will also look at the advantageous properties of a data warehouse.

Obviously, data sources have a lot of technical characteristics, which we will look at in *Chapter 5 – The Origin of Data*.

3.2 Operational or Dispositive Data Systems?

Computer systems that process data can be classified into two main categories: operational (or transactional) systems and dispositive systems. There are other possible categorizations; however, we will look only at the properties of these two types because they are important for data preparation considerations.

3.2.1 Operational Systems

Operational systems are designed to assist business operations. These systems allow enterprises to run their daily businesses by providing the technical infrastructure to process data from their business processes. Therefore, these systems are (or should be) tightly integrated with the business processes and workflows of the companies or organizations that use them. Operational systems that deal with business transactions are also referred to as *transactional systems*.

Examples of operational systems include the following:

- Sales force automation systems that help salespeople administer their offers, orders, and contacts.

- Call center systems that maintain customer databases where address changes, customer complaints, or inquires are handled.

- Booking systems at hotels, airlines, or travel agencies that allow the entering of reservations and bookings, and that have methods implemented so different agents are not able to simultaneously book the same seat on the same airplane.

- Legacy systems in banks that maintain customer and account data and process account transactions.

- Clinical databases, where data are entered and maintained in adherence to good clinical practice regulations. These databases allow double data entry, data correctness or plausibility verification, and maintain audit trails of data modifications.

- Enterprise Resource Planning (ERP) systems perform tasks such as human resource management, stock keeping, or accounting.

The following are important characteristics of these systems:

- Because the underlying business processes are time critical, quick response times are a key issue.

- The focus is in most cases on a single transaction; a bank customer who wants to make a few financial transactions or withdrawals, or an agent in the call center who needs the data for a calling customer.

- Having the actual version of the data available, rather than providing time histories of data value changes. This does not mean that these systems have no focus on historic information, just the immediate need to provide a transaction history for a customer's bank account. However, with customer data for example, the focus is more on the actual version of the address for sending the bill rather than on providing the information, such as where a customer lived 12 months ago. In many cases the decision to maintain a lot of historic data is dropped in favor of improving the performance and response times of these systems.

3.2.2 Dispositive Systems

Different from transactional systems, *dispositive systems* are designed to provide information to the user. Their name is derived from the fact that information in the form of reports, tables, graphs, or charts is at the consumer's disposal. These systems are also referred to as reporting systems, information retrieval systems, or data warehouse systems.

Examples of dispositive systems include the following:

- Sales reporting systems. These systems provide sales measures such as the number of sold pieces, price, or turnover. These measures are made available to the user in the form of tables, summaries, or charts, allowing subgroups by regions, product categories, or others.

- Customer data warehouses. With the target to provide a 360° view of the customer, this kind of system holds customer data from sources such as a call center database, billing system, product usage database, contract database, or socio-demographics.

- Data warehouse systems in general. These systems hold information from various data sources of the enterprise or organization. A data warehouse is subject-oriented and integrates data from those data sources by a data model. It is usually a read-only database to the warehouse (information) consumer and is time-variant in the sense that its data model is built to provide historic views (or snapshots) of the data.

The following are important characteristics of these systems:

- Dispositive systems or data warehouse systems are mostly read-only. They have to handle read operations over the entire table (full table scans) very quickly, whereas they do not have to provide updates of single records.

- Data warehouse systems handle time histories. They provide historic information so that the information consumer can retrieve the status of the data at a certain historic point in time. As previously mentioned, operational systems deal more with the actual version of data.

  - For example, customers that canceled their contracts with the company two years ago will no longer appear in the operational system. The customers and their related data, however, will still be available in a data warehouse (with a record-actuality indicator).

  - Data warehouses also track changes of subject attributes over time and allow the creation of a snapshot of the data for a given date in the past.

  - Data warehouse environments provide aggregations of several measures in predefined time intervals. Depending on the data warehouse design and the business requirement, the time granularity is monthly, weekly, or daily. Also note that due to the huge space requirements, the underlying detail records are very seldom maintained in the data warehouse for a long time in online mode. Older versions of the data might exist on tapes or other storage media.

- Extending the time granularity. Billing systems create balances with each bill. The bill cycles, however, will very often not coincide with calendar months. Therefore, sums from the bill cannot be used to exactly describe the customer's monthly usage. Data warehouses store additional information in the billing data, and monthly usage by calendar months, weeks, and days. The granularity itself depends on the data warehouse design.

The following table summarizes the properties of operational and dispositive systems:

Table 3.1: Comparison of operational and dispositive systems

Operational System	Dispositive System
Serves as a transaction system in order to perform specified business tasks.	Serves as an information system in order to provide businesses a basis for decision making.
Those business tasks include billing, customer data maintenance, customer care tracking, and sales assistance.	Historic snapshots of data are available or can be re-created from history logs of the tables.
Importance is on the reliability of actual data.	Many transactions are run in batch, often in overnight jobs.
Performance is needed to guarantee short response times.	Performance is needed to quickly update tables in one pass.
Allows performance of business transactions.	Provides information such as reports or forecasts.
Has a transactional orientation.	Is the data basis for reporting, monitoring, and analytics.

3.3 Data Requirement: Periodic Availability

In the previous chapter we learned that for re-run analyses we need data periodically, i.e., more than once. In many cases this is not difficult because we can prepare a more recent snapshot of the data at a later time, again and again, in order to prepare our analysis table.

There are, however, cases where not all data are periodically available. This happens if different data sources are used in the analysis that have different reasons for their existence. Data from some sources, such as surveys and special analyses, are gathered only once (e.g., to perform a market research analysis). These are examples of typical one-shot analysis.

Data that were gathered for a one-shot analysis should be included with care in a re-run analysis. If it cannot be assured that the data can be provided in the same or similar quality and actuality at a later time for re-running the analysis or scoring, then the analytic data mart will contain outdated or irrelevant information.

The following situations are important in this context:

- When using data that were acquired from an external data provider, care has to be taken whether these data will be available in the future. Negotiation and contractual status with the external provider have to be taken into account.

- Surveys such as market research are performed on samples and not on the whole customer base. The resulting variables are only available for a subset of the customer base. Analysis results for the samples can hardly be reproduced or generalized to the whole customer base.

- Analyses with inherent dependencies that use the scores of other scorecards as input variables. If model B depends on model A in the sense that the scores derived from model A are input variables for model B, then model B can be applied only as long as model A is used and periodically applied. The awareness of those dependencies is important because a retraining of model A will force an immediate retraining of model B.

- Analyses that are based on classifications that periodically change. If the model uses input variables that are based on lookup tables or definitions that change over time, such as price plans or tariffs, two problems can arise:

  - Changes in the definition of the classifications force a retraining of the model(s).

  - Care has to be taken that the actual and valid version of the lookup table or classification is used for analysis. Again we are faced with the case where the analyst tries to gather as many variables as possible and oversees the data maintenance effort and responsibility for these data.

3.4 Wording: Analysis Table or Analytic Data Mart?

When we write a book on analytic data preparation, it is not easy to decide which name to use for the final product that comes out of the data preparation process. Looking at it from a technical point of view, we see it as a table with rows and columns. Most likely the rows logically represent subjects, repeated measurements on subjects, or both. The columns in most cases represent attributes of the subjects.

In data mining it has become usual to call this table a *data mart*. This name is not necessarily correct, because a data mart is not always only one table; it can also be a set of tables that hold data for a certain business domain. For example, tables of a data mart might need to be joined together in order to have the appropriate format. However, data mart has become a synonym for the table that holds the data for data mining.

In non-data mining areas, the term data mart is almost unknown. Names such as analysis table, data table, or data matrix are common here and mean the same thing as data mart in the preceding example, which is a table with data for analysis.

We have therefore tried to use the global name *analysis table* as often as possible. In some chapters, the term data mart is used. If not separately specified, this term also means a table with data for analysis.

Our definition of the final product of the data preparation process is that we have a single table that can be used for analysis.

3.5 Quality of Data Sources for Analytics

In the first section of this chapter we compared the properties of a data warehouse to a transactional database system. We saw that the quality of data from a data warehouse for analysis is more appropriate than data directly from an operational system.

The following figure displays data inputs from a logical point of view:

Figure 3.1: Data inputs in the analytic data mart from a logical point of view

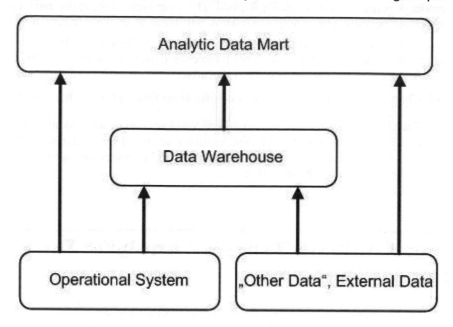

We see that there are two main sources where data for analysis originally reside:

- Data in operational systems, which were described in the first section of this chapter.
- Other data such as external data, survey data, ad-hoc data, lookup tables, and categorizations that also reside in databases, spreadsheets, PC databases, or text files.

Data from operational systems or other data can either flow directly into the analytic data mart or can be retrieved from a data warehouse.

3.5.1 Advantages of Data Warehouses

It is obvious that if a data warehouse is in place that holds the required information, the data would be accessed from there. Data from a data warehouse have the following advantages:

- Data are usually quality checked.
- Maintenance procedures are in place to keep lookup tables and definitions up-to-date.
- Data are periodically refreshed.
- Data are historicized, and historic snapshots of data can be retrieved.
- Changes in the operational system, such as tables, entities, column names, codes, ranges, and formats are handled by the data warehouse, and data marts are consistent in their definition.

Again, the feature of data historization in a data warehouse is emphasized here, because it is very important for analysts to have historic snapshots of data in order to study the influence of input variables on a target variable in different periods. *Data historization* means that historic versions or historic snapshots of the data can be reproduced.

Operational systems, in contrast, hold historic information—for example, accounts that were opened seven years ago, or transactions that were made last year. However, *historic data* and *historic versions of data* are not the same. An operational system may provide information about accounts that were opened seven years ago, but will not necessarily provide information about how the conditions on this account, such as the interest rate, looked seven, six, or five years ago.

3.5.2 The Analyst's Dilemma

The analyst's dilemma starts if he cannot retrieve all the data he needs from the data warehouse. A data warehouse might not provide all necessary data sources that are needed for a certain analysis, i.e., data from a recent market survey are not (yet) available in the data warehouse.

The analyst must now decide whether or not to use data sources that have not been loaded into the data warehouse, or to access these data sources directly. Technically, SAS is of great help in this case, because external data or other additional tables for analysis can be quickly imported, data can be cleansed and quality checked, and complex data structures can be decoded.

Given this, the analyst assumes all duties and responsibilities, including data quality checking, lookup table maintenance, and interface definitions. He might encounter trouble with every change in the operational system, including renaming variables, and redefining classifications or interfaces to external data.

This is not problematic with one-shot analysis because once the data are extracted from whatever system, the analysis can be performed and there is no need to reload the data at a later time. However, if the analysis is a re-run analysis and analysis results need to be refreshed at a later time, problems in the data interface can be encountered. This is also true for scoring at a later time, because the scoring table needs to be rebuilt as well.

This does not mean that the analyst can always decide against including data directly from some systems. In many cases this is necessary to access the relevant data for an analysis. However, analysts should not work independently of IT departments and build their own analytic data warehouses if they do not have the resources to maintain them.

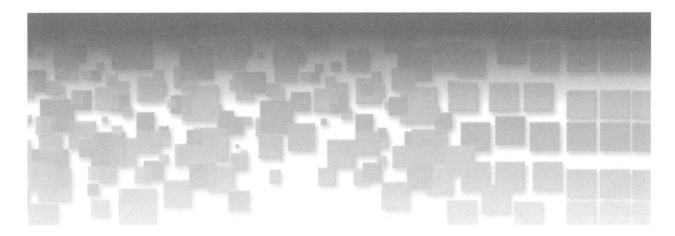

Chapter 4

Different Points of View on Analytic Data Preparation

4.1 Introduction

In *Chapter 2 – Characteristics of Analytic Business Questions,* we discussed the complexity of the analysis team as a characteristic of a business question. If the business question and the related analysis are simple and only one or a few people are involved in this analysis, there will be little interaction between people.

If, however, more people from different disciplines with different backgrounds and interests are involved in the analysis process, different points of view on this topic come up. In this chapter we will not deal with project management and personnel management theory. Instead we will focus on typical roles that are frequently encountered in analysis projects and will look at the interests, objectives, and concerns of the people in those roles.

4.2 Simon, Daniele, and Elias: Three Different Roles in the Analysis Process

General

In this section we will define three roles we usually encounter in the analysis process. We will define these roles by their responsibilities and job descriptions in an organization. We will make the acquaintance of three people who work in our organization.

- Simon—the business analyst
- Daniele—the quantitative expert
- Elias—the IT and data expert

These roles are not necessarily filled by three people: the roles can be divided more granularly and involve more than three people. It is also possible that one person fills more than one role.

This separation of roles is suitable for our purposes to illustrate different viewpoints of different people who are involved in data preparation.

Note that in the following sections we will list some of these people's attitudes and opinions. For illustration we do this in an intentionally extreme manner, although we know that not many people are so extreme. We also want to mention that we had no particular Simon, Daniele, or Elias in mind when writing this chapter.

4.3 Simon—The Business Analyst

His role

Simon, the business analyst, is the person who needs results for his business. He usually has a business question that needs to be answered. There are business analysts like Simon in almost any organization. He is the one who approaches the statistician with a request for an analysis.

- In marketing, Simon might be a campaign, segment, or product manager who is responsible for his product line or segment.
- In a sales department, Simon might be a team leader of salespeople or a key account manager who is in charge of preparing sales campaigns.
- In medical statistics, Simon might be a medical doctor who wants to have his medical trial results evaluated. Here he is specifically interested in significant p-values.
- Simon can also be a decision maker who needs results on which to base his decisions.

In general he does not have vast knowledge of statistics. He has already heard that data mining methods include regression, decision trees, and neural networks. However, he has limited experience, such as some statistics courses in his college education.

His working environment

- Simon's working environment can be quite hectic. There is time pressure when the next campaign will go out and a target group is needed. There is success pressure when the sales figures have to be raised. There is also time pressure when a decision has to be made.
- His job also involves some creativity and flexibility. It is not possible to foresee and plan all analysis requests, and he cannot always plan for when a certain result is needed.

His interests

- Simon wants results. For example, Simon needs the results for an analysis in order to make a business decision about a target group for a campaign. He wants to see and understand the results, possibly graphically illustrated.

- In conversations with Daniele, the quantitative expert, he likes simple explanations. He knows that Daniele can deliver very valuable results to him in order to improve his business.

- In his role as a medical doctor, he does not like dealing with problems such as multiplicity and the fact that multiple choice responses to questions on the case record form or multiple possible diagnoses can't be entered into the same table column for a patient.

What he dislikes

- Not understanding why an analysis takes so much preparation time, whereas another analysis, such as building a predictive model, can be done much more quickly

- Being involved in too much data acquisition discussion

- Getting things explained in a complex statistical language

- Being told that things are too complicated, questions cannot be answered with these types of data, and more time is needed

4.4 Daniele—The Quantitative Expert

Her role

Daniele is the statistician in the organization. She is in charge of models that help the company answer its business questions and be more productive. She is confronted with requirements from the business departments and she needs data from the IT department. She tries to explain her points of view with as few statistical terms as possible, but sometimes she needs to rely on her statistical arguments.

Her working environment

From an organizational point of view, Daniele can be found in different departments:

- in the business department, she assists the business directly with results

- in the IT department, if analysis is seen as a service to other (business) departments

- in a separate analysis department, if this is especially foreseen in the respective organization

- as an external consultant who assists the organization with the analysis

Other people expect Daniele to deliver correct and beneficial results. Sometimes, she finds herself wondering whether she should leave the path of "correct" statistical analysis and be creative with a workaround in order to get desired results.

She knows that analysis cannot be completely separated from data management, and she often finds herself facing the difficulty of why she cannot name the necessary attributes.

Her interests

- Have an analytic database rich in attributes
- Perform data management on her own
- Respond to requests from the business side very quickly
- See her results being used in practice
- Need for analysis of not only active customers but also of inactive customers (or patients that have already left the clinical trial)
- Understand that historic data are not only data from the past but historic snapshots of the data
- Investigate longitudinal data herself—which trends, triggers, or aggregations best fits the business problem
- Create from perhaps hundreds of input variables a final prediction model with between 10 and 20 predictors
- Make it clear that there is a difference between multivariate statistical suitable data and univariate cleansed data for reporting purposes

What she dislikes

- Long and confrontational discussions with the IT department about why a certain variable is needed
- Doing too much ad-hoc analysis
- Getting data too late or of bad quality
- Getting data that are not quality checked and discovering only during analysis that data are implausible
- Being under time pressure because of late data delivery, which in many cases means that there is not enough time for analysis

4.5 Elias—The IT and Data Expert

His role

Elias is the IT and data expert. He manages a lot of databases, operational systems, and the reporting data warehouse. He knows the data sources, the tables, and the data model very well. He also knows about the effort and resources needed to create additional variables.

His working environment

Elias is working in a department that is very often confronted with requests from the business department that cannot be fulfilled in the planned time frame. The communication level between him and the business people is not always the best because the business people are frustrated by the long time delays. Elias knows that other departments do not understand that providing data and IT resources is not only the process of creating a table—it's also a lot of necessary documentation and process considerations.

His interests

- That data necessary for analysis can be regularly loaded from the data warehouse or other source systems and not too much external ad-hoc data are used
- That the data preparation process runs smoothly, automatically, and securely
- That the data preparation job can run in batch and can be scheduled
- That no errors in the data preparation process occur
- Perform scoring on the scoring data mart
- Balance the cost of data acquisition with the benefit of these data
- Consider data to be clean, if they are clean for reporting or clean in a univariate way

What he dislikes

- Transferring data to the analytic platform just to be scored there
- Being confronted with a large SAS program with many (dubious) macro calls, where it is hard to understand what a program does
- Looking for error messages in the SAS log
- Having an unsuccessful scoring run because of new categories or other reasons
- Having an unsuccessful scoring run that he cannot interpret because of errors in the SAS log
- Daniele's need for historic snapshots of data that are hard to produce from the source systems
- Loading times that are very long because many attributes, historic snapshots, and aggregations need to be made
- Providing data on the detail or record level instead of aggregated values
- The fact that the cost for only one column (e.g., the purchase probability) is very high in terms of programming, monitoring, and data loading
- Daniele's need to explore so many attributes instead of providing him a final list of attributes that she will need in the modeling
- Not understanding why Daniele cannot specify the final shape of the analysis table and why she wants to perform data management or access source data during data preparation
- Not understanding why an analysis should still be performed when a desired variable is missing

4.6 Who Is Right?

General

As you might expect there is no universal truth concerning the preceding points. And there is no judge to define who is allowed to like or to dislike something. We always have to bear in mind that everyone in an organization plays a role and understands and copes with expectations that are posed to that role, given their differential needs and responsibilities.

Therefore, the following sections do not judge but rather explain the background of certain wishes, fears, and attitudes of people.

4.6.1 Performing Statistical Data Management

Simon wants to perform data management because he needs to prepare the data for statistical analysis. This is not because he wants to compete with Elias, but this is a task that cannot be predefined because the statistical properties of the data have to be analyzed in advance. Outliers have to be checked, missing values have to be evaluated, and statistical transformations such as normalization and standardization have to be performed.

Simon also wants to check the data quality. This does not mean that statisticians have stronger requirements than IT people. But for statistical analysis, statistical and multivariate properties of the data have to be checked.

Sometimes the respective analysis is the first time these data are analyzed in detail. Therefore, data management is also needed during the analysis.

These types of data management and checks require statistical functions such as median or standard deviation, which are not always found in the SQL language of relational databases. SAS is very well suited for these types of analyses.

4.6.2 Running Analysis within Half a Day or Four Weeks

There is very often a lot of confusion about how long analyses usually take. Business people often do not understand why analyses can take several weeks, whereas a simple clustering or decision tree takes only a few hours to create. Here we have to consider when a statistical method can be applied to data. The data are usually prepared, which means that a lot of work— not only programming work—has already been performed. Data preparation, including the conceptual work, is a task that can be very time-intensive.

This is something that has to be made clear to business people. Their view of matters is only part of the game.

4.6.3 Statistical and Non-statistical Explanation

Business people do not like to get explanations in a language that they do not understand. If arguments to IT or business people are too theoretical or too mathematical, they will not understand. This is why Daniele is asked to make her results very easily interpretable.

On the other hand, Simon and Elias have to accept that there are arguments that cannot be broken down into a few simple words or sentences. And they need to be patient and try to understand why statistical things are as they are.

It is also a Daniele's responsibility to explain the results to Simon in an illustrative way. For example, what are the reasons that a customer buys a product or leaves a company? Often this responsibility would require Daniele to prepare statistical results.

4.6.4 Not All Business Questions Can Be Answered within a Day

Organizations usually prepare analytical databases for the questions they are most often confronted with. These data are available for the dedicated analyses, and many similar analyses can also be answered with these data.

It is however, inefficient and impractical to create a stock analysis database on the chance that someone might require a statistic on these data. In this case, resources such as time and disk storage would be needed.

Simon needs to understand that not all data can be made available so that at any time an analysis can be started if it is suddenly decided to be a high priority from a business point of view.

4.6.5 "Old Data" and Many Attributes

From an IT point of view Daniele will need data that are in many cases hard to provide. Historic snapshots are needed, not only for the last period, but for a series of prior periods. If in sales analysis the influence of price on the sold quantity will be analyzed, the sale for each historic period has to be compared with the historic price in the same time period.

We will discuss in *Chapter 12 – Considerations for Predictive Modeling*, the case of latency windows. We will discuss that a prediction for period k sometimes has to ignore data from period k-1 because this is the period where the preparation of the business action will take place. So in many cases data from period k-2 and earlier are needed. For statistical estimation, more than one period should be taken because of stability concerns. If we are talking about years it is likely that a long time history is needed.

4.6.6 When Is an Attribute Really Important?

There is very often confusion in business and IT departments when Daniele is asked to list the attributes she needs for the analysis. In many cases some attributes from this wish list cannot be provided, but Daniele can still run the analysis. Was the attribute important or not?

The answer is yes and no. The attribute is important or Daniele believed in its importance. However, the project's success does not rest on these attributes, because other correlative attributes can compensate for their effect or predictability. Therefore, it makes sense to deliver the attributes if possible; however, Daniele did not want to bother Simon to provide an unnecessary attribute.

4.6.7 Automation and Clear Interfaces in Data Acquisition Make Sense

Daniele often does not understand why Elias insists so much on clear interfaces of data retrieval and why he dislikes ad-hoc external data. From an analysis point of view, ad-hoc external data that are not retrieved via regular database interface from the source system make sense as they allow us to retrieve important data. However, Elias will want to have his data interfaces as clean as possible, in case he needs to prepare the same data at a later time for scoring or for another analysis. Elias is concerned with controlling his environment with best practices that are not easily conducive to analytical efforts.

4.7 The Optimal Triangle

When looking at the different points of view, we might get the impression that the analysis process with people from different departments and interests can produce conflicts. In fact there are analysis projects where people from different departments have heated debates on these topics. However, it doesn't always have to be that dramatic. It is more a personal and organizational culture concerning how different points of view are handled.

Moreover, the presence of different points of view also means that people have a different background, knowledge, and understanding of the respective analysis and they can benefit from each other. Analysis is not always a very straightforward process. We need to be curious about various details and creative in order to uncover flexible solutions.

The people or roles we have described in the preceding sections are experts in their respective domains. They have very specific knowledge, which others probably don't have. Therefore, the cooperation of these people is an important success factor for a project.

- For example, Daniele might want to include a certain attribute in the data, but Elias might tell her that this attribute is not well maintained and that the content, if not missing, is more subjective textual input than real categorical data.

- Because Elias knows the data very well, he might suggest different aggregation levels or data sources that haven't occurred to Daniele. For historical data or historical snapshots of the data, he might also give good advice on the respective data sources and their definitions at a historic point in time.

- Daniele might be a good "sparring partner" for Simon in questioning his business practices for more details. Very often points come up that Simon had not considered.

- In return, Simon might have valuable input in the selection of data and data sources from a business point of view.

- In consulting projects it often happens that an external consultant in the persona of Daniele comes to an organization and encounters Simon and Elias on the business and technical sides. Again the exchange of suggestions and statistical requirements of Daniele, the business process and requirements, and the technical possibilities are very crucial parts of each project.

We also see here that many of these conflicts and discussions take place long before the first analysis table is built or the first derived variable is calculated. We also want to mention that these discussions should not be seen as counterproductive. They can provide momentum and positive energy for the entire project's successful completion.

Part 2

Data Structures and Data Modeling

Introduction

Part 2 deals with data structures and data models of the data. Data are stored in the source systems in a data model, which can range from a simple flat table data model to a complex relational model. Based on these data from the source system, an analysis data mart is built.

The structure of the analysis data mart depends on the definition of the analysis subject and the handling of repeated observations.

The possibilities for the technical source of our data will be discussed in *Chapter 5 – The Origin of Data*, where we will consider data sources by their technical platform and business application.

In *Chapter 6 – Data Models,* we will provide an introduction to data models such as the relational data model and the star schema and show the possibilities of the graphical representation of data models with an entity relationship diagram.

In *Chapter 7 – Analysis Subjects and Multiple Observations,* the influence of the analysis subject and multiple-observations-per-analysis subject will be discussed.

Chapters 8, 9, and 10 will go into detail about analytic data structures.

In *Chapter 8 – The One-Row-per-Subject Data Mart,* we will discuss the properties of the one-row-per-subject paradigm and the resulting data table.

In *Chapter 9 – The Multiple-Rows-per-Subject Data Mart,* we will show examples and properties of tables that can have more than one row per subject.

In *Chapter 10 – Data Structures for Longitudinal Analysis,* we will cover application of data structures in time series analysis.

In *Chapter 11 – Considerations for Data Marts,* will discuss the properties of variables in data marts, such as their measurement scale, type, and role in the analysis process.

Finally, in *Chapter 12 – Considerations for Predictive Modeling,* we will look at selected properties of data marts in predictive modeling such as target windows and overfitting.

Chapter 5

The Origin of Data

5.1 Introduction

In *Chapter 3 – Characteristics of Data Sources*, we started classifying data sources in respect to their relationship to the analytic business question and identified properties that are important for analytic data preparation.

In this chapter, we will deal with the origin of data that we will use in analytic data preparation. We will do this from a technical viewpoint where we separate data sources by their technical platform and storage format.

5.2 Data Origin from a Technical Point of View

Data that are used to build an analysis table can have different technical origins. The options range from manual data entry directly into the analysis table to a complex extraction from a hierarchical database. In this section we will look at the most common technical data origins for analysis data.

They include the following:

- directly entered data
- simple text files or spreadsheets
- relational database systems
- Enterprise Resource Planning (ERP) systems
- hierarchical databases
- large text files

These data sources can overlap. For example, data in an Enterprise Resource Planning system can be accessed directly at the application layer, the data layer where the underlying database is accessed, or from a table dump into a large text file.

5.3 Application Layer and Data Layer

In an *application layer* data are accessed from a system using its business logic and application functions. When data are accessed at the *data layer*, data are directly imported from the underlying tables.

In some systems accessing data for an entity called CUSTOMER ORDER can involve joining a number of tables on the data level. Accessing this information on the application level can use the functions that are built into the application to combine this information. These functions can use the metadata and the business data dictionary of the respective application.

5.4 Simple Text Files or Spreadsheets

Simple text files or spreadsheets hold data in a rectangular form (subjects in the rows, variables in the columns). These files are simple in that they have around 100 rows rather than 100,000 rows and a few columns. We call them *simple* here and differentiate them from large text files, which we will discuss later in this chapter.

The technical formats of simple text files or spreadsheets include delimited text files, comma-separated files, Microsoft Excel and Microsoft Access tables, or other formats that frequently occur on personal computers.

Data from these data sources are frequently used in small or simple analyses, where a researcher enters data on his PC and transfers or imports these data to SAS. External data such as geo-demographics from data providers in marketing, lookup tables such as a list of product codes and product names, or randomization lists in clinical trials are also frequently imported from this type of data source.

This type of data source is very widespread, and it can be created and opened on nearly any personal computer.

Text files and spreadsheets allow users to enter data however they want. The rules for data structures are not set by the software itself, but have to be considered during the data entry process.

5.5 Relational Database Systems

In a relational database system, data are organized and accessed according to the relationships between data items. Relational database systems store data in separate tables for each entity. Entities can be thought of as the objects for which data are stored. In a system that processes customer data where customers can place orders, the entities will be CUSTOMER and ORDER. The relationship between these tables is represented by special columns in the tables—the primary key and foreign key columns.

Data that are accessed from relational database systems are usually accessed table by table, which afterward are merged together corresponding to the primary and foreign keys, or they are accessed from a view in the database that already represents the merge of these tables. The relationships between tables are also called the relational model or relational data model, which we will discuss in the next section.

Structured Query Language (SQL) is used for data definition, data management, and data access and retrieval in relational database systems. This language contains elements for the selection of columns and subsetting of rows, aggregations, and joins of tables.

Strictly speaking, SQL refers to a specific collection of data, but it is often used synonymously with the software that is used to manage that collection of data. That software is more correctly called a relational database management system, or RDBMS.

Relational database systems also provide tables with metadata on tables and columns. Leading relational database systems include Oracle, Microsoft SQL Server, and DB2. Microsoft Access does not provide the same database administration functions as a relational database system, but it can be mentioned here because data can be stored in a relational model and table views can be defined.

A data warehouse, which we discussed in *Chapter 3 – Characteristics of Data Sources*, is also relational from a logical point of view. The storage platform for a data warehouse can be SAS or a relational database management system.

5.6 Enterprise Resource Planning Systems

Enterprise Resource Planning (ERP) systems support business processes by providing them with application functionality. Typical functional areas of ERP systems include materials management, finance and accounting, controlling, human resource management, research and development, sales and marketing, and master data administration.

The functionality of ERP systems is tightly linked to the business process. They are typically operational systems as we discussed in Chapter 3. Their data model is also optimized for the operative process and not for information retrieval. In the background, most of these systems store their data within a relational database management system.

The complexity of the data model, however, makes it difficult to retrieve data directly from the relational database. The access to information is mostly done on the application layer. Here the data are retrieved by using certain programmatic functions of the ERP system. In this case the ERP system provides functions that retrieve and join data from several tables and translates technical table and variable names into names with business meaning.

5.7 Hierarchical Databases

Hierarchical databases are based on an information model that was created in the 1960s. Only in the '70s did relational databases start to evolve. Still, hierarchical databases continue to play a role. They are frequently found on host platforms in large insurance companies, banks, or public and governmental organizations.

Data retrieval from hierarchical databases is not as easy as data access from a relational database system because the underlying data are not stored in rectangular tables that can be joined by a primary key. Data access does not follow table relationships; rather it follows so-called access hierarchies. Host programming and detailed knowledge of the respective hierarchical data structure are needed to decode the data files and to retrieve data.

The output dump of a hierarchical database is characterized by the fact that rows can belong to different data objects and can therefore have different data structures. E.g., customer data can be followed by account data for that customer (see section 13.5 for an illustration).

5.8 Large Text Files

Different from simple text files are *large text files*, which represent either the dump of a table from a relational database system with thousands of rows or an output structure from a hierarchical database system that needs to be logically decoded. Log file data from Web or application servers are also usually output as large text files.

Transferring and importing text files is very common and typically fast and practical. Usually an additional file describing the variable's formats, lengths, and positions in a table are delivered. For data import, a data collector has to be defined that reads the data from the correct positions in the text file.

In contrast to a relational database where additional or new columns can easily be imported without additional definition work, every change of the structure of the text file needs a respective change in the definition of the data collector.

5.9 Where Should Data Be Accessed From?

After listing the possible data sources from a technical point of view, we see that the same data can be made available via different technical methods.

- Data from an operational system can be accessed directly via the application itself, from the underlying relational database, or via a dump into a text file.

- External data can be loaded directly from a spreadsheet, or they can already be in a data warehouse where they can be accessed from the relational table.

Selecting the technical method for providing data for an analysis is influenced by the following criteria:

- In the case of a one-shot analysis a manual output dump from database data can be sufficient. For re-run analyses, however, it has to be certain that the data in the same format can be provided on a regular basis.

- Authorization policies sometimes do not allow the direct access of data from a system itself, even if it would be technically possible. In these cases the system that holds data usually exports its data to another source, text file, or relational database, where they can be retrieved for analysis.

- Performance considerations can also play a role in the selection of the data delivery method.

Now that we have seen the technical origin of data, we will look at possible data models and data structures of the input data in the next chapter.

C h a p t e r **6**

Data Models

6.1 Introduction

The source systems, which we discussed in *Chapter 5 – The Origin of Data*, and which we will use to access data, store the data in a certain data model. Except in the case of a flat table, where we have only a single table to import, in most situations we will incorporate a number of tables. These tables have a certain relationship to each other, which is described by a data model.

We will consider the possible data models that are frequently encountered, namely the flat table, the relational model, and the star schema. We will go into detail about entity relationship diagrams and will look at logical and physical data models.

This chapter builds a bridge between the input data sources that come into our analytic data preparation environment and the final analytic data table.

6.2 Relational Model and Entity Relationship Diagrams

The relational model is the most common database model. It follows the paradigm that the business context for which data are stored is represented by entities and *relationships*. When describing a business context in sentences, the nouns usually represent entities, and the verbs represent relationships.

The business fact "One customer can have one or more accounts/One account is owned by exactly one customer" results in two entities, CUSTOMER and ACCOUNT, and the relationship between them is "can have"/"is owned." We also say that CUSTOMER and ACCOUNT have a one-to-many relationship because one customer can have many accounts. The restrictions "one or more" and "exactly one" represent cardinalities in the relationships.

Cardinalities are important for business applications, which in our preceding example have to check that an account without a customer can't be created. In data preparation for analytics cardinalities form the basis of data quality checks. Different from business applications, which check analytics for each transaction when an account is opened, in data quality checks for analytics, we check these rules on the entire table and create a list of non-conformities.

The visual representation of a relational model is called an *entity relationship diagram*. The following is an entity relationship diagram for this example.

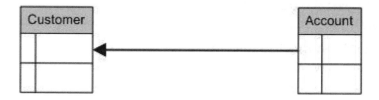

We see that each entity is represented by a rectangle, and the relationship is represented by an arrow. In our diagram in a one-to-many relationship, the direction of the arrow points from the 'many' to the 'one' entity. This relationship is sometimes also called a parent-child relationship. We will use this term in the book. Note that there are different conventions for the representation of relationships, their direction, and their cardinality.

6.3 Logical versus Physical Data Model

In our description so far, we have not distinguished between logical and physical data models. The logical data model represents the logical relationships between entities. Entities are described by attributes. The logical data model does not describe which attribute is used to represent a certain relationship.

In the physical data model entities are converted to tables, and attributes are converted to columns. Relationships are represented by the appropriate columns or additional table.

- One-to-many relationships are represented by a foreign key column. In this case the 'many' table holds the primary key of the 'one' table as a foreign key. In our preceding example the 'many' table ACCOUNT holds the customer ID for the corresponding customer.

- A many-to-many relationship requires its own table that resolves the relationship to two one-to-many relationships. An example of a many-to-many relationship in the preceding context is that not only can one customer have one account, but also one account can be owned by one or more customers. In this case we need a separate table that resolves the many-to-many relationship between customers and accounts by holding the combinations of CustID and AccountID with one row for each combination.

The following diagram shows the physical data model for our customer-account example.

We see that column CustID is stored in the ACCOUNT table as a foreign key. These columns are important because they are needed to join the tables together in order to create an analysis table.

6.4 Star Schema

A *star schema* is a special form of a relational data structure. Star schemas are composed of one fact table and a number of dimension tables. The dimension tables are linked to the fact table with a one-to-many relationship. In the following example of a star schema in retail, the fact table is the central table, holding foreign keys for each of the dimensions.

The fact table POINTOFSALE holds the sales data per DATE, CUSTOMER, PRODUCT, and PROMOTION.

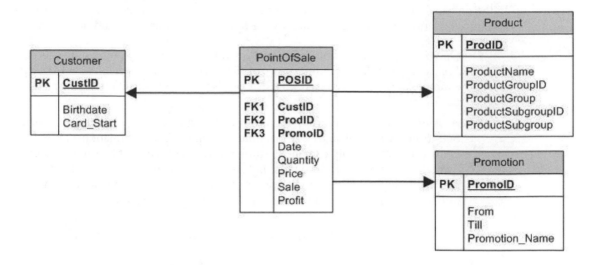

It is apparent that the name star schema comes from its shape. This type of data storage is efficient for querying and reporting tasks. For example, the purchases of a certain customer in a certain period of time can easily be queried, as well as a report per branches, showing the monthly purchase amount for each article group.

Note that more elaborate data models such as the snowflake schema exist. We will, however, leave details about this to the data modeling literature for data warehouses.

Star schemas are very common in data warehouse environments, especially for reporting. Multidimensional data structures, so-called cubes, are built for OLAP reporting, mostly on the basis of the star schema. In data warehouses, star schemas are also important for the historization of attributes of certain points in time.

For analytics, however, this structure needs to be converted into one single table. For example, if we want to generate a table that shows the purchase behavior per customer, we will use the dimension table CUSTOMER as its basis and include all relevant attributes that belong directly to the customer. Then we will start aggregating or merging data from the POINTOFSALE table, potentially by subgrouping per PROMOTION, PRODUCTGROUP, and TIME to the CUSTOMER table (derived from DATE in POINTOFSALE).

In *Chapter 7 – Analysis Subjects and Multiple Observations*, we will give an overview of possible data structures with repeated observations per subject. In Part 3 of this book, we will show which derived variables can be created in various data structures. And in *Chapter 26 – Case Study 2— Deriving Customer Segmentation Measures from Transactional Data*, we will show a comprehensive example of the star schema data model.

6.5 Normalization and De-normalization

General

In our preceding example for customers and accounts we saw that one customer with many accounts is represented by a CUSTOMER table and an ACCOUNT table.

- The CUSTOMER table holds a unique customer identifier and all relevant information that is directly related to the customer.

- The ACCOUNT table holds a unique account identifier, all information about accounts, and the customer key.

- The unique account identifier in the account table and the unique customer identifier in the customer table are called *primary key*s. The customer identifier in the account table denotes which customer the account belongs to and is called the *foreign key*.

The process of combining information from several tables based on the relationships expressed by primary and foreign keys is called *joining* or *merging*.

Normalization

In a *normalized relational model* no variables, aside from primary and foreign keys, are duplicated among tables. Each piece of information is stored only once in a dedicated table. In data modeling theory this is called the second normal form. Additional normal forms exist, such as third normal form and the Boyce Codd normal form, but we will leave details about these to the data modeling theory literature and will not discuss them in this book.

Normalization is important for transactional systems. The rationale is that certain information is stored in a single table only, so that updates on data are done in only one table. These data are stored without redundancy.

De-normalization

The opposite of normalization is de-normalization. *De-normalization* means that information is redundantly stored in the tables. This means that the same column appears in more than one table. In the case of a one-to-many relationship, this leads to the fact that values are repeated.

There are two reasons for de-normalization of data:

- De-normalization is necessary for analytics. All data must be merged together into one single table.

- De-normalization can be useful for performance and simple handling of data. In reporting, for example, it is more convenient for the business user if data are already merged together. For performance reasons, in an operational system, a column might be stored in de-normalized form in another table in order to reduce the number of table merges.

Example

We return to our preceding example and see the content of the CUSTOMER and ACCOUNT tables.

Table 6.1: Content of CUSTOMER table

CustID	Birthdate	Gender
1	16.05.1970	Male
2	19.04.1964	Female

Table 6.2: Content of ACCOUNT table

AccountID	CustID	Type	OpenDate
1	1	Checking	05.12.1999
2	1	Savings	12.02.2001
3	2	Savings	01.01.2002
4	2	Checking	20.10.2003
5	2	Savings	30.9.2004

Tables 6-1 and 6-2 represent the normalized version. Besides column CustID, which serves as a foreign key in the ACCOUNT table, no column is repeated.

Merging these two tables together creates the de-normalized CUSTOMER_ACCOUNT table.

Table 6.3: Content of the de-normalized CUSTOMER_ACCOUNT table

CustID	Birthdate	Gender	AccountID	Type	OpenDate
1	16.05.1970	Male	1	Checking	05.12.1999
1	16.05.1970	Male	2	Savings	12.02.2001
2	19.04.1964	Female	3	Savings	01.01.2002
2	19.04.1964	Female	4	Checking	20.10.2003
2	19.04.1964	Female	5	Savings	30.09.2004

In the de-normalized version, the variables BIRTHDATE and GENDER appear multiple times per customer. This version of the data can directly be used for analysis of customers and accounts because all information is stored in one table.

In the remaining chapters in Part 2, we will examine how to efficiently create de-normalized data tables for statistical analysis. The process of moving from a relational data structure to a single table is also often called the de-normalizing or flat-making of a table.

Chapter 7

Analysis Subjects and Multiple Observations

7.1 Introduction

In *Chapter 5 – The Origin of Data*, we explored possible data sources from a technical point of view, and in *Chapter 6 – Data Models*, we discussed the data models and data structures that we might encounter when accessing data for analytics.

In this chapter we will cover the basic structure of our analysis table.

In the following sections we will look at two central elements of analytic data structures:

- the identification and definition of the analysis subject
- the determination of whether multiple observations per analysis subject exist and how they will be handled

Finally, we will also see that in some analysis tables, individual analysis subjects are not present, but aggregates of these subjects are analyzed.

7.2 Analysis Subject

Definition

Analysis subjects are entities that are being analyzed, and the analysis results are interpreted in their context. Analysis subjects are therefore the basis for the structure of our analysis tables.

The following are examples of analysis subjects:

- Persons: Depending on the domain of the analysis, the analysis subjects have more specific names such as patients in medical statistics, customers in marketing analytics, or applicants in credit scoring.

- Animals: Piglets, for example, are analyzed in feeding experiments; rats are analyzed in pharmaceutical experiments.

- Parts of the body system: In medical research analysis subjects can also be parts of the body system such as arms (the left arm compared to the right arm), shoulders, or hips. Note that from a statistical point of view, the validity of the assumptions of the respective analysis methods has to be checked if dependent observations per person are used in the analysis.

- Things: Such as cash machines in cash demand prediction, cars in quality control in the automotive industry, or products in product analysis.

- Legal entities: Such as companies, contracts, accounts, and applications.

- Regions or plots in agricultural studies, or reservoirs in the maturity prediction of fields in the oil and gas industry.

Analysis subjects are the heart of each analysis because their attributes are measured, processed, and analyzed. In deductive (inferential) statistics the features of the analysis subjects in the sample are used to infer the properties of the analysis subjects of the population. Note that we use feature and attribute interchangeably here.

Representation in the data set

When we look at the analysis table that we want to create for our analysis, the analysis subjects are represented by rows, and the features that are measured per analysis subject are represented by columns. See Table 7.1 for an illustration.

Table 7.1: Results of ergonometric examinations for 21 runners

	PersonNr	Age in years	Weight in kg	Oxygen consumption	Min. to run 1.5 miles	Heart rate while resting	Heart rate while running	Maximum heart rate	Experimental group
1	1	44	89.47	44.609	11.37	62	178	182	2
2	2	40	75.07	45.313	10.07	62	185	185	2
3	3	44	85.84	54.297	8.65	45	156	168	2
4	4	42	68.15	59.571	8.17	40	166	172	2
5	5	38	89.02	49.874	9.22	55	178	180	2
6	6	47	77.45	44.811	11.63	58	176	176	2
7	7	40	75.98	45.681	11.95	70	176	180	2
8	8	43	81.19	49.091	10.85	64	162	170	2
9	9	44	81.42	39.442	13.08	63	174	176	2
10	10	38	81.87	60.055	8.63	48	170	186	2
11	11	44	73.03	50.541	10.13	45	168	168	2
12	12	45	87.66	37.388	14.03	56	186	192	1
13	13	45	66.45	44.754	11.12	51	176	176	1
14	14	47	79.15	47.273	10.6	47	162	164	1
15	15	54	83.12	51.855	10.33	50	166	170	1
16	16	49	81.42	49.156	8.95	44	180	185	1
17	17	51	69.63	40.836	10.95	57	168	172	1
18	18	51	77.91	46.672	10	48	162	168	1
19	19	48	91.63	46.774	10.25	48	162	164	1
20	20	49	73.37	50.388	10.08	67	168	168	1
21	21	57	73.37	39.407	12.63	58	174	176	1

In this table 21 runners have been examined, and each one is represented by one row in the analysis table. Features such as age, weight, and runtime, have been measured for each runner, and each feature is represented by a single column. Analyses, such as calculating the mean age of our population or comparing the runtime between experimental group 1 and 2, can directly start from this table.

Analysis subject identifier

A column PersonNr has been added to the table to identify the runner. Even if it is not used for analysis, the presence of an ID variable for the analysis subjects is important for the following reasons:

- data verifications and plausibility checks, if the original data in database queries or data forms have to be consulted
- the identification of the analysis subject if additional data per subject has to be added to the table
- if we work on samples and want to refer to the sampled analysis subject in the population

Also note that in some cases it is illegal, and in general it is against good analysis practice to add people's names, addresses, social security numbers, and phone numbers to analysis tables. The statistician is interested in data on analysis subjects, not in the personal identification of analysis subjects. If an anonymous subject number is not available, a surrogate key with an arbitrary numbering system has to be created for both the original data and the analysis data. The statistician in that case receives only the anonymous analysis data.

7.3 Multiple Observations

General

The analysis table in Table 7.1 is simple in that we have only one observation per analysis subject. It is therefore straightforward to structure the analysis table in this way.

There are, however, many cases where the situation becomes more complex; namely, when we have multiple observations per analysis subject.

Examples

- In the preceding example we will have multiple observations when each runner does more than one run, such as a second run after taking an isotonic drink.
- A dermatological study in medical research where different creams are applied to different areas of the skin.
- Evaluation of clinical parameters before and after surgery.
- An insurance customer with insurance contracts for auto, home, and life insurance.
- A mobile phone customer with his monthly aggregated usage data for the last 24 months.
- A daily time series of overnight stays for each hotel.

In general there are two reasons why multiple observations per analysis subject can exist:

- repeated measurements over time
- multiple observations because of hierarchical relationships

We will now investigate the properties of these two types in more detail.

7.3.1 Repeated Measurements over Time

Repeated measurements over time are obviously characterized by the fact that for the same analysis subject the observation is repeated over time. From a data model point of view, this means that we have a one-to-many relationship between the analysis subject entity and a time-related entity.

Note that we are using the term repeated measurement where observations are recorded repeatedly. We are not necessarily talking about measurements in the sense of numeric variables per observation of the same analysis subject—only the presence or absence of an attribute (yes or no) would be noted on each of X occasions.

The simplest form of repeated measurements is the two-observations-per-subject case. This case happens most often when comparing observations before and after a certain event and we are interested in the difference or change in certain criteria (pre-test and post-test). Examples of such an event include the following:

- giving a certain treatment or medication to patients
- execution of a marketing campaign to promote a certain product

If we have two or more repetitions of the measurement, we will get a measurement history or a time series of measurements:

- Patients in a clinical trial make quarterly visits to the medical center where laboratory values and vital signs values are collected. A series of measurement data such as the systolic and diastolic blood pressure can be analyzed over time.

- The number and duration of phone calls of telecommunications customers are available on a weekly aggregated basis.

- The monthly aggregated purchase history for retail customers.

- The weekly total amount of purchases using a credit card.

- The monthly list of bank branches visited by a customer.

The fact that we do not have only multiple observations per analysis subject, but ordered repeated observations, allows us to analyze their course over time such as by looking at trends. In Chapters 18–20 we will explore in detail how this information can be described and retrieved per analysis subject.

7.3.2 Multiple Observations because of Hierarchical Relationships

If we have multiple observations for an analysis subject because the subject has logically related child hierarchies, we call this *multiple observations because of hierarchical relationships*. The relation to the entity relationship diagrams that we introduced in Chapter 6 is that here we have so-called one-to-many relationships between the analysis subject and its child hierarchy. The following are examples:

- One insurance customer can have several types of insurance contracts (auto insurance, home insurance, life insurance). He can also have several contracts of the same type, e.g., if he has more than one car.

- A telecommunications customer can have several contracts; for each contract, one or more lines can be subscribed. (In this case we have a one-to-many relationship between the customer and contract and another one-to-many relationship between the contract and the line.)

- In one household, one or more persons can each have several credit cards.
- Per patient both eyes are investigated in an ophthalmological study.
- A patient can undergo several different examinations (laboratory, x-ray, vital signs) during one visit.

7.3.3 Multiple Observations Resulting from Combined Reasons

Multiple observations per analysis subject can also occur as a result of a combination of repeated measures over time and multiple observations because of hierarchical relationships:

- Customers can have different account types such as a savings account and a checking account. And for each account a transaction history is available.
- Patients can have visits at different times. At each visit, data from different examinations are collected.

If from a business point of view a data mart based on these relationships is needed, data preparation gets more complex, but the principles that we will see in "Data Mart Structures" remains the same.

7.3.4 Redefinition of the Analysis Subject Level

In some cases, the question arises whether we have useful multiple observations per analysis subject or whether we have to (are able to) redefine the analysis subject. Redefining the analysis subject means that we move from a certain analysis subject level, such as patient, to a more detailed one, such as shoulder.

The problem of redefining the analysis subject is that we then have dependent measures that might violate the assumptions of certain analysis methods. Think of a dermatological study where the effect of different creams applied to the same patient can depend on the skin type and are therefore not independent of each other. The decision about a redefinition of the analysis subject level requires domain-specific knowledge and a consideration of the statistically appropriate analysis method.

Besides the statistical correctness, the determination of the correct analysis subject level also depends on the business rationale of the analysis. The decision whether to model telecommunication customers on the customer or on the contract level depends on whether marketing campaigns or sales actions are planned and executed on the customer or contract level.

Note that so far we have only identified the fact that multiple observations can exist and the causal origins. We have not investigated how they can be considered in the structure of the analysis table. We will do this in the following section.

7.4 Data Mart Structures

In order to decide how to structure the analysis table for multiple observations, we will introduce the two most important structures for an analysis table, the one-row-per-subject data mart and the multiple-rows-per-subject data mart. In Chapters 8 and 9, respectively, we will discuss their properties, requirements, and handling of multiple observations in more detail.

7.4.1 One-Row-per-Subject Data Mart

In the one-row-per-subject data mart, all information per analysis subject is represented by one row. Features per analysis subject are represented by a column. When we have no multiple observations per analysis subject, the creation of this type of data mart is straightforward—the value of each variable that is measured per analysis subject is represented in the corresponding column. We saw this in the first diagram in Chapter 6. The one-row-per-subject data mart usually has only one ID variable, namely that of identifying the subjects.

Table 7.2: Content of CUSTOMER table

CustID	Birthdate	Gender
1	16.05.1970	Male
2	19.04.1964	Female

Table 7.3: Content of ACCOUNT table

AccountID	CustID	Type	OpenDate
1	1	Checking	05.12.1999
2	1	Savings	12.02.2001
3	2	Savings	01.01.2002
4	2	Checking	20.10.2003
5	2	Savings	30.09.2004

In the case of the presence of multiple observations per analysis subject, we have to represent them in additional columns. Because we are creating a one-row-per-subject data mart, we cannot create additional rows per analysis subject. See the following example.

Table 7.4: One-row-per-subject data mart for multiple observations

CustID	Birthdate	Gender	Number of Accounts	Proportion of Checking Accounts	Opendate of oldest account
1	16.05.1970	Male	2	50 %	05.12.1999
2	19.04.1964	Female	3	33 %	01.01.2002

Table 7.4 is the one-row-per-subject representation of Tables 7.2 and 7.3. We see that we have only two rows because we have only two customers. The variables from the CUSTOMER table have simply been copied to the table. When aggregating data from the ACCOUNT table, however, we experience a loss of information. We will discuss that in Chapter 8. Information from the underlying hierarchy of the ACCOUNT table has been aggregated to the customer level by completing the following tasks:

- counting the number of accounts per customer
- calculating the proportion of checking accounts
- identifying the open date of the oldest account

We have used simple statistics on the variables of ACCOUNT in order to aggregate the data per subject. More details about bringing all information into one row will be discussed in detail in *Chapter 8 – The One-Row-per-Subject Data Mart*.

7.4.2 The Multiple-Rows-per-Subject Data Mart

In contrast to the one-row-per-subject data mart, one subject can have multiple rows. Therefore, we need one ID variable that identifies the analysis subject and a second ID variable that identifies multiple observations for each subject. In terms of data modeling we have the child table of a one-to-many relationship with the foreign key of its master entity. If we also have information about the analysis subject itself, we have to repeat this with every observation for the analysis subject. This is also called de-normalizing.

- In the case of multiple observations because of hierarchical relationships, ID variables are needed for the analysis subject and the entities of the underlying hierarchy. See the following example of a multiple-rows-per-subject data mart. We have an ID variable CUSTID for CUSTOMER and an ID variable for the underlying hierarchy of the ACCOUNT table. Variables of the analysis subject such as birth date and gender are repeated with each account. In this case we have a de-normalized table as we explained in Chapter 6.

Table 7.5: Multiple-rows-per-subject data mart as a join of the CUSTOMER and ACCOUNT tables

CustID	Birthdate	Gender	AccountID	Type	OpenDate
1	16.05.1970	Male	1	Checking	05.12.1999
1	16.05.1970	Male	2	Savings	12.02.2001
2	19.04.1964	Female	3	Savings	01.01.2002
2	19.04.1964	Female	4	Checking	20.10.2003
2	19.04.1964	Female	5	Savings	30.09.2004

- In the case of repeated observations over time the repetitions can be enumerated by a measurement variable such as a time variable or, if we measure the repetitions only on an ordinal scale, by a sequence number. See the following example with PATNR as the ID variable for the analysis subject PATIENT. The values of CENTER and TREATMENT are repeated per patient because of the repeated measurements of CHOLESTEROL and TRIGLYCERIDE at each VISITDATE.

Table 7.6: Multiple-rows-per-subject data mart as a join of the CUSTOMER and ACCOUNT tables

PATNR	CENTER	TREATMENT	MEASUREMENT	VISITDATE	CHOLESTEROL	TRIGLYCERIDE
1	VIENNA	A	1	15JAN2002	220	220
1	VIENNA	A	2	20JUL2002	216	216
1	VIENNA	A	3	07JAN2002	205	205
2	SALZBURG	B	1	15APR2001	308	308
2	SALZBURG	B	2	01OCT2001	320	320

7.4.3 Summary of Data Mart Types

Table 7.7 summarizes how different data mart structures can be created, depending on the structure of the source data.

Table 7.7: Data mart types

	Data mart structure that is needed for the analysis	
Structure of the source data: "Multiple observations per analysis subject exist?"	**One-row-per-subject data mart**	**Multiple-rows-per-subject data mart**
NO	Data mart with one row per subject is created.	(Key-value table can be created.)
YES	Information of multiple observations has to be aggregated per analysis subject (see also Chapter 8).	Data mart with one-row-per-multiple observations is created. Variables at the analysis subject level are duplicated for each repetition (see also Chapter 9).

7.4.4 Using Both Data Mart Structures

There are analyses where data need to be prepared in both versions: the one-row-per-subject data mart and the multiple-rows-per subject data mart.

Consider the case where we have measurements of credit card activity per customer on a monthly basis.

- In order to do a segmentation analysis or prediction analysis on customer level we need the data in the form of a one-row-per-subject data mart.
- In order to analyze the course of the monthly transaction sum per customer over time, however, we need to prepare a multiple-rows-per-analysis subject data mart. In this data mart we will create line plots and calculated trends.
- The visual results of the line plots are input for the analyst to calculate derived variables for the one-row-per-subject data mart. Trends in the form of regression coefficients on the analysis subject level are calculated on the basis of the multiple-rows-per-subject data mart. These coefficients are then added to the one-row-per-subject data mart.

In the next two chapters we will take a closer look at the properties of one- and multiple-rows-per subject data marts. In *Chapter 14 – Transposing One- and Multiple-Rows-per-Subject Data Structures*, we will see how we can switch between different data mart structures.

7.5 No Analysis Subject Available?

General

In the preceding sections we dealt with cases where we were able to identify an analysis subject. We saw data tables where data for patients or customers were stored.

There are, however, analysis tables where we do not have an explicit analysis subject. Consider an example where we have aggregated data on a monthly level—for example, the number of airline travelers, which can be found in the SASHELP.AIR data set. This table is obviously an analysis table, which can be used directly for time series analysis. We do not, however, find an analysis subject in our preceding definition of it.

	DATE	international airline travel (thousands)
1	JAN49	112
2	FEB49	118
3	MAR49	132
4	APR49	129
5	MAY49	121
6	JUN49	135
7	JUL49	148
8	AUG49	148
9	SEP49	136
10	OCT49	119
11	NOV49	104
12	DEC49	118

We, therefore, have to refine the definition of analysis subjects and multiple observations. It is possible that in analysis tables we consider data on a level where information of analysis subjects is aggregated. The types of aggregations are in most cases counts, sums, or means.

Example

We will look at an example from the leisure industry. We want to analyze the number of overnight stays in Vienna hotels. Consider the following three analysis tables:

- Table that contains the monthly number of overnight stays per HOTEL
- Table that contains the monthly number of overnight stays per CATEGORY (5 stars, 4 stars ...)
- Table that contains the monthly number of overnight stays in VIENNA IN TOTAL

The first table is a typical multiple-rows-per-subject table with a line for each hotel and month. In the second table we have lost our analysis subjects because we have aggregated, or summed, over them. The hotel category could now serve as a new "analysis subject," but it is more an analysis level than an analysis subject. Finally, the overnight stays in Vienna in total are on the virtual analysis level 'ALL,' and we have only one analysis subject, 'VIENNA'.

Longitudinal Data Structures

We have seen that as soon we start aggregating over an analysis subject, we come to analysis levels where the definition of an analysis subject is not possible or does not make sense. The aggregated information per category is often considered over time, so we have multiple observations per category over time.

These categories can be an aggregation level. We have seen the example of hotel categorization as 5 stars, 4 stars, and so on. The category at the highest level can also be called the ALL group, similar to a grand total.

Note that categorizations do not need to necessarily be hierarchical. They can also be two alternative categorizations such as hotel classification, as we saw earlier, and regional district in the preceding example. This requires that the analysis subject HOTEL have the properties classification and region, which allow aggregation of their number of overnight stays.

Data structures where aggregated data over time are represented either for categories or the ALL level are called *longitudinal data structures* or *longitudinal data marts*.

Strictly speaking the multiple-rows-per-subject data mart with repeated observations over time per analysis subject can also be considered a longitudinal data mart. The analytical methods that are applied to these data marts do not differ. The only difference is that in the case of the multiple-rows-per-subject data mart we have dependent observations per analysis subject, and in the case of longitudinal data structures we do not have an analysis subject in the classic sense.

C h a p t e r 8

The One-Row-per-Subject Data Mart

8.1 Introduction

In this chapter we will concentrate on the one-row-per-subject data mart. This type of data mart is frequently found in classic statistical analysis. The majority of data mining marts are of this structure. In this chapter, we will work out general properties, prerequisites, and tasks of creating a one-row-per-subject data mart. In Chapters 18–20 we will show how this type of data mart can be filled from various data sources.

8.2 The One-Row-per-Subject Paradigm

8.2.1 Importance and Frequent Occurrence of This Type of Data Mart

The one-row-per-subject data mart is very important for statistical and data mining analyses. Many business questions are answered on the basis of a data mart of this type. See the following list for business examples:

- Prediction of events such as the following:
  - Campaign response
  - Contract renewal
  - Contract cancellation (attrition or churn)
  - Insurance claim
  - Default on credit or loan payback
  - Identification of risk factors for a certain disease

- Prediction of values such as the following:
  - Amount of loan that can be paid back
  - Customer turnover and profit for the next period
  - Time intervals (e.g., remaining lifetime of customer or patient)
  - Time until next purchase
  - Value of next insurance claim

- Segmentation and clustering of customers
  - Customer segmentation
  - Clustering of text documents
  - Clustering of counties based on sociodemographic parameters

The following analytical methods require a one-row-per-subject data mart:

- regression analysis
- analysis-of-variance (ANOVA)
- neural networks
- decision trees
- survival analysis
- cluster analysis
- principal components analysis and factor analysis

8.2.2 Putting All Information into One Row

The preceding application examples coupled with the underlying algorithm require that all the data source information be aggregated into one row. Central to this type of data mart is the fact that we have to put all information per analysis subject into one row. We will, therefore, call this the one-row-per-subject data mart paradigm. Multiple observations per analysis subject must not appear in additional rows; they have to be converted over into additional columns of the single row.

Obviously, putting all information into one row is very simple if we have no multiple observations per analysis subject. The values are simply read into the analysis data mart and derived variables are added.

If we have multiple observations per analysis subject it is more effort to put all the information into one row. Here we have to cover two aspects:

- the technical aspect of exactly how multiple-rows-per-subject data can be converted into one-row-per-subject data
- the business aspect—which aggregations, derived variables—make the most sense to condense the information from multiple rows into columns of the single row

The process of taking data from tables that have one-to-many relationships and putting them into a rectangular one-row-per-subject analysis table has many names: transposing, de-normalizing, "making flat," and pivoting, among others.

Table 8.1 illustrates the creation of a one-row-per-subject data mart in terms of an entity relationship diagram.

Table 8.1: Creation of a one-row-per-subject data mart

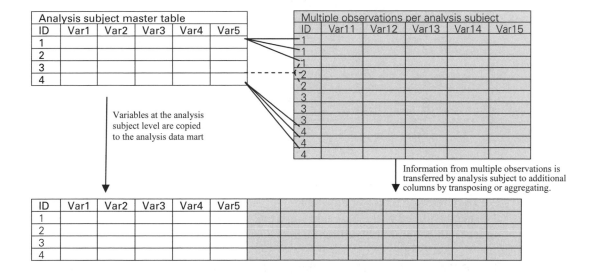

8.3 The Technical Point of View

From a technical point of view we can solve the task of putting all information into one row by using two main techniques:

- Transposing: Here we transpose the multiple rows per subject into columns. This technique can be considered the "pure" way because as we take all data from the rows and represent them in columns.

- Aggregating: Here we aggregate the information from the columns into an aggregated value per analysis subject. We perform information reduction by trying to express the content of the original data in descriptive measures that are derived from the original data.

Transposing

We look at a very simple example with three multiple observations per subject.

Table 8.2: Base table with static information per patient (= analysis subject)

PatNr	Gender
1	Male
2	Female
3	Male

And we have a table with repeated measurements per patient.

Table 8.3: Table with multiple observations per patient (= analysis subject)

PatNr	Measurement	Cholesterol
1	1	212
1	2	220
1	3	240
2	1	150
2	2	145
2	3	148
3	1	301
3	2	280
3	3	275

We have to bring data from the table with multiple observations per patient into a form so that the data can be joined to the base table on a one-to-one basis. We therefore transpose Table 8.3 by patient number and bring all repeated elements into columns.

Table 8.4: Multiple observations in a table structure with one row per patient

PatNr	Cholesterol 1	Cholesterol 2	Cholesterol 3
1	212	220	240
2	150	145	148
3	301	280	275

This information is then joined with the base table and we retrieve our analysis data mart in the requested data structure.

Table 8.5: The final one-row-per-subject data mart

PatNr	Gender	Cholesterol 1	Cholesterol 2	Cholesterol 3
1	Male	212	220	240
2	Female	150	145	148
3	Male	301	280	275

Note that we have brought all information on a one-to-one basis from the multiple-rows-per-subject data set to the final data set by transposing the data. This data structure is suitable for analyses such as repeated measurements analysis of variance.

Aggregating

If we aggregate the information from the multiple-rows-per-subject data set, e.g., by using descriptive statistics such as the median, the minimum and maximum, or the interquartile range, we do not bring the original data on a one-to-one basis to the one-row-per-subject data set. Instead we analyze a condensed version of the data. The data might then look like Table 8.6.

Table 8.6: Aggregated data from tables 8.2 and 8.3

PatNr	Gender	Cholesterol_Mean	Cholesterol_Std
1	Male	224	14.4
2	Female	148	2.5
3	Male	285	13.8

We see that aggregations do not produce as many columns as transpositions because we condense the information. The forms of aggregations, however, have to be carefully selected because the omission of an important and, from a business point of view, relevant aggregation means that information is lost. There is no ultimate truth for the best selection aggregation measures over all business questions. Domain-specific knowledge is key when selecting them. In predictive analysis, for example, we try to create candidate predictors that reflect the properties of the underlying subject and its behavior as accurately as possible.

8.4 The Business Point of View: Transposing or Aggregating Original Data

In the preceding section we saw two ways of bringing multiple observation data into a one-row-per-subject structure: transposing and aggregating. The selection of the appropriate method depends on the business question and the analytical method.

We mentioned in the preceding section that there are analyses where all data for a subject are needed in columns, and no aggregation of observations makes sense. For example, the repeated measurement analysis of variance needs the repeated measurement values transposed into columns. In cases where we want to calculate derived variables, describing the course of a time series, e.g., the mean usage of months 1 to 3 before cancellation divided by the mean usage in months 4 to 6, the original values are needed in columns.

There are many situations where a one-to-one transfer of data is technically not possible or practical and where aggregations make much more sense. Let's look at the following questions:

- If the cholesterol values are expressed in columns, is this the information we need for further analysis?

- What will we do if we have not only three observations for one measurement variable per subject, but 100 repetitions for 50 variables? Transposing all these data would lead to 5,000 variables.

- Do we really need the data in the analysis data mart on a one-to-one basis from the original table, or do we more precisely need the information in concise aggregations?

- In predictive analysis, don't we need data that have good predictive power for the modeling target and that suitably describe the relationship between target and input variables, instead of the original data?

The answers to these questions will depend on the business objectives, the modeling technique(s), and the data itself. However, there are many cases, especially in data mining analysis, where clever aggregations of the data make much more sense than a one-to-one transposition. How to best aggregate the data requires business and domain knowledge and a good understanding of the data.

So we can see that putting all information into one row is not solely about data management. We will show in this book how data from multiple observations can be cleverly prepared to fit our analytic needs. Chapters 19–21 will extensively cover this topic. Here we will give an overview of methods that allow the extraction and representation of information from multiple observations per subject.

When aggregating measurement (quantitative) data for a one-row-per-subject data mart we can use the following:

- simple descriptive statistics, such as sum, minimum, or maximum

- frequency counts and number of distinct values

- measures of location or dispersion such as mean, median, standard deviation, quartiles, or special quantiles

- concentration measures on the basis of cumulative sums

- statistical measures describing trends and relationships such as regression coefficients or correlation coefficients

When analyzing multiple observations with categorical (qualitative) data that are aggregated per subject we can use the following:

- total counts

- frequency counts per category and percentages per category

- distinct counts

- concentration measures on the basis of cumulative frequencies

8.5 Hierarchies: Aggregating Up and Copying Down

General

In the case of repeated measurements over time we are usually aggregating this information by using the summary measurements described earlier. The aggregated values such as means or sums are then used at the subject level as a property of the corresponding subject.

In the case of multiple observations because of hierarchical relationships, we are also aggregating information from a lower hierarchical level to a higher one. But we can also have information at a higher hierarchical level that has to be available to lower hierarchical levels. It can also be the case that information has to be shared between members of the same hierarchy.

Example

Consider the example where we have the following relationships:

- In one HOUSEHOLD one or more CUSTOMERs reside.
- One CUSTOMER can have one or more CREDIT CARDs.
- On each CREDIT CARD, a number of TRANSACTIONs are associated.

For marketing purposes, a segmentation of customers will be performed; therefore, CUSTOMER is chosen as the appropriate analysis subject level. When looking at the hierarchies, we will aggregate the information from the entities CREDIT CARD and TRANSACTION to the CUSTOMER level and make the information that is stored on the HOUSEHOLD level available to the entity CUSTOMER.

We will now show how data from different hierarchical levels are made available to the analysis mart. Note that for didactic purposes, we reduce the number of variables to a few.

Information from the same level

At the customer level we have the following variables that will be made available directly to our analysis data mart.

- Customer birthday, from which the derived variable AGE is calculated
- Gender
- Customer term or duration, from which the basis of the duration of the customer relationship is calculated

Aggregating up

In our example, we will create the following variables on the CUSTOMER level that are aggregations from the lower levels, CREDIT CARD and TRANSACTION.

- Number of cards per customer
- Number of transactions over all cards for the last three months
- Sum of transaction values over all cards for the last three months
- Average transaction value for the last three months, calculated on the basis of the values over all cards

Copying down

At the HOUSEHOLD level the following information is available, which is copied down to the CUSTOMER level.

- Number of persons in the household
- Geographic region type of the household with the values RURAL, HIGHLY RURAL, URBAN, and HIGHLY URBAN.

Graphical representation

Table 8.7 shows the graphical representation in the form of an entity relationship diagram and the respective variable flows.

Table 8.7: Information flows between hierarchies

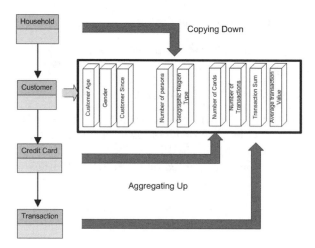

8.6 Conclusion

We see that information that is finally used at the analysis level CUSTOMER is retrieved from several levels. We used simple statistical aggregations to illustrate this. In Chapters 18–20 we will go into more detail about these processes and will also present their respective coding examples.

Chapter 9

The Multiple-Rows-per-Subject Data Mart

9.1 Introduction

As we saw in Table 7.6 of *Chapter 7 – Analysis Subjects and Multiple Observations*, the multiple-rows-per-subject data mart can be created only if we have multiple observations per subject. We also identified the two reasons for the multiple observations, namely, repeated measurements over time and multiple observations because of hierarchical relationships.

A one-to-many relationship exists between the analysis subject entity and the entity of the multiple observations. In the multiple-rows-per-subject data mart the number of observations is determined by the number of observations in the entity with the multiple observations. This is different from the one-row-per-analysis subject data mart, where the number of observations is determined by the number of observations in the table of the analysis subject.

The columns in the multiple-rows-per-subject data mart are derived from the table with the multiple observations. Different from the one-row-per-subject data mart, the columns from the table with the multiple observations can be copied directly to the multiple-rows-per-subject data mart.

Aggregations, in the sense of putting multiple observations information into additional columns, are not needed. Therefore, in the case of multiple observations, building multiple-rows-per-subject data marts is a bit more straightforward than building a one-row-per-subject data mart. The reason is that multiple observation data are usually stored in a multiple-rows-per-subject structure by data processing systems.

The columns can also come from the table that holds non-repeated or static information for the analysis subject. In that case, this information is repeated per subject as often as observations for the respective subject exist in the analysis subject table. These data are then de-normalized.

Multiple-rows-per-subject data marts can also consist of data only from the 'many' table of a one-to-many relationship. This means that static information on the analysis subject level does not necessarily have to be copied to the multiple-rows-per-subject data mart.

9.2 Using Multiple-Rows-per-Subject Data Marts

Multiple-rows-per-subject data marts are used, for example, to answer the following representative business questions:

- What is the demand forecast for a certain product for the next 18 months?
- Are there products that are frequently bought together (market basket analysis)?
- What are the typical buying sequences of my customers?
- What are the Web paths of Internet users (clickstream analysis)?
- Is there a visual trend of cholesterol values over time?

The following analytic methods require a multiple-rows-per-subject data mart:

- time series analysis
- association analysis
- sequence analysis
- link analysis
- line plots for longitudinal analysis

9.3 Types of Multiple-Rows-per-Subject Data Marts

General

We will look at common types of multiple-rows-per-subject data marts and also present an example. The types are distinguished by the following two questions:

- Does the table contain data only from multiple observations (besides the subject ID), or are de-normalized data from the subject level in the table also present?

- How are multiple rows per subject enumerated? The table can have the following features:

  - an interval scaled variable such as a time variable

  - an ordinal numerator variable or sequence variable

  - no numeration of the multiple rows

9.3.1 No Numeration of Multiple Rows

Table 9.1 contains a list of products that each customer has purchased. This table format is the typical input for a market basket analysis. In this example no de-normalized information on a customer basis is available, and no variable for the numeration of the multiple units is present. The lack of the sequence variable means that only a market-basket analysis but not a purchase-sequence analysis can be performed on the basis of these data.

Note that sometimes from the ordering of records in the table the sequence can be assumed. However, this would not be the usual case, and a sequence number, if available, should be provided because the record sequence can be easily changed during data import or table creation or for other reasons.

Table 9.1: Market basket data for two customers without a buying sequence variable

	CUSTOMER	PRODUCT
1	0	hering
2	0	comed_b
3	0	olives
4	0	ham
5	0	turkey
6	0	bourbon
7	0	ice_crea
8	1	baguette
9	1	soda
10	1	hering
11	1	cracker
12	1	heineken
13	1	olives
14	1	comed_b

Table 9.2 contains the same data as in Table 9.1. However, in this case, additional information for the analysis subject level has been added—the value segment. The value of this variable is repeated as many times as observations exist per subject. This table allows a so-called BY-category analysis, where the market basket analysis is performed per value segment.

Table 9.2: Market basket data for two customers with additional data on CUSTOMER level

	CUSTOMER	PRODUCT	Segment
1	213	baguette	SILVER
2	213	hering	SILVER
3	213	avocado	SILVER
4	213	artichok	SILVER
5	213	heineken	SILVER
6	213	chicken	SILVER
7	213	coke	SILVER
8	217	baguette	GOLD
9	217	hering	GOLD
10	217	avocado	GOLD
11	217	artichok	GOLD
12	217	heineken	GOLD
13	217	apples	GOLD
14	217	peppers	GOLD
15	221	soda	SILVER
16	221	olives	SILVER
17	221	bourbon	SILVER
18	221	cracker	SILVER
19	221	heineken	SILVER
20	221	turkey	SILVER
21	221	steak	SILVER

9.3.2 Key-Value Table

The key-value table is another version of a multiple-rows-per-subject data mart without numeration of the multiple rows. Here we have multiple rows per analysis subject where each row represents a certain characteristic of the analysis subject, such as GENDER = MALE. The name *key-value table* is derived from the fact that besides the subject ID we have a *key variable,* which holds the name of the characteristic, and a *value variable*, which holds the value of the characteristic. Table 9.3 shows an example of a key-value table.

Table 9.3: Key-value table for SASHELP.CLASS

	ID	Name	Key	Value
1	1	Alice	Sex	F
2	1	Alice	Age	13
3	1	Alice	Height	56.5
4	1	Alice	Weight	84
5	2	Barbara	Sex	F
6	2	Barbara	Age	13
7	2	Barbara	Height	65.3
8	2	Barbara	Weight	98
9	3	Carol	Sex	F
10	3	Carol	Age	14
11	3	Carol	Height	62.8
12	3	Carol	Weight	102.5
13	4	Jane	Sex	F
14	4	Jane	Age	12
15	4	Jane	Height	59.8
16	4	Jane	Weight	84.5

This type of data representation is sometimes found in storage systems. We will see in *Chapter 12 – Considerations for Predictive Modeling*, that the restructuring of a key-value table to a one-row-per-subject data mart is very simple to accomplish.

9.3.3 Ordinal Numeration of Observations

Table 9.4 shows the same data as in Table 9.3 but with a sequence variable added. This allows the identification of the buying sequence.

Table 9.4: Market basket data for two customers with a buying sequence variable

	CUSTOMER	TIME	PRODUCT
1	0	0	hering
2	0	1	comed_b
3	0	2	olives
4	0	3	ham
5	0	4	turkey
6	0	5	bourbon
7	0	6	ice_crea
8	1	0	baguette
9	1	1	soda
10	1	2	hering
11	1	3	cracker
12	1	4	heineken
13	1	5	olives
14	1	6	comed_b

Table 9.5 gives a further example of an ordinal-scaled variable for the repeated observations. The data are an extract from a Web log and can be used for Web path analysis. The subject in that case is the session, identified by the SESSION IDENTIFIER; the variable SESSION SEQUENCE enumerates the requested files in each session.

Table 9.5: Web log data with a session sequence variable

	Session Identifier	requested_file	session_sequence
1	43d0a4da826149b5 2002-02-17 08:38:12	/Home.jsp	1
2	43d0a4da826149b5 2002-02-17 08:38:12	/Cookie_Check.jsp	2
3	43d0a4da826149b5 2002-02-17 08:38:12	/Home.jsp	3
4	43d0a4da826149b5 2002-02-17 08:38:12	/Corporate_Relations.jsp	4
5	43d0a4da826149b5 2002-02-17 08:38:12	/Retail_Store.jsp	5
6	43d0a4da826149b5 2002-02-17 08:38:12	/Store/Store_Locations.jsp	6
7	43d639ebce6c73d8 2002-02-17 23:43:16	/Home.jsp	1
8	43d639ebce6c73d8 2002-02-17 23:43:16	/Cookie_Check.jsp	2
9	43d639ebce6c73d8 2002-02-17 23:43:16	/Home.jsp	3
10	43d639ebce6c73d8 2002-02-17 23:43:16	/Department.jsp	4

9.3.4 Time Series Data

The data basis for Table 9.6 is a one-to-many relationship between a customer and monthly profit. The monthly profit for each customer for the months September 2004 through December 2004 are provided. This is an example where we have data only from the 'many' table of the one-to-many relationship.

These types of multiple-rows-per-subject data are usually applicable in time series analysis. The MONTH variable defines equidistant intervals.

Table 9.6: Monthly profit data

	CustomerID	Month	Profit
1	1001	200409	120
2	1001	200410	123
3	1001	200411	158
4	1001	200412	219
5	1043	200409	90
6	1043	200410	64
7	1043	200411	65
8	1043	200412	68
9	1049	200409	223
10	1049	200410	198
11	1049	200411	191
12	1049	200412	185

Table 9.7 shows the same data as in Table 9.6. However, the variables GENDER and TARIFF have been added and repeated for each customer.

Table 9.7: Monthly profit data, enhanced with GENDER and TARIFF

	CustomerID	Gender	Tariff	Month	Profit
1	1001	MALE	Tariff_A	200409	120
2	1001	MALE	Tariff_A	200410	123
3	1001	MALE	Tariff_A	200411	158
4	1001	MALE	Tariff_A	200412	219
5	1043	FEMALE	Tariff_B	200409	90
6	1043	FEMALE	Tariff_B	200410	64
7	1043	FEMALE	Tariff_B	200411	65
8	1043	FEMALE	Tariff_B	200412	68
9	1049	MALE	Tariff_B	200409	223
10	1049	MALE	Tariff_B	200410	198
11	1049	MALE	Tariff_B	200411	191
12	1049	MALE	Tariff_B	200412	185

9.4 Multiple Observations per Time Period

We will also show an example of multiple observations per period and a case in which we have two possible analysis subjects. In Table 9.8 MACHINE and OPERATOR are potential analysis subjects. Depending on the analysis question, we can decide to have multiple observations per MACHINE or multiple observations per OPERATOR.

Additionally, we see that we have multiple observations per MACHINE and OPERATOR for a given date. So in our example we have a one-to-many relationship between MACHINE/OPERATOR and DAY and a one-to-many-relationship between DAY and MEASUREMENT. The variable MEASUREMENT is not included in this table. MEASUREMENT in this case does not mean that the observations are taken in an order. They can also be non-ordered repetitions.

Such occurrences are common in analysis tables for ANOVA or quality control chart analysis (see PROC SHEWHART in SAS/QC). In these analyses, a measurement is often repeated for a combination of categories.

These table structures are also multiple-rows-per-subject tables. The focus, however, is not on the analysis at the subject level but on the comparison of the influences of various categories.

Table 9.8: Table with two possible analysis subjects

	machine	operator	day	diameter
23	A386	MKS	04MAY1990	4.49874
24	A386	MKS	04MAY1990	4.53749
25	A386	RMM	05MAY1990	4.89581
26	A386	RMM	05MAY1990	3.68588
27	A386	RMM	05MAY1990	3.60865
28	A386	RMM	05MAY1990	4.1402
29	A386	RMM	05MAY1990	4.23453
30	A386	RMM	05MAY1990	4.12907
31	A386	MKS	06MAY1990	4.74748
32	A386	MKS	06MAY1990	4.68869
33	A386	MKS	06MAY1990	4.50394
34	A386	MKS	06MAY1990	4.68806
35	A386	MKS	06MAY1990	4.4782
36	A386	MKS	06MAY1990	4.37201
37	A386	CMB	07MAY1990	4.52382
38	A386	CMB	07MAY1990	4.63532
39	A386	CMB	07MAY1990	4.03783
40	A386	CMB	07MAY1990	4.71058
41	A386	CMB	07MAY1990	4.3612
42	A386	CMB	07MAY1990	4.35975
43	A386	CMB	08MAY1990	4.61969
44	A386	CMB	08MAY1990	4.65951
45	A386	CMB	08MAY1990	4.56661
46	A386	CMB	08MAY1990	4.45578
47	A386	CMB	08MAY1990	4.4117
48	A386	CMB	08MAY1990	5.09301
49	A455	DRJ	09MAY1990	4.6881
50	A455	DRJ	09MAY1990	4.53631
51	A455	DRJ	09MAY1990	3.92734
52	A455	DRJ	09MAY1990	4.6213
53	A455	DRJ	09MAY1990	4.23637
54	A455	DRJ	09MAY1990	4.34124
55	A455	DRJ	10MAY1990	4.69474

9.5 Relationship to Other Data Mart Structures

9.5.1 The One-Row-per-Subject Data Mart

As mentioned in Chapter 7, both structures (the one-row-per-subject data mart and the multiple-rows-per-subject data mart) can be used for analysis. Final analysis, for example, is conducted on the one-row-per-subject data mart. For exploratory analyses in the form of line plots over time, the multiple-rows-per-subject data mart is required.

9.5.2 Longitudinal Data Structures

In Chapter 7 we mentioned that the multiple-rows-per-subject data mart with multiple observations over time for a subject can also be considered a longitudinal data mart. With the multiple-rows-per-subject data mart, however, we have dependent observations per analysis subject, whereas in the case of longitudinal data structures, we have observations without an individual analysis subject.

Chapter **10**

Data Structures for Longitudinal Analysis

10.1 Introduction

In this chapter we will cover *longitudinal data marts*. In longitudinal data marts we have observations over time. This means that time or a sequential ordering is included in these data marts by a time or sequence variable.

Longitudinal data marts do not have an analysis subject such as in multiple-rows-per-subject data marts. They can represent one or more variables measured on several points in time, as we see in the following examples.

Table 10.1: Example of longitudinal data marts

	Period	Actual Sales
1	1993.01	$29,813.00
2	1993.02	$29,584.00
3	1993.03	$29,873.00
4	1993.04	$30,581.00
5	1993.05	$31,617.00
6	1993.06	$33,605.00
7	1993.07	$33,578.00
8	1993.08	$31,160.00
9	1993.09	$28,696.00
10	1993.10	$31,355.00
11	1993.11	$27,659.00
12	1993.12	$31,956.00

	Period	Actual Sales	Predicted Sales
1	1993.01	$29,813.00	$32,385.00
2	1993.02	$29,584.00	$29,163.00
3	1993.03	$29,873.00	$31,818.00
4	1993.04	$30,581.00	$27,429.00
5	1993.05	$31,617.00	$30,263.00
6	1993.06	$33,605.00	$27,634.00
7	1993.07	$33,578.00	$33,220.00
8	1993.08	$31,160.00	$28,874.00
9	1993.09	$28,696.00	$28,470.00
10	1993.10	$31,355.00	$30,262.00
11	1993.11	$27,659.00	$31,434.00
12	1993.12	$31,956.00	$29,259.00

Or longitudinal data marts can have measurements over time for each category. In Table 10.2 the category is COUNTRY.

Table 10.2: Longitudinal data mart with time series per country

	Period	Country	Actual Sales
1	1993.01	CANADA	$8,115.00
2	1993.02	CANADA	$10,431.00
3	1993.03	CANADA	$10,458.00
4	1993.04	CANADA	$9,695.00
5	1993.05	CANADA	$9,916.00
6	1993.06	CANADA	$11,204.00
7	1993.07	CANADA	$12,313.00
8	1993.08	CANADA	$8,047.00
9	1993.09	CANADA	$10,686.00
10	1993.10	CANADA	$10,739.00
11	1993.11	CANADA	$8,139.00
12	1993.12	CANADA	$11,277.00
13	1993.01	GERMANY	$11,239.00
14	1993.02	GERMANY	$9,297.00
15	1993.03	GERMANY	$9,772.00
16	1993.04	GERMANY	$11,490.00
17	1993.05	GERMANY	$12,495.00
18	1993.06	GERMANY	$10,895.00
19	1993.07	GERMANY	$10,557.00
20	1993.08	GERMANY	$12,843.00
21	1993.09	GERMANY	$9,233.00
22	1993.10	GERMANY	$11,303.00
23	1993.11	GERMANY	$9,110.00
24	1993.12	GERMANY	$9,170.00

Note that the borders of longitudinal and multiple-rows-per-subject data marts are very fluid. As soon as we define COUNTRY as an analysis subject, we have a multiple-rows-per-subject data mart. We will, however, keep this distinction for didactic purposes.

Analyses that require longitudinal data structures will be called longitudinal analysis in this book. Time series analyses or quality control analyses are the main types of longitudinal analyses.

Examples of longitudinal analyses include the following:

- predicting the number of tourist overnight stays in Vienna on a daily basis for a forecasting horizon of three months
- predicting the daily cash demand in ATMs, depending on factors such as region or day of week

- monitoring the number of failures of produced goods in a production line
- analyzing whether the filling quantity in soda bottles stays within specification limits over production lots

Many SAS procedures from the SAS/ETS and the SAS/QC module require data to be arranged in the longitudinal data structure. For more details, see Appendix A.

10.2 Data Relationships in Longitudinal Cases

In the "Introduction" section we saw some examples of longitudinal data marts. In this section, we will elucidate the underlying data relationships. All longitudinal data marts have at least a time variable and a value variable present. Therefore, a TIME and a VALUE entity will be present when we consider entity relationship diagrams. These two entities comprise the simple form of a longitudinal data structure.

10.2.1 The Simple Form

In the simple form of a longitudinal relationship we have a value that is measured at different points in time. The graphical representation is shown in Figure 10.1.

Figure 10.1: Entity relationship diagram for the simple form

Month	Actual Sales
1993.01	$10,887.00
1993.02	$12,315.00
1993.03	$12,235.00
1993.04	$13,018.00
1993.05	$12,064.00
1993.06	$14,196.00

Note that the VALUE entity here can represent one or more values. Both tables in Table 10.1 are represented by this entity relationship diagram.

10.2.2 Extension of the Simple Form

The simple form can be extended if the series of values is given for each category (or subgroup). We saw an example in Figure 10.1. In this case we have an additional variable for COUNTRY, which represents the category. The entity relationship diagram for the data in Figure 10.1 is given in Figure 10.2.

Figure 10.2: Entity relationship diagram for values per month and country

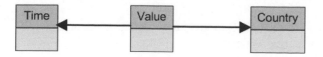

Month	Country	Actual Sales
1993.01	CANADA	$2,809.00
1993.02	CANADA	$4,064.00
1993.03	CANADA	$4,395.00
1993.01	GERMANY	$5,598.00
1993.02	GERMANY	$4,322.00
1993.03	GERMANY	$4,021.00
1993.01	U.S.A.	$2,480.00
1993.02	U.S.A.	$3,929.00
1993.03	U.S.A.	$3,819.00

If we add another categorization such as PRODUCT, our entity relationship diagram will have an additional entity, PRODUCT.

Figure 10.3: Entity relationship diagram for categories COUNTRY and PRODUCT

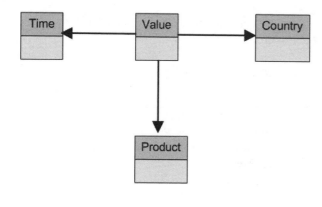

Month	Country	Product	Actual Sales
1993.01	CANADA	BED	$856.00
1993.02	CANADA	BED	$1,581.00
1993.03	CANADA	BED	$1,900.00
1993.01	CANADA	SOFA	$1,953.00
1993.02	CANADA	SOFA	$2,483.00
1993.03	CANADA	SOFA	$2,495.00
1993.01	GERMANY	BED	$1,875.00
1993.02	GERMANY	BED	$1,929.00
1993.03	GERMANY	BED	$1,222.00
1993.01	GERMANY	SOFA	$3,723.00
1993.02	GERMANY	SOFA	$2,393.00
1993.03	GERMANY	SOFA	$2,799.00

If we refer to *Chapter 6 – Data Models*, we see that here we have a classic star schema. In fact, data models for longitudinal data marts are a star schema where we have the VALUES in the fact table and at least one dimension table with the TIME values. For each category, a dimension table will be added; in this case, COUNTRY and PRODUCT are added.

The categories themselves can have additional hierarchical relationships, such as the following:

- One country is in a sales region, and each sales region belongs to one continent.
- One product is in a product group, and product groups are summarized in product main groups.

10.2.3 The Generalized Picture

We will, for didactic purposes, summarize the categorizing entities into a logical CATEGORY entity. This includes all entities except for the VALUE and TIME entities. This results in the following entity relationship diagram:

Figure 10.4: Generalized picture of an entity relationship diagram

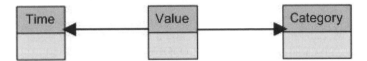

In terms of the star schema, we have put all possible dimensions except the TIME dimension into CATEGORY, resulting in a star schema with the fact table VALUE and two dimension tables, TIME and CATEGORY. This representation is useful for us in this chapter because we will investigate which data mart structures represent these relationships. Before we look at the different data mart structures, we will summarize properties of our generic data relationship model.

TIME

The TIME entity contains the ordering dimension of the data. In most cases it is a time variable, but it can also be an ordinal sequence or some other type of ordering that is not time-related. The time information can be stored in one variable, or it can be given in a set of variables such as YEAR, MONTH, and DAY. In many cases the time information is available at different aggregation levels such as YEAR, QUARTER, MONTH, WEEK, or DAY, allowing analysis at different time granularities. This entity corresponds to the TIME dimension in a star schema.

CATEGORY

The CATEGORY entity is a logical entity that combines classification information from different categorizing entities except for the TIME entity. In terms of the star schema, CATEGORY entity corresponds to the different dimensions (except for the TIME dimension). The single categories can have hierarchical orderings, such as branch, region, and country, allowing analysis at different hierarchical aggregation levels.

Note that if no categorization is present, as we saw in the examples in Table 10.1, we can speak of a virtual ALL category in order to fit our definition.

VALUE

The VALUE entity can contain one or more values that are given for combinations of the other entities TIME and CATEGORY. The values can be either interval-scaled or nominal-scaled. The analysis method will be chosen based on the level of measurement for the variables in question. This entity corresponds to the fact table in a star schema.

10.3 Transactional Data, Finest Granularity, and Most Appropriate Aggregation Level

When we look at the TIME dimension from the preceding section we can define the finest time granularity according to how the time values are measured. If we have daily data, the finest time granularity is DATE; if we have weekly data, the finest time granularity is WEEK. This data can then be aggregated to higher aggregation levels.

Looking at the logical CATEGORY dimension, we can have different dimensions with hierarchies. For example, we can have a CUSTOMER and a PRODUCT dimension. The PRODUCT dimension can have hierarchies such as PRODUCT, PRODUCT_SUBGROUP, and PRODUCT_GROUP.

Consider an example where we have daily data at the CUSTOMER and PRODUCT level. In this case the finest granularity is DATE, CUSTOMER, and PRODUCT. Table 10.3 shows an example of these data.

Table 10.3: Transactional data

Date	CustomerID	Product	Quantity
15.06.2004	1002	A	43
15.06.2004	1002	D	36
15.06.2004	1007	E	91
15.06.2004	1018	A	31
15.06.2004	1018	C	54
15.06.2004	1018	E	11
16.06.2004	1007	C	54
16.06.2004	1007	D	10
16.06.2004	1008	A	5
16.06.2004	1008	B	20
16.06.2004	1008	A	7
16.06.2004	1018	C	12
16.06.2004	1018	F	9
16.06.2004	1021	A	22
16.06.2004	1021	D	54

We see that in our data we can have more than one row per DATE, CUSTOMER, and PRODUCT. This can happen, for example, if a customer purchases the same product more than once on a certain day. This representation of data is frequent if data are retrieved from an ordering system. In *Chapter 3 – Characteristics of Data Sources*, we also saw that these systems are called transactional systems. In the ordering system the orders will have a unique order ID.

If we now aggregate the data to their finest aggregation levels, DATE, CUSTOMER, and PRODUCT, the result is Table 10.4.

Table 10.4: Transactional data, aggregated to the finest granularity

Date	CustomerID	Product	Quantity
15.06.2004	1002	A	43
15.06.2004	1002	D	36
15.06.2004	1007	E	91
15.06.2004	1018	A	31
15.06.2004	1018	C	54
15.06.2004	1018	E	11
16.06.2004	1007	C	54
16.06.2004	1007	D	10
16.06.2004	1008	A	12
16.06.2004	1008	B	20
16.06.2004	1018	C	12
16.06.2004	1018	F	9
16.06.2004	1021	A	22
16.06.2004	1021	D	54

In practice, Tables 10.3 and 10.4 are often referred to as *transactional data*. We want to mention here that *real* transactional data from an operational system is usually in the finest time granularity, but they can also contain duplicate rows per TIME and CATEGORY. The duplicates per TIME and CATEGORY are usually identified by a unique order ID. Usually these data are aggregated in a next step to a higher aggregation level for analysis.

This higher aggregation level can also be called the most appropriate aggregation level. The most appropriate aggregation level depends on the business questions. If forecasts are produced on a monthly basis, then data will be aggregated at the TIME dimension on a monthly basis. The finest granularity can be seen as something that is technically possible with the available data. However, the level, either TIME or CATEGORY, to which data are finally aggregated depends on the business questions and the forecasting analysis method.

In *Chapter 21 – Data Preparation for Multiple-Rows-per-Subject and Longitudinal Data Marts*, we will go into more detail about aggregating data. A comprehensive example will be shown in *Chapter 27 – Case Study 3—Preparing Data for Time Series Analysis*.

10.4 Data Mart Structures for Longitudinal Data Marts

General
Let's look again at a simple version of a longitudinal data mart.

Table 10.5: Simple longitudinal data mart

	Period	Actual Sales
1	1993.01	$29,813.00
2	1993.02	$29,584.00
3	1993.03	$29,873.00
4	1993.04	$30,581.00
5	1993.05	$31,617.00
6	1993.06	$33,605.00
7	1993.07	$33,578.00
8	1993.08	$31,160.00
9	1993.09	$28,696.00
10	1993.10	$31,355.00
11	1993.11	$27,659.00
12	1993.12	$31,956.00

At the end of the previous section we discussed that we can have more than one VALUE, and we can have values not only for the virtual ALL group but for subgroups or CATEGORIES. This means that the table will grow in order to accommodate the additional VALUES or CATEGORIES.

Not surprisingly, the table can grow by adding more rows or columns, or both. There are three main data mart structures for longitudinal data:

- the standard form for longitudinal data
- the cross-sectional dimension data mart
- the interleaved longitudinal data mart

10.4.1 The Standard Form of a Longitudinal Data Set

A time series data set in standard form has the following characteristics:

- The data set contains one variable for each time series.
- The data set contains exactly one observation for each time period.
- The data set contains an ID variable or variables that identify the time period of each observation.
- The data set is sorted by the ID variables associated with datetime values, so the observations are in time sequence.
- The data are equally spaced in time. That is, successive observations are a fixed time interval apart, so the data set can be described by a single sampling interval such as hourly, daily, monthly, quarterly, yearly, and so forth. This means that time series with different sampling frequencies are not mixed in the same SAS data set.

We can see that the tables in Table 10.1 are in the standard form of a longitudinal data set. Table 10.2 is not in the standard form because we have more than one row per time interval. In order to bring this table into standard form, we have to arrange it as in Table 10.6.

Table 10.6: Standard form of a longitudinal data set with values per country

	Period	CANADA	GERMANY	U_S_A_
1	1993.01	$8,115.00	$11,239.00	$10,459.00
2	1993.02	$10,431.00	$9,297.00	$9,856.00
3	1993.03	$10,458.00	$9,772.00	$9,643.00
4	1993.04	$9,695.00	$11,490.00	$9,396.00
5	1993.05	$9,916.00	$12,495.00	$9,206.00
6	1993.06	$11,204.00	$10,895.00	$11,506.00
7	1993.07	$12,313.00	$10,557.00	$10,708.00
8	1993.08	$8,047.00	$12,843.00	$10,270.00
9	1993.09	$10,686.00	$9,233.00	$8,777.00
10	1993.10	$10,739.00	$11,303.00	$9,313.00
11	1993.11	$8,139.00	$9,110.00	$10,410.00
12	1993.12	$11,277.00	$9,170.00	$11,509.00

Note that the standard form of a longitudinal data set has some similarity to the one-row-per-subject paradigm. In the one-row-per-subject paradigm, we are not allowed to have multiple rows per subject, but here we cannot have more than one row per period. As a consequence we have to represent the information in additional columns.

10.4.2 The Cross-Sectional Dimension Data Structure

When we look at Table 10.2 and consider each COUNTRY separately, we see that we have a table in the standard form of a longitudinal data set for each COUNTRY. The cross-sectional dimension structure is a concatenation of standard form tables for each category.

If we have more than one category in our logical entity CATEGORY, each additional category results in an additional cross-dimension variable. If, for example, we have data as in Figure 10.3, we will have a cross-sectional dimension for COUNTRY and one for PRODUCT. The corresponding table would look like Table 10.7.

Table 10.7: Cross-sectional dimension data mart for COUNTRY and PRODUCT (truncated)

	Country	Product	Period	Actual Sales
1	CANADA	BED	1993.01	$856.00
2	CANADA	BED	1993.02	$1,581.00
3	CANADA	BED	1993.03	$1,900.00
4	CANADA	BED	1993.04	$2,744.00
5	CANADA	BED	1993.05	$1,104.00
6	CANADA	BED	1993.06	$2,611.00
7	CANADA	BED	1993.07	$2,095.00
8	CANADA	BED	1993.08	$863.00
9	CANADA	BED	1993.09	$2,241.00
10	CANADA	BED	1993.10	$1,484.00
11	CANADA	BED	1993.11	$1,326.00
12	CANADA	BED	1993.12	$3,039.00
13	CANADA	CHAIR	1993.01	$979.00
14	CANADA	CHAIR	1993.02	$2,323.00
15	CANADA	CHAIR	1993.03	$1,813.00
16	CANADA	CHAIR	1993.04	$1,139.00

10.4.3 Interleaved Time Series

In the standard form of a longitudinal data set we can have more than one measurement variable per row. In the second example in Table 10.1, these variables are ACTUAL SALES and PREDICTED SALES. If we represent additional measurement variables not in additional columns but in additional rows, we speak of an interleaved time series data mart.

Table 10.8: Interleaved time series data mart

	Period	Variable	Value
1	1993.01	ACTUAL	$29,813.00
2	1993.01	PREDICT	$32,385.00
3	1993.02	ACTUAL	$29,584.00
4	1993.02	PREDICT	$29,163.00
5	1993.03	ACTUAL	$29,873.00
6	1993.03	PREDICT	$31,818.00
7	1993.04	ACTUAL	$30,581.00
8	1993.04	PREDICT	$27,429.00
9	1993.05	ACTUAL	$31,617.00
10	1993.05	PREDICT	$30,263.00
11	1993.06	ACTUAL	$33,605.00
12	1993.06	PREDICT	$27,634.00
13	1993.07	ACTUAL	$33,578.00
14	1993.07	PREDICT	$33,220.00
15	1993.08	ACTUAL	$31,160.00
16	1993.08	PREDICT	$28,874.00
17	1993.09	ACTUAL	$28,696.00
18	1993.09	PREDICT	$28,470.00
19	1993.10	ACTUAL	$31,355.00
20	1993.10	PREDICT	$30,262.00
21	1993.11	ACTUAL	$27,659.00
22	1993.11	PREDICT	$31,434.00
23	1993.12	ACTUAL	$31,956.00
24	1993.12	PREDICT	$29,259.00

In the diction of our data structures, we have in this case two variables from the fact table observed for each time period. Representing the second fact variable in additional rows, instead of additional columns, results in the interleaved data structure.

10.4.4 Combination of the Interleaved and the Cross-Sectional Data Structure

If we put all information into rows, i.e., if we repeat the standard form for each subgroup combination and variable in the fact table, we have a combination of the interleaved and cross-sectional data structure. This data structure is relevant if we want to analyze the data by CATEGORIES and VALUES with a BY statement in one pass.

10.4.5 Comparison: Interleaved or Cross-Sectional versus Standard Form

The standard form has some similarity with the one-row-per-subject data mart; we have only one row for each time period. Information for underlying categories or different fact variables is represented by columns. This type of data representation allows for the calculation of correlations between time series.

In the interleaved or cross-sectional data mart structures we multiply the number of observations of the corresponding standard form by repeating the rows as often as we have categories or repeated values. This type of data representation is convenient for analysis because we can use BY statements for BY-group processing and WHERE statements for filtering variables. And this type of data representation allows for the creation of graphs for BY groups on the data.

We will encounter the three forms of longitudinal data marts again in *Chapter 15 – Transposing Longitudinal Data*, where we will demonstrate methods of converting among these structures by using SAS code.

Chapter 11

Considerations for Data Marts

11.1 Introduction

In Chapters 6–10 we discussed possible data models and data structures for our data mart. Before we move to the creation of derived variables and coding examples we will examine definitions and properties of variables in a data mart. We will look at attribute scales, variable types, and variable roles. And we will see different ways that derived variables can be created.

11.2 Types and Roles of Variables in a Data Mart

Variables are usually represented by the columns of a table and are the basis for analysis. In this section we will define variables in order to reference them in the following chapters. We will look at the following:

- attribute scales
- variable types
- variable roles

11.2.1 Attribute Scales

Attributes that are measured on the analysis subject can be referred to by a certain scale. We call this an *attribute scale*. The following attribute scales are important considerations for attributes in a data mart:

- Binary-scaled attributes contain two discrete values, such as PURCHASE: Yes, No.

- Nominal-scaled attributes contain a discrete set of values that do not have a logical ordering such as PARTY: Democrat, Republican, other.

- Ordinal-scaled attributes contain a discrete set of values that do have a logical ordering such as GRADE: A, B, C, D, F.

- Interval-scaled attributes contain values that vary across a continuous range such as AGE: 15, 34, 46, 50, 56, 80, ..., 102.

In some cases binary, nominal, and ordinal scaled attributes are also referred to as *categorical* or *qualitative attributes* (or categorical variables) because their values are categories. Interval measures are considered quantitative attributes.

Note that the preceding classification equals the classification of variable types in SAS Enterprise Miner, which is important for all statistical considerations that we will deal with. The only difference in statistical theory is that we do not distinguish between interval-scaled and ratio-scaled attributes but treat them as one type. A ratio-scaled attribute is defined by the fact that ratios of its values make sense; for example,

- "Duration in years" is ratio-scaled because we can say that 10 years is twice as much as 5 years.

- "Temperature in degrees Celsius" is (only) interval-scaled because ratios on this scale do not make sense.

11.2.2 Variable Types

From a table columns definition point of view, we can differentiate two main categories:

- numeric variables
- character variables

Numeric variables can contain interval-scaled values and categorical values. In the case of interval-scaled measures the value itself is stored in the variable. The value of binary, nominal, or ordinal attributes needs to be numerically coded for interval variables.

Character variables usually hold binary, nominal, and ordinal scaled values. It is also technically possible to store interval-scaled values in character variables, but in most cases this does not allow for performing calculations on them.

11.2.3 Numeric or Character Type Variables

The decision of which variable type will be used for which attributes is influenced by the following facts:

- Numeric variables have at least a length of 3 bytes in a SAS data set. If numeric codes of a few digits are to be stored, using a character type variable can save disk space.

- Storing binary values in numeric variables can make sense in calculations, as the mean of a binary variable with values 0 and 1 equals the proportion of the observations with a value of 1.

- SAS Enterprise Miner, SAS/INSIGHT, and SAS/IML derive the metadata information—whether a variable is an interval or categorical variable—from their variable type. Storing categorical information in character type variables automates the correct definition of the variable's metadata type.

- SAS formats can be used to assign meaningful labels to values and reduce disk space requirements.

11.2.4 Variable Roles

In the analysis data mart each variable can have one or more roles. *Role*, in this context, means the function the variable has in the analyses. See the following list for possible variable roles. Note that we use a part of the definition of roles in SAS Enterprise Miner here. These definitions also apply to other analyses not involving SAS Enterprise Miner.

- ID variables hold the ID values of subjects and the ID values of entities in underlying hierarchies. Examples are Customer Number, Account ID, Patient Number, and so forth. Subject identifiers such as primary and foreign keys are candidates for ID variables.

- TARGET variables are predicted or estimated by input variables in predictive modeling. In the y=$\mathbf{X}$b equation, target variables stand on the left side of the equation and represent the y's. Note that target variables can also be a set of variables. In PROC LOGISTIC, for example, the y can also be given in the form of count data, where the number of events and the number of trials are in two variables.

- INPUT variables are all non-target variables that contain information that is used in analyses to classify, estimate, or predict a target or output variable. In predictive modeling, for example, INPUT variables are used to predict the target variables (those that are represented by the $\mathbf{X}$ in the y=$\mathbf{X}$b equation). In cluster analyses, where there is no target variable, INPUT variables are used to calculate the cluster assignments.

- TIMEID variables can be found in longitudinal data marts and multiple-rows-per-subject data marts only and contain the value of the TIME dimension.

- CROSSID variables can be found in longitudinal data marts only and contain the categories for cross-sectional analyses in longitudinal data marts. The CROSSID variables are used to subgroup or subset an analysis.

- SEQUENCE variables can be found in longitudinal and multiple-rows-per-subject data marts. They are similar to the TIMEID variables, but can also contain ordinal or non-time-related information.

Note that one variable can have more than one role—a variable can be a target variable in analysis A and an input variable in analysis B.

11.3 Derived Variables

From their source, variables in a data mart can be roughly divided into two groups:

- variables that are taken on a one-to-one basis from the source data and are copied to the analysis table
- derived variables that are created from other variables by mathematical or statistical functions or by aggregations

Variables that are taken on a one-to-one basis from the source data are often subject identifiers or values that do not need further transformation or calculation such as the "number of children" or "gender."

The variable BIRTHDATE, for example, is taken from the operative system on a one-to-one basis into the data mart and the derived variable AGE is calculated from the actual date and BIRHTDATE.

Analysis tables almost never exclusively contain variables that are taken on a one-to-one basis from the source data. In many cases derived variables play an important role in statistical analysis. We will, therefore, have a more detailed look at them and how they are created.

11.3.1 Categorical Variables That Need to Be Dummy Coded

Categorical information with more than two categories, such as region or contract type cannot always be processed as a categorical value in one variable. The categorical information needs to be distributed into a set of dummy variables. *Dummy variables* are binary variables with the values 0 or 1. We will look at the rationale for dummy variables and their creation in detail in *Chapter 16 – Transformations of Interval-Scaled Variables*.

11.3.2 Derived Variables That Depend Only on the Subject's Values

The values of these variables for a certain subject are calculated only on the basis of the subject's own values. No aggregations over all subjects or comparisons to other subjects are included in these variables.

For example, the calculated variable AGE is calculated from 'SNAPSHOT_DATE – BIRHTDATE'. The values of AGE at a certain point in time for a customer depend only on the customer's date of birth. Values in this group do not depend on the values of other subjects such as the age of other patients.

11.3.3 Derived Variables That Depend on the Values of Other Subjects

This type of derived variable has the property that its values are not calculated only on values of the same subject. The values of other observations also influence its values.

- For example, we want to group our customers by their income into the groups LOW, MEDIUM, and HIGH. In this case we will set up rules such as INCOME < 15000 = LOW. Thus, the derived variable INCOME_GROUP for a subject depends not only on its INCOME value but also on the class limits, which are derived from business rules or the quantiles of the distribution of overall customers.

- An aggregated value over all observations, such as the sum or the mean, is used in the calculation of the derived variable. For example, the mean over all subjects of a variable is subtracted from the subject's value of a variable in order to have positive values in the derived variable for values higher than the mean and negative values for values lower than the mean.

These types of variables are usually very important in order to describe specific properties of a subject. In data mining, most variables in an analysis data mart are created by comparing the subject's value to the values of other subjects and trying to create meaningful characteristics for the analysis. In Chapters 16–19 we will discuss potential methods used to calculate those types of variables.

Note that these types of derived variables are mostly found with numerical variables but are not limited to those. Think of a simple indicator variable that defines whether a customer lives in a district where most other customers live or where the most car insurance claims have been recorded.

11.3.4 Multiple Uses of an Original Variable

Note that one original variable can lead to a number of derived variables that might belong to different groups that we described earlier.

For example, consider the variable NUMBER OF CHILDREN:

- The variable itself belongs to a group "Variables that are taken on a one-to-one basis from the source data."
- Creating an indicator variable that tells whether a subject has children or not, we have a derived variable that depends only on the subject's values themselves.
- Creating a variable that tells us whether the subject has fewer or more children than the average number of children for all customers, results in a derived variable that also depends on the values of other subjects.

11.4 Variable Criteria

When creating an analysis table from a data source or a number of data sources we potentially have a lot of variables in the source tables. Whether all these variables are transferred to the analysis table depends on the business question and the data structure.

For example, in the case of the creation of a one-row-per-subject analysis table, where we have data sources with multiple observations per analysis subject, we have to represent the repeated information in additional columns. We discussed this in Chapters 7 and 8.

The following four criteria characterize properties of the resulting derived variables that are created in an analysis table from the data source(s):

- sufficiency
- efficiency
- relevance
- interpretability

Sufficiency means that all potentially relevant information from the available source systems is also found in the analysis table.

Efficiency means that we try to keep the number of variables as small as possible. For example, in the case of transposing repeated observations, we can easily get thousands of variables in the analysis table.

Relevance means that data are aggregated in such a form that the derived variables are suitable for the analysis and business question. For example, in predictive modeling, we want to have input variables with high predictive power.

Interpretability means the variables that are used for analysis can be interpreted and are meaningful from a business point of view. In predictive modeling, however, variables that have high predictive power but are not easy to interpret are often used in the analysis. In this case, other variables are used for model interpretation.

Example

For customers we have a measurement history for the account balance over 20 months. When we have to condense this information into a one-row-per-customer data mart, we are faced with the challenge of creating meaningful variables from the measurement history.

Simply creating 20 variables M1–M20 for the measurements would fulfill the sufficiency requirement. However, creating these variables would not be efficient, nor would it usually provide relevant variables for analysis.

In this case data preparation has to provide meaningful derived values of the measurements over time, which does not necessarily mean that we consider each value itself. *Chapter 18 – Multiple Interval-Scaled Observations per Subject* will go into detail about this topic and discuss the applications of indicators, trend variables, and moving averages, among others.

Chapter 12

Considerations for Predictive Modeling

12.1 Introduction

Predictive modeling is a discipline of data mining that plays a very important role. Because voluminous literature can be found concerning the training and tuning of predictive models, we will not consider these topics here. However, because the analysis data mart is a very important success factor for the predictive power of models, we will consider some design details for predictive modeling. In *Chapter 20 – Coding for Predictive Modeling*, we will deal with the creation of derived variables with high predictive power.

The points that we discuss here mostly concern one-row-per-subject data marts in data mining analyses but are not restricted to them. In this section we will consider the following:

- target windows and observation windows
- multiple target windows
- overfitting

12.2 Target Windows and Observation Windows

Consider an example of event prediction: we want to predict whether a customer is going to cancel his contract in the next month. To predict this event, we need to have data for the past—whether a customer has canceled his contracts within a certain time interval.

This time interval is called the *target window* because the target event—the cancellation—has to happen within it. Note that the length of the target window has to be within the context of the business question. In our case it is one month, because we want to predict customers who cancel in the next month. Assume our target window is January 2005.

We will then consider only customers who were active customers at the beginning of the target window on January 1, 2005. If a customer cancels his contract at any time from January 1 until January 31, 2005, he will be considered an event; otherwise, he will be considered a non-event. Note that customers who cancelled before January 1, 2005, are not considered at all and customers who cancelled after January 31, 2005, are considered as customers who did not cancel.

The end of the *observation window* can then for example, be December 31, 2004. This means that a data snapshot after December 31, 2004, is used to predict an event one month in the future. Obviously, we cannot take a data snapshot after January 31, 2005, to predict an event that has already happened. We would then use data that are measured after the event has happened and would likely get a good but useless prediction that variables such as customer status are good predictors.

In many cases the end of the observation window in our preceding example is set to one month earlier, at November 30, 2004. This trains the model to predict events that occur one month after the end of the observation window instead of events that occur immediately in the month after the end of the observation window. This additional time interval accounts for the time that might be needed to make data available for scoring and to prepare a campaign for customer retention. In many cases the data from the last month are available at, for example, the fifth of the following month in the data warehouse. To execute a retention campaign might take another three to five days. This means that in practice we might reach the customer only eight days after the start of the target window. This additional time interval is also called the *offset*.

Figure 12.1: Sampling and target interval with an offset

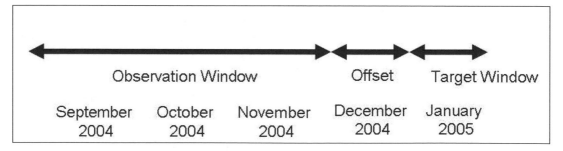

We see that the observation window can span several months—in order to have customer account transactions for a longer time period so that we can detect patterns of time courses that correlate with the cancellation event.

In *Chapter 3 – Characteristics of Data Sources*, we mentioned that the need for snapshots of the data at a historic point in time can be a challenge if a data warehouse is not in place. We see that for a prediction for January 2005 we have to go back to September 2004 in our simple example.

If only a small number of target events exist, the target window will need to be extended by setting it to one year instead of one month, for example. Sometimes the context in which a business question is analyzed also suggests a one-year target window. In credit scoring, for example, defaults of credit customers are cumulated over one calendar year for analysis.

Note that in this case or rare incidence of target events, we need a data history extending over years if we want to use the events in one target window as training data and the events in another target window as validating data.

12.3 Multiple Target Windows

If the number of events is small for one period or if we want to consider seasonal effects of the target events, we might use *multiple target windows*. In this case we consider different months as the target window, prepare one data mart for each target window, and then stack the data marts together. Figure 12.2 illustrates this.

Figure 12.2: Multiple target windows

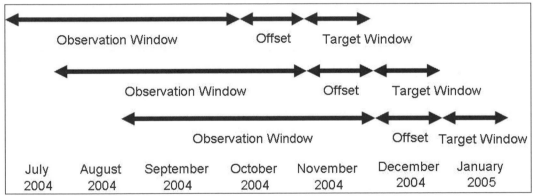

We see that we used the months November 2004, December 2004, and January 2005 as target windows in the respective data marts. The final data mart is then stacked together by aligning the data marts with their target windows over each other. The first row of Table 12.1 shows that relative month, as it is used in the analysis table. Rows 1–3 show the calendar months from which those data are taken in each of the three analysis data marts that are stacked together.

Table 12.1: Analysis table with multiple target windows

Observation Month 1	Observation Month 2	Observation Month 3	Offset	Target Windows
July 2004	August 2004	September 2004	October 2004	November 2004
August 2004	September 2004	October 2004	November 2004	December 2004
September 2004	October 2004	November 2004	December 2004	January 2005

In our example the target window spans only one month. This case is frequently found in marketing analyses where the behavior of customers in the next month will be predicted. There are, however, many other cases where the target window can span several months or a year. In credit scoring, for example, target windows with a span over a calendar year are frequently found.

Note that the selection of the sampling months also provides differential (and perhaps biasing) contributions to the model structure.

In this case we violated the assumption of independent observations in the input data because we used the same analysis subject more than once. The benefit, however, is a more stable prediction because we have more events and also events from different periods, which can offset seasonal effects of the event behavior.

12.4 Overfitting

12.4.1 Reasons for Overfitting

When preparing data for predictive analysis the danger of overfitting in the analysis has to be considered. With *overfitting* we mean that the resulting model fits perfectly or nearly perfectly to the data used for modeling, but the model poorly generalizes an data of other subjects or data in the next time period.

- For categorical data, overfitting often occurs if a high number of categories with a possible low number of observations per category exist. Considering all these categories individually leads to a predictive model on the training data, but will probably not generalize on other data if, for example, some of the categories with low frequency show other behavior.

- In the case of interval data, overly complex mathematical functions that describe the relationship between the input variable and the target variable can also lead to overfitting, if the relationship cannot be reproduced on other data.

- The most dramatic case of overfitting would be to use the individual subject ID to measure its relationship to the target variable. This would perfectly model the available data, but it would only by chance model other data. Of course, this example is more of didactic importance than a real-life mistake, but it illustrates very well what happens if highly specific but inappropriate information from the training data is used during modeling.

Whereas overfitting describes the fact that we model the noise in the data in our prediction model, the term *generalization* is used for the fact that a model makes good predictions for cases that are in the training data. A *generalization error* describes the fact that a prediction model loses its predictive power on non-training data. More details about this can also be found in the SAS Enterprise Miner Help.

12.4.2 Validation Methods

In order to allow estimation of the generalization error or to control the effect of overfitting, several validation methods exist. These methods include the following:

- *Split sample validation* is very popular and easy to implement. The available data are split into training and validation data. The split is done either by simple random sampling or by stratified sampling. The training data mart is used to train the model and to learn the relationships. It is used for preliminary model fitting. The analyst attempts to find the best model weights using this data set. The validation data mart is used to assess the adequacy of the model that was built on the training data.

  With some methods the validation data are already used in the model creation process such as the pruning of leaves in a decision tree. In these cases, the generalization error based on the validation data is biased. Therefore, a third data set, the test data, is created by random or stratified sampling and assessment statistics are calculated for this data set.

- Cross-validation is preferable for small data sets. Here the data are split in different ways, models are trained for each split, and then the validation results are combined across the splits.

- Bootstrap aggregation (bagging) works similarly to cross-validation. However, instead of subsets of the data, subsamples of the data are analyzed, where each subsample is a random sample with a replacement from the original sample of data.

Part 3

Data Mart Coding and Content

Introduction

After we introduce different data models and data structures for analytic data marts, we will start to fill the tables with content in this part of the book. *Part 3 – Data Mart Coding and Content,* is fully dedicated to the creation of the appropriate data mart structure and the generation of meaningful variables for the analysis.

This part of the book is essential for powerful data preparation for analytics. We will show how we can fill the tables with content in order to answer our business questions.

Whereas in Parts 1 and 2 we only introduced the concepts and rationale for certain elements in analytic data preparation, we will start to use SAS code in this part of the book. Part 3 is a good balance between business explanations for the importance and rationale of certain derived variables and data mart structures and the SAS coding to create these variables and structures.

In total, Part 3 introduces 11 SAS macros for data preparation and is filled with numerous SAS code examples that explain step by step how derived variables can be created or data mart structures can be changed. All SAS code examples also show the underlying data tables in order to illustrate the effect of the respective SAS code. Note that all SAS macros and code examples in this book can be downloaded from the companion Web site at http://support.sas.com/publishing/bbu/companion_site/60502.html.

Part 3 contains nine chapters:

- In *Chapter 13 – Accessing Data,* we will deals with how data can be loaded into SAS and be loaded into SAS data sets.

- The next two chapters deal with the transposition of data to the appropriate data mart structure:

  □ In *Chapter 14 – Transposing One-and Multiple-Rows-per-Subject Data Structures*, we will show how we can transpose data marts in order to change between one-row-per-subject structures and multiple-rows-per-subject structures.

  □ In *Chapter 15 – Transposing Longitudinal Data*, we will show how to change between different data structures for longitudinal data marts.

- Chapters 16 and 17 will show how derived variables for interval-scaled data and categorical data can be created. The methods that are presented here apply to one-row-per-subject tables and multiple-rows-per-subject tables.

  □ In *Chapter 16 – Transformations of Interval-Scaled Variables*, we will consider topics such as standardization, the handling of time intervals, and the binning of observations.

  □ In *Chapter 17 – Transformations of Categorical Variables*, we will cover typical topics for categorical variables such as dummy variables, the definition of an OTHER group, and multidimensional categorical variables.

- The treatment of data of one-to-many relationships in the creation of a one-row-per-subject table is very important and poses a number of challenges for effective data management. Chapters 18 and 19 will show how meaningful derived variables can be created from one-to-many relationships.

  □ In *Chapter 18 – Multiple Interval-Scaled Observations per Subject*, we will cover those derived variables based on interval-scaled data such as static aggregations, correlation, concentrations and courses over time.

  □ In *Chapter 19 – Multiple Categorical Observations per Subject*, we will show how derived variables can be created considering absolute and relative frequencies and total or distinct counts for categories.

  □ Because predictive modeling is a very important task in data mining, in *Chapter 20 – Coding for Predictive Modeling*, we will specifically deal with data preparation for predictive modeling where we show how we can create meaningful input variables such as means or proportions of the target variable or the transformation or binning of interval variables.

- Finally, in *Chapter 21 – Data Preparation for Multiple-Rows-per-Subject and Longitudinal Data Marts*, we will see how we can prepare data for association or sequence analyses and how we can enhance and aggregate time series data. We will also show how to use SAS functions for this type of data preparation.

Part 3 of this book will serve as a pool of ideas for data preparation and the creation of derived variables. You are invited to read this part of the book, to learn from it, and to get ideas and come back to these chapters during data preparation work in order to look for coding examples.

Chapter 13

Accessing Data

13.1 Introduction

In *Chapter 5 – Data Structures and Data Mining*, we investigated different technical origins of data such as simple text files, spreadsheets, relational database systems, Enterprise Resource Planning systems, large text files, and hierarchical databases.

When investigating methods of data preparation we also need to look at the methods that allow accessing data and importing data to SAS. After data are available in SAS we can start to change the data mart structure, to create derived variables, and to perform aggregations.

In this chapter we will look at data access from a technical point of view. We will investigate with code examples how to access data from relational databases using SAS/ACCESS modules, then we will see how data can be accessed from Microsoft Office and from rectangular and hierarchical text files.

13.2 Accessing Data from Relational Databases Using SAS/ACCESS Modules

General

SAS offers access modules for the following relational databases: DB2, Informix, Microsoft SQL Server, MySQL, Oracle, Sybase, and Teradata. Access to ODBC and OLEDB is also available from relational databases.

13.2.1 Specifying a Libref to a Relational Database

With SAS/ACCESS modules a libref can be established that points to database tables and views in a database schema. A libref can be established with the following LIBNAME statement.

```
LIBNAME RelDB oracle USER     = 'scott'
                     PASSWORD = 'tiger'
                     PATH     = 'unix'
                     SCHEMA   = 'production';
```

In this example we specify a libref to an Oracle database. SAS/ACCESS to Oracle is needed to run this statement.

The advantage of specifying a libref is that the data in the database can be used in the syntax, as they were in the SAS tables. The library is displayed in the SAS Explorer window or the libraries listing in SAS Enterprise Guide. Tables can be opened for browsing.

13.2.2 Accessing Data

Data can now be imported from the relational database to SAS by using the appropriate procedures or a DATA step.

```
DATA data.CustomerData;
 SET RelDB.CustomerData;
RUN;
```

This DATA step imports the data from the Oracle database to SAS. An alternative is to use PROC COPY.

```
PROC COPY IN  = RelDB OUT = data;
 SELECT CustomerData UsageData;
RUN;
```

13.2.3 Implicit Pass-Through

The SAS/ACCESS modules to relational database allow the implicit pass-through of operations of statements to the source database. This means that the access module decides which statement(s) of the SAS syntax can be passed to the relational database for processing.

The following statement, for example, causes the filtering of the observations for Segment 1 in the source database and only the import of the filtered data.

```
DATA data.CustomerData_S1;
 SET RelDB.CustomerData;
 WHERE Segment = 1;
RUN;
```

In this case, the WHERE statement is passed through to the database.

The libref to the relational database can also be used to access data for analysis directly.

```
PROC MEANS DATA - RelDB.CustomerData;
 VAR age income;
RUN;
```

In the preceding statement, instead of the whole table, only the variables AGE and INCOME are imported from the table CUSTOMERDATA.

13.2.4 Explicit Pass-Through

PROC SQL can be used to perform an explicit pass-through of statements to the relational database. We see from the following statements that we specify the connection to the Oracle database in the SQL procedure. Statements that will be passed through to the database are shown in bold.

```
PROC SQL;
 CONNECT TO ORACLE (USER     = 'scott'
                    PASSWORD = 'tiger'
                    PATH     = 'unix');
 %PUT &sqlxmsg;
 CREATE TABLE CustomerData_S1
 AS
 SELECT *
 FROM CONNECTION TO ORACLE
   (SELECT *
    FROM dwh.CustomerData
    WHERE Segment = 1)
 %PUT &sqlxmsg;
 DISCONNECT FROM ORACLE;
QUIT;
```

The pass-through statements can also perform an aggregation in the relational database itself by using the appropriate group functions such as SUM or AVG and the respective GROUP BY clause.

13.3 Accessing Data from Microsoft Office

13.3.1 Using LIBNAME Statements

Access to data from Microsoft Office products such as Excel and Access can be performed using the SAS/ACCESS to PC File Formats module. In this module a LIBNAME to an Excel worksheet or an Access database can be established with the following statements.

```
LIBNAME xlslib 'c:\data\lookup.xls';

LIBNAME mdblib 'c:\data\productlist.mdb';
```

The LIBNAME XLSLIB allows accessing of the available sheets from the Excel worksheet LOOKUP.XLS, e.g., by specifying the following statements.

```
DATA lookup;
 SET xlslib.'Sheet1$'n;
RUN;
```

Note that we have used 'Sheet1$'n because we have to mask the specific character '$'.

The LIBNAME MDBLIB allows accessing of the tables and views of the Microsoft Access database.

13.3.2 Using PROC IMPORT

With the following statements, data from Microsoft Excel can be imported to SAS using PROC IMPORT.

```
PROC IMPORT OUT= lookup
            DATAFILE= "c:\data\lookup.xls"
            REPLACE;
            GETNAMES=YES;
RUN;
```

Note that we have specified GETNAMES=YES in order to read the column names from the first row.

13.4 Accessing Data from Text Files

Various formats

Data in text files can have numerous formats.

- Values can be separated by tabs, commas, or blanks, or they can be aligned at a certain column position. One row in the text file can represent exactly one row in the table.
- One row can also represent more than one table row, or one table row can extend over more than one line.

- All rows in the text file can hold data from the same entity, or the text file can also represent data from a hierarchical database where a row with customer attributes is followed by rows with the corresponding account data.
- The first row in the table can already hold data, or it can hold the variable names.

SAS can handle all of these types of text data for import. The SAS DATA step offers numerous ways to decode simple and complex text files. We will not show all options here because that would go beyond the scope of this book. We will deal with an example of rectangular-oriented data in a comma-separated file and we will show an example with hierarchical data in the next section.

Using the SAS DATA Step and the INFILE Statement

Suppose we have a text file with the following content and we want to import this file to a SAS data set.

```
Date;CustomerID;Duration;NumberOfCalls;Amount
010505;31001;8874.3;440;32.34
010505;31002;0;0;0
010505;31003;2345;221;15.99
010505;31004;1000.3;520;64.21
010505;31005;5123;50;77.21
020505;31001;1887.3;640;31.34
020505;31002;0;0;0
020505;31003;1235;221;27.99
020505;31004;1100.3;530;68.21
020505;31005;1512;60;87.21
```

The following statements allow accessing of the data from the text file by specifying the appropriate informats, and they output the data to the table CDR_DATA by specifying the respective formats.

```
DATA CDR_Data;
  INFILE 'C:\DATA\CDR.csv' DELIMITER = ';'
         MISSOVER DSD LRECL=32767 FIRSTOBS=2 ;

  INFORMAT Date         ddmmyy.
           CustomerID   8.
           Duration     8.1
           NumberOfCalls 8.
           Amount       8.2 ;

  FORMAT Date         date9.
         CustomerID   8.
         Duration     8.2
         NumberOfCalls 8.
         Amount       8.2 ;

  INPUT Date CustomerID Duration NumberOfCalls Amount;
RUN;
```

Note the following from the code:

- We use an INFILE statement in order to reference the text file.
- We specify the DELIMITER and the row number where the first observation starts (FIRSTOBS=2).

- We specify the LRECL option for the record length in order to overwrite the default of 256 characters. Note that this is not necessary for this example, but it has been included for didactic purposes because it is necessary in many cases when one row in the text file is longer than 256 characters.
- We specify INFORMAT date=MMDDYY and FORMAT date=DATE9 to convert the representation of the date format from 010505 to 01MAY2005.

The result is shown in Table 13.1.

Table 13.1: CDR_DATA table

	Date	CustomerID	Duration	NumberOfCalls	Amount
1	01MAY2005	31001	8874.30	440	32.34
2	01MAY2005	31002	0.00	0	0.00
3	01MAY2005	31003	234.50	221	15.99
4	01MAY2005	31004	1000.30	520	64.21
5	01MAY2005	31005	512.30	50	77.21
6	02MAY2005	31001	1887.30	640	31.34
7	02MAY2005	31002	0.00	0	0.00
8	02MAY2005	31003	123.50	221	27.99
9	02MAY2005	31004	1100.30	530	68.21
10	02MAY2005	31005	151.20	60	87.21

13.5 Accessing Data from Hierarchical Text Files

The data

The output dump of a hierarchical database is characterized by the fact that different rows can belong to different data objects and therefore can have different data structures. For example, customer data can be followed by account data for that customer.

An example of a hierarchical text file is shown next. We have rows for CUSTOMER data, and we have rows for USAGE data.

The CUSTOMER data rows hold the following attributes: CUSTID, BIRTH DATE, GENDER, and TARIFF. The USAGE data rows hold the following attributes: DATE, DURATION, NUMBEROFCALLS, and AMOUNT. At the first character a flag is available that indicates whether the data come from the CUSTOMER or USAGE data.

```
C;31001;160570;MALE;STANDARD
U;010505;8874.3;440;32.34
U;020505;1887.3;640;31.34
U;030505;0;0;0
U;040505;0;0;0
C;31002;300748;FEMALE;ADVANCED
U;010505;2345;221;15.99
U;020505;1235;221;27.99
U;030505;1000.3;520;64.21
C;31003;310850;FEMALE;STANDARD
U;010505;1100.3;530;68.21
U;020505;5123;50;77.21
U;030505;1512;60;87.21
```

The code

In order to input this data structure to a SAS data set the following code can be used.

```
DATA CustomerCDR;
    DROP check;
    RETAIN CustID BirthDate Gender Tariff;
    INFORMAT
            CustID          8.
            BirthDate       ddmmyy.
            Gender          $6.
            Tariff          $10.
            Date            ddmmyy.
            Duration        8.1
            NumberOfCalls   8.
            Amount          8.2 ;
    FORMAT  CustID          8.
            BirthDate       date9.
            Gender          $6.
            Tariff          $10.
            Date            date9.
            Duration        8.2
            NumberOfCalls   8.
            Amount          8.2 ;
    INFILE 'C:\data\hierarch_DB.csv' DELIMITER = ';'
            DSD LRECL=32767 FIRSTOBS=1 ;
    LENGTH check $ 1;
    INPUT @@ check $;
    IF check = 'C' THEN INPUT  CustID BirthDate Gender $ Tariff $;
    ELSE if check = 'U' THEN INPUT Date Duration NumberOfCalls Amount;
    IF check = 'U' THEN OUTPUT;
RUN;
```

Note the following from the code:

- We use a RETAIN statement to retain the values from the CUSTOMER hierarchy and make them available at the USAGE hierarchy.
- We input the first character from the data in order to decide which INPUT statement will be used to input the data.
- Finally, we output only those rows that correspond to usage data because they now also hold the de-normalized values from the customer data.

The result

The resulting data set is shown in Table 13.2.

Table 13.2: Results in a de-normalized data structure

	CustID	BirthDate	Gender	Tariff	Date	Duration	NumberOfCalls	Amount
1	31001	16MAY1970	MALE	STANDARD	01MAY2005	8874.30	440	32.34
2	31001	16MAY1970	MALE	STANDARD	02MAY2005	1887.30	640	31.34
3	31001	16MAY1970	MALE	STANDARD	03MAY2005	0.00	0	0.00
4	31001	16MAY1970	MALE	STANDARD	04MAY2005	0.00	0	0.00
5	31002	30JUL1948	FEMALE	ADVANCED	01MAY2005	234.50	221	15.99
6	31002	30JUL1948	FEMALE	ADVANCED	02MAY2005	123.50	221	27.99
7	31002	30JUL1948	FEMALE	ADVANCED	03MAY2005	1000.30	520	64.21
8	31003	31AUG1950	FEMALE	STANDARD	01MAY2005	1100.30	530	68.21
9	31003	31AUG1950	FEMALE	STANDARD	02MAY2005	512.30	50	77.21
10	31003	31AUG1950	FEMALE	STANDARD	03MAY2005	151.20	60	87.21

We have seen that a SAS DATA step can be used in a very flexible way to decode a hierarchical text file and to load the data into a SAS data set.

Output separate data sets

In order to output separate data sets for a normalized data structure, we can amend the preceding code to output a CUSTOMER table and a CDR table.

```
data CustomerCDR
     Customer(keep= CustID BirthDate Gender Tariff)
     CDR(keep = CustID Date Duration NumberOfCalls Amount);
< more DATA step code lines, see above example >
   if check = 'U' then do;  output CDR;
                            output CustomerCDR; end;
   if check = 'C' then output Customer;
RUN;
```

This outputs, in addition to the CUSTOMERCDR table, the CUSTOMER and CDR tables.

Table 13.3: CUSTOMER table

	CustID	BirthDate	Gender	Tariff
1	31001	16MAY1970	MALE	STANDARD
2	31002	30JUL1948	FEMALE	ADVANCED
3	31003	31AUG1950	FEMALE	STANDARD

Table 13.4: CDR table

	CustID	Date	Duration	NumberOfCalls	Amount
1	31001	01MAY2005	8874.30	440	32.34
2	31001	02MAY2005	1887.30	640	31.34
3	31001	03MAY2005	0.00	0	0.00
4	31001	04MAY2005	0.00	0	0.00
5	31002	01MAY2005	234.50	221	15.99
6	31002	02MAY2005	123.50	221	27.99
7	31002	03MAY2005	1000.30	520	64.21
8	31003	01MAY2005	1100.30	530	68.21
9	31003	02MAY2005	512.30	50	77.21
10	31003	03MAY2005	151.20	60	87.21

We see that we have output separate tables in a relational structure from the preceding hierarchical text file.

13.6 Other Access Methods

General

The preceding methods allow access of data at the data layer. The data are accessed directly from their source and no application logic is applied. In *Chapter 5 – Data Structures and Data Modeling*, we also mentioned the access of data from the application layer. In this case the data are not directly accessed from their underlying database tables or data dump into text files, but application logic is used to derive the data with their semantic relationships.

13.6.1 Accessing Data from Enterprise Resource Planning Applications

SAS provides interfaces, the so-called data surveyors, to enterprise resource planning applications such as SAP, SAP-BW, PeopleSoft, or Oracle e-Business Suite. These data surveyors and the underlying SAS/ACCESS modules operate on the application layer. Business logic of the respective source system is used to access data and to pre-join data that dispersed over a number of tables.

13.6.2 Enterprise Application Integration

Another possibility to access data is *enterprise application integration*. According to its name data storage and data processing applications are integrated and exchange data in real time rather than in batch mode.

A new customer record that is entered into a customer database, for example, also notifies the sales force automation system, the campaign management system, and the data warehouse in real time. This means that data are not batch-loaded between systems at certain time points, but data are exchanged on a real-time basis.

The underlying data exchange techniques are, for example, message-queuing techniques that represent the data exchange logic and data exchange techniques to disperse data to other systems. SAS Integration Technologies provide the techniques to integrate SAS into enterprise application integration.

We have now imported our data and can start bringing the data into the appropriate structure.

Chapter **14**

Transposing One- and Multiple-Rows-per-Subject Data Structures

14.1 Introduction

Changing the data structure of data sets is a significant and common task in preparing data sets. In this book, we call this *transposing* of data sets. The structure of data sets might need to be changed due to different data structure requirements of certain analyses, and data might need to be transposed in order to allow a join of different data sets in an appropriate way.

Here are the most common cases of the need for transposing:

- The preparation of a one-row-per-subject data set might need the transposition of data sets with multiple observations per subject. Other names for this task are pivoting or flattening of the data. See "Putting All Information into One Row" for more details about this.

- Longitudinal or multiple-rows-per-subject data sets are converted to a one-row-per-subject structure and vice versa depending on the requirements of different types of analysis. We will discuss data structure requirements for SAS analytic procedures in Appendix B.

- Longitudinal data sets need to be switched from one type to another (standard form of a time series, the cross-sectional data sets, or the interleaved time series) for various types of longitudinal analysis. See *Chapter 9 – The Multiple-Rows-per-Subject Data Mart* for details about this topic.

- Bringing the categories of a table with transactional data into columns for further analyses on a one-row-per-subject level.

- Intermediate results or the output from SAS procedures need to be rearranged in order to be joined to the analysis data set.

SAS is an excellent tool for transposing data. The functionality of the DATA step and PROC TRANSPOSE provides data management power that exceeds the functionality of various RDBMSs.

In this chapter we will look at different types of transpositions of data sets:

- the change from a multiple-rows-per-subject data set with repeated measurements to a one-row-per-subject data set

- the change from a one-rows-per-subject data set to a multiple-rows-per-subject data set

- the change from a multiple-rows-per-subject data set with categorical entries to a one-row-per-subject data set

This chapter details how various transpositions can be done with PROC TRANSPOSE. In Appendix C we will discuss the performance of PROC TRANSPOSE with 10,000 observations and a number of variables. As an alternative we will introduce SAS DATA step programs for transpositions.

Terminology

When looking at the shape of the data sets, we are working with "long" data sets (many rows) and "wide" data sets (many columns). In practice it has proven to be very intuitive to use the term LONG for a multiple-rows-per-subject data set and WIDE for a one-row-per-subject data set in this context. We can also use LONG to refer to univariate data sets and WIDE to refer to multivariate data sets. See Table 14.1 for an illustration.

Table 14.1: LONG and WIDE data sets

	ID	TIME	WEIGHT
1	1	1	77
2	1	2	79
3	1	3	83
4	2	1	62
5	2	2	58
6	2	3	59
7	3	1	99
8	3	2	97
9	3	3	92

Multiple-rows-per-subject data mart

Univariate data set

LONG data set

	ID	weight1	weight2	weight3
1	1	77	79	83
2	2	62	58	59
3	3	99	97	92

One-row-per-subject data set

Multivariate data set

WIDE data set

We will use the terminology LONG and WIDE in the following sections of this chapter.

Macros

In this chapter we will use the following macros:

- %MAKEWIDE creates a one-row-per-subject data mart from a multiple-rows-per-subject data mart.
- %MAKELONG creates a multiple-rows-per-subject data mart from a one-row-per-subject data mart.
- %TRANSP_CAT creates a one-row-per-subject data mart from a multiple-rows-per-subject data mart with categories per subject.
- %REPLACE_MV replaces missing values to zero for a list of variables.

14.2 Transposing from a Multiple-Rows-per-Subject Data Set to a One-Row-per-Subject Data Set

Overview

In *Chapter 8 – The One-Row-per-Subject Data Mart*, we discussed two different ways to include multiple-rows-per-subject data into a one-row-per-subject data set, namely *transposing* and *aggregating*. We also saw that in some cases it makes sense to aggregate data. There are, however, a lot of cases where the original data have to be transposed on a one-to-one basis per subject into one row.

- Repeated measurement analysis of variance can be performed with the REPEATED statement in PROC GLM of SAS/STAT. In this case data need to be in the format of one-row-per-subject.
- Indicators, describing the course of a time series, will be created. Examples are the mean usage of months 1 to 3 before cancellation divided by the mean usage in months 4 to 6 (we will discuss different indicators to describe such a course in *Chapter 18 – Multiple Interval-Scaled Observations per Subject*).

The simple case

In our terminology we are starting from a LONG data set and are creating a WIDE data set.

If we consider our two data sets in Table 14.1 the change from a LONG to a WIDE data set can be done simply with the following code:

```
PROC TRANSPOSE DATA = long
               OUT = wide_from_long(DROP = _name_)
               PREFIX = weight;
BY id ;
VAR weight;
ID time;
RUN;
```

The macro version

This code can be encapsulated simply in the following macro %MAKEWIDE:

```
%MACRO MAKEWIDE (DATA=,OUT=out,COPY=,ID=,
                 VAR=, TIME=time);

PROC TRANSPOSE DATA   = &data
               PREFIX = &var
               OUT    = &out(DROP = _name_);
 BY  &id &copy;
 VAR &var;
 ID  &time;
RUN;
%MEND;
```

Note that the data must be sorted for the ID variable. The legend to the macro parameters is as follows:

DATA and OUT

The names of the input and output data sets, respectively.

ID

The name of the ID variable that identifies the subject.

COPY

A list of variables that occur repeatedly with each observation for a subject and will be copied to the resulting data set. Note that the COPY variable(s) must not be used in the COPY statement of PROC TRANSPOSE for our purposes, but are listed in the BY statement after the ID variable. We assume here that COPY variables have the same values within one ID.

VAR

The variable that holds the values to be transposed. Note that only one variable can be listed here in order to obtain the desired transposition. See the following section for an example of how to deal with a list of variables.

TIME

The variable that numerates the repeated measurements.

Note that the TIME variable does not need to be a consecutive number; it can also have non-equidistant intervals. See the following data set from an experiment with dogs.

Table 14.2: Transposing with a non-equidistant interval numeration

	ID	Drug	Depleted	Histamine	Measurement
1	1	Morphine	N	0.04	0
2	1	Morphine	N	0.2	1
3	1	Morphine	N	0.1	3
4	1	Morphine	N	0.08	5
5	2	Morphine	N	0.02	0
6	2	Morphine	N	0.06	1
7	2	Morphine	N	0.02	3
8	2	Morphine	N	0.02	5
9	3	Morphine	N	0.07	0
10	3	Morphine	N	1.4	1
11	3	Morphine	N	0.48	3
12	3	Morphine	N	0.24	5
13	4	Morphine	N	0.17	0
14	4	Morphine	N	0.57	1
15	4	Morphine	N	0.35	3
16	4	Morphine	N	0.24	5
17	5	Morphine	Y	0.1	0
18	5	Morphine	Y	0.09	1
19	5	Morphine	Y	0.13	3
20	5	Morphine	Y	0.14	5

DOGS_LONG data set

	ID	Drug	Depleted	Histamine0	Histamine1	Histamine3	Histamine5
1	1	Morphine	N	0.04	0.2	0.1	0.08
2	2	Morphine	N	0.02	0.06	0.02	0.02
3	3	Morphine	N	0.07	1.4	0.48	0.24
4	4	Morphine	N	0.17	0.57	0.35	0.24
5	5	Morphine	Y	0.1	0.09	0.13	0.14

DOGS_WIDE data set

```
%MAKEWIDE(DATA=dogs_long,
         OUT=dogs_wide,
         ID=id,
         COPY=drug depleted,
         VAR=Histamine,
         TIME=Measurement);
```

Transposing more than one measurement variable

If more than one measurement variable will be transposed, providing a list of variables to the macro or to the VAR statement in PROC TRANSPOSE does not generate the desired results. We have to generate a transposed data set for each variable and merge the resulting data sets afterwards.

If we have, in addition to HISTAMINE, a variable HEAMOGLOBIN in our DOGS data set from the preceding example, the following statements will result in a data set with variables HISTAMINE1–HISTAMINE3 and HEAMOGLOBIN1–HEAMOGLOBIN3.

```
%MAKEWIDE(DATA=dogs_long_2vars,OUT=out1,ID=id, COPY=drug depleted,
         VAR=Histamine, TIME=Measurement);
%MAKEWIDE(DATA=dogs_long_2vars,OUT=out2,ID=id,
         VAR=Heamoglobin, TIME=Measurement);
DATA dogs_wide_2vars;
 MERGE out1 out2;
 BY id;
RUN;
```

Table 14.3: DOGS_WIDE_2VARS with two transposed variables

	ID	Drug	Depleted	Histamine0	Histamine1	Histamine3	Histamine5	Heamoglobin0	Heamoglobin1	Heamoglobin3	Heamoglobin5
1	1	Morphine	N	0.04	0.2	0.1	0.08	14.7	14	14.2	14.1
2	2	Morphine	N	0.02	0.06	0.02	0.02	14.4	14.5	14.2	14.2
3	3	Morphine	N	0.07	1.4	0.48	0.24	14.4	14.2	14.9	14.2
4	4	Morphine	N	0.17	0.57	0.35	0.24	15	14.9	14.3	14.3
5	5	Morphine	Y	0.1	0.09	0.13	0.14	14.5	14.7	14	14.2
6	6	Morphine	Y	0.12	0.11	0.1	.	14.4	14.5	14.9	15
7	7	Morphine	Y	0.07	0.07	0.06	0.07	14.3	14.5	14	14.1
8	8	Morphine	Y	0.05	0.07	0.06	0.07	14.3	14.1	14.7	14.2
9	9	Trimethaphan	N	0.03	0.62	0.31	0.22	14.1	14	14.1	14.4
10	10	Trimethaphan	N	0.03	1.05	0.73	0.6	14.1	14.7	14.5	14.3
11	11	Trimethaphan	N	0.07	0.83	1.07	0.8	14.6	15	14.2	14
12	12	Trimethaphan	N	0.09	3.13	2.06	1.23	14.5	14.4	14.3	14.1
13	13	Trimethaphan	Y	0.1	0.09	0.09	0.08	14.7	14.3	14.2	14.6
14	14	Trimethaphan	Y	0.08	0.09	0.09	0.1	14.9	14.2	14.4	14.1
15	15	Trimethaphan	Y	0.13	0.1	0.12	0.12	14.7	14.7	15	14.5
16	16	Trimethaphan	Y	0.06	0.05	0.05	0.05	14.8	14.9	14.7	14.5

Note that in the second invocation of the %MAKEWIDE macro no COPY variables were given, because they were transposed with the first invocation. Also note that every variable requires its separate transposition here. If there are a number of variables to transpose another way can be of interest.

14.2.1 Transposing the Data Set Twice

An alternative to calling macro %MAKEWIDE for each variable is to transpose the data set twice. We will transpose the data set twice and concatenate the variable name and the repetition number in between in order to be able to transpose more than one variable.

```
PROC TRANSPOSE DATA = dogs_long_2vars
               OUT  = dogs_tmp;
 BY id measurement drug depleted;
RUN;

DATA dogs_tmp;
 SET dogs_tmp;
 Varname = CATX("_",_NAME_,measurement);
RUN;

PROC TRANSPOSE DATA = dogs_tmp
               OUT  = dogs_wide_2vars_twice(drop = _name_);
 BY id drug depleted;
 VAR col1;
 ID Varname;
RUN;
```

It is up to you to decide which version best suits your programs. The last version has the advantage because it can be easily converted into a SAS macro, as the code itself does not change with the number of variables that will be transposed. To create a macro, the bold elements in the preceding code have to be replaced by macro variables in a macro.

Note the use of the CATX function, which allows an elegant concatenation of strings with a separator. For details about the CAT function family, see SAS OnlineDoc.

14.3 Transposing from a One-Row-per-Subject Data Set to a Multiple-Rows-per-Subject Data Set

Overview

In the following cases multiple observations per subject need to be in multiple rows:

- when using PROC GPLOT in order to produce line plots on the data
- when using PROC REG or other statistical procedures to calculate, for example, trend or correlation coefficients
- when calculating aggregated values per subject such as means or sums with PROC MEANS or PROC SUMMARY

The simple case

In our terminology we are starting from a WIDE data set and are creating a LONG data set.

We will again use our simple example from Table 14.1. The change from a WIDE to a LONG data set can be done simply with the following code:

```
PROC TRANSPOSE DATA = wide
               OUT = long_from_wide(rename = (col1 = Weight))
               NAME = _Measure;
BY id ;
RUN;
```

Note that with these statements we will receive a variable MEASURE that contains for each observation the name of the respective variable name of the WIDE data set (HISTAMINE1, HISTAMINE2, ...).

Table 14.4: LONG table with variable names in column MEASURE

	ID	_measure	weight
1	1	weight1	77
2	1	weight2	79
3	1	weight3	83
4	2	weight1	62
5	2	weight2	58
6	2	weight3	59
7	3	weight1	99
8	3	weight2	97
9	3	weight3	92

The following statements convert this to a variable with only the measurement numbers in a numeric format.

```
*** Create variable with measurement number;
DATA long_from_wide;
SET long_from_wide;
 FORMAT Time 8.;
 time = INPUT(TRANWRD(_measure,"weight",''),$8.);
DROP _measure;
RUN;
```

This creates the table we already saw in Table 14.1.

Table 14.5: LONG_FROM_WIDE table with the time sequence in variable TIME

	ID	weight	time
1	1	77	1
2	1	79	2
3	1	83	3
4	2	62	1
5	2	58	2
6	2	59	3
7	3	99	1
8	3	97	2
9	3	92	3

The macro

This code can be encapsulated simply in the following macro %MAKELONG:

```
%MACRO MAKELONG(DATA=,OUT=,COPY=,ID=,ROOT=,MEASUREMENT=Measurement);
PROC TRANSPOSE DATA = &data(keep = &id &copy &root.:)
               OUT = &out(rename = (col1 = &root))
               NAME = _measure;
 BY &id &copy;
RUN;
*** Create variable with measurement number;
DATA &out;
 SET &out;
 FORMAT &measurement 8.;
 &Measurement = INPUT(TRANWRD(_measure,"&root",''),$8.);
 DROP _measure;
RUN;
%MEND;
```

Note that the data must be sorted for the ID variable. The legend to the macro parameters is as follows:

DATA and OUT

The names of the input and output data sets, respectively.

ID

The name of the ID variable that identifies the subject.

COPY

A list of variables that occur repeatedly with each observation for a subject and will be copied to the resulting data set. Note that the COPY variable(s) must not be used in the COPY statement of PROC TRANSPOSE for our purposes, but should be listed in the BY statement after the ID variable. We assume here that COPY variables have the same values within one ID.

ROOT

The part of the variable name (without the measurement number) of the variable that will be transposed. Note that only one variable can be listed here in order to obtain the desired transposition. See below for an example of how to deal with a list of variables.

MEASUREMENT

The variable that numerates the repeated measurements.

Note that the TIME variable does not need to be a consecutive number, but can also have non-equidistant intervals. See the following data set from an example with dogs.

Table 14.6: Transposing with a non-equidistant interval numeration (DOGS_WIDE)

ID	Drug	Depleted	Histamine0	Histamine1	Histamine3	Histamine5
1	1 Morphine	N	0.04	0.2	0.1	0.08
2	2 Morphine	N	0.02	0.06	0.02	0.02
3	3 Morphine	N	0.07	1.4	0.48	0.24
4	4 Morphine	N	0.17	0.57	0.35	0.24
5	5 Morphine	Y	0.1	0.09	0.13	0.14

DOGS_WIDE data set

```
%MAKELONG(DATA=dogs_wide,
          OUT=
Dogs_long_from_wide,
          ID=id,
          COPY=drug
Depleted,
          ROOT=Histamine,
          MEASUREMENT=
             Measurement);
```

Table 14.7: Transposing with a non-equidistant interval numeration
(DOGS_LONG_FROM_WIDE)

ID	Drug	Depleted	Histamine	Measurement
1	1 Morphine	N	0.04	0
2	1 Morphine	N	0.2	1
3	1 Morphine	N	0.1	3
4	1 Morphine	N	0.08	5
5	2 Morphine	N	0.02	0
6	2 Morphine	N	0.06	1
7	2 Morphine	N	0.02	3
8	2 Morphine	N	0.02	5
9	3 Morphine	N	0.07	0
10	3 Morphine	N	1.4	1
11	3 Morphine	N	0.48	3
12	3 Morphine	N	0.24	5
13	4 Morphine	N	0.17	0
14	4 Morphine	N	0.57	1
15	4 Morphine	N	0.35	3
16	4 Morphine	N	0.24	5
17	5 Morphine	Y	0.1	0
18	5 Morphine	Y	0.09	1
19	5 Morphine	Y	0.13	3
20	5 Morphine	Y	0.14	5

DOGS_LONG_FROM_WIDE
data set

14.3.1 Transposing More Than One Group of Measurement Variables

If more than one group of measurement variables will be transposed, providing a list of variables to the macro or to the VAR statement in PROC TRANSPOSE does not generate the desired results. We have to generate a transposed data set for each variable and merge the resulting data sets afterwards.

If we have, in addition to the variable group HISTAMINE1–HISTAMINE3, another variable group HEAMOGLOBIN1–HEAMOGLOBIN3 in our DOGS data set from the preceding example, the following statements will result in a data set with variables HISTAMINE and HEAMOGLOBIN.

```
%MAKELONG(DATA=dogs_wide_2vars,OUT=out1,ID=id, COPY=drug depleted,
          ROOT=Histamine, MEASUREMENT=Measurement);
%MAKELONG(DATA= dogs_wide_2vars,OUT=out2,ID=id,
          ROOT=Heamoglobin, MEASUREMENT =Measurement);
DATA dogs_long_2vars;
 MERGE out1 out2;
 BY id measurement;
RUN;
```

Note that in the second invocation of the %MAKELONG macro no COPY variables were given because they were transposed with the first invocation. Also note that every variable requires its separate transposition here.

In Appendix C.2 we introduce a macro that performs transpositions from a WIDE to a LONG data set by using a SAS DATA step instead of PROC TRANSPOSE.

14.4 Transposing a Transactional Table with Categorical Entries

In the preceding sections we had multiple observations in the sense that we had repeated measurements over time (or any sequential indicator). In this section we will cover the case where we have transactional data and a list of categorical entries for each subject. We will give an example of such a table structure.

This type of data structure is typical in the following cases:

- In market basket analysis a list of purchased products is given per customer.
- In Web path analysis a list of visited pages is given for each Web site visitor.

Note that in the second case a sequence variable, such as a timestamp, is available that allows ordering the entries per subject. We will, however, ignore the sequential information and will consider only the categories, i.e., the products or the pages in the preceding cases.

Simple example

In our simple example we will use the data set SAMPSIO.ASSOCS. This data sets holds market basket data in transactional form. Each PRODUCT a CUSTOMER purchased is represented on a separate line. The variable TIME gives the purchasing sequence for each customer.

In order to transpose this data set into a one-row-per-subject data set, PROC TRANSPOSE will be used.

Table 14.8: Transactional data

	CUSTOMER	TIME	PRODUCT
1	0	0	hering
2	0	1	comed_b
3	0	2	olives
4	0	3	ham
5	0	4	turkey
6	0	5	bourbon
7	0	6	ice_crea
8	1	0	baguette
9	1	1	soda
10	1	2	hering
11	1	3	cracker
12	1	4	heineken
13	1	5	olives
14	1	6	comed_b
15	2	0	avocado
16	2	1	cracker
17	2	2	artichok
18	2	3	heineken
19	2	4	ham
20	2	5	turkey
21	2	6	sardines

```
PROC TRANSPOSE DATA = sampsio.assocs(obs=21)
               OUT = assoc_tp (DROP = _name_);
  BY customer;
  ID Product;
RUN;
```

Table 14.9: Transposed products per customer

	CUSTOMER	bourbon	corned_b	ham	hering	ice_crea	olives	turkey	baguette
1	0		1	1	1	1	1	1	
2	1			1		1		1	1
3	2				1			1	
4	3	1			1		1	1	1
5	4				1	1		1	1
6	5				1		1		
7	6	1				1	1	1	
8	7	1	1			1			1
9	8	1					1		1
10	9	1	1		1				
11	10							1	1
12	11		1		1				1
13	12		1		1		1		

The big advantage of PROC TRANSPOSE as part of SAS compared with other SQL-like languages is that the list of possible products itself is not part of the syntax. Therefore, PROC TRANSPOSE can be used flexibly in cases where the list of products is extended.

14.4.1 Duplicate Entries per Subject

You might have noticed that in the preceding example we used the data set option OBS=21 in the PROC TRANSPOSE syntax. This option caused only the first 21 rows of the data set SAMPOSIO.ASSOCS to be processed. We did this for didactic purposes here because omitting this option in our example and processing all the data would have caused the following error:

```
ERROR: The ID value "ice_crea" occurs twice in the same BY
group.
NOTE: The above message was for the following by-group:
      CUSTOMER=45
ERROR: The ID value "avocado" occurs twice in the same BY
group.
NOTE: The above message was for the following by-group:
      CUSTOMER=57
```

Printing the data for customer 45 into the output windows gives the following:

```
Obs     CUSTOMER     TIME      PRODUCT

316        45          0        corned_b
317        45          1        peppers
318        45          2        bourbon
319        45          3        cracker
320        45          4        chicken
321        45          5        ice_crea
322        45          6        ice_crea
```

We see that customer 45 has an "ice_crea" entry in times 5 and 6. PROC TRANSPOSE however, does not allow duplicate values per subject. We will therefore introduce the macro %TRANSP_CAT, which uses PROC FREQ before PROC TRANSPOSE in order to compress duplicate entries per subject.

```
%MACRO TRANSP_CAT(DATA = , OUT = TRANSP, VAR = , ID =);

PROC FREQ DATA  = &data NOPRINT;
 TABLE &id * &var / OUT = tmp(DROP = percent);
RUN;

PROC TRANSPOSE DATA = tmp
               OUT  = &out (DROP = _name_);
 BY &id;
 VAR count;
 ID &var;
RUN;


%MEND;
```

Note that unlike our simple example we use a VAR statement in PROC TRANSPOSE, which holds the frequency for each category per subject. This frequency is then inserted into the respective column, as we can see from the transposed data set for customer 45.

Table 14.10: Transposed products per customer

	CUSTOMER	bourbon	corned_b	ham	hering	ice_crea
46	45	1	1	.	.	2
47	46	.	.	.	.	.
48	47	.	.	.	1	.
49	48	1	1	.	.	1
50	49	.	.	1	1	.
51	50	1	.	.	.	.
52	51	.	.	.	.	1
53	52	.	.	.	1	.
54	53	.	1	.	1	.
55	54	.	1	1	1	.
56	55	.	1	1	1	.
57	56	.	1	.	.	1
58	57	.	.	.	1	.
59	58	1	.	1	.	.

The description of the macro syntax is as follows:

DATA and OUT

The names of the input and output data sets, respectively.

ID

The name of the ID variable that identifies the subject.

VAR

The variable that holds the categories, e.g., in market basket analysis the products a customer purchased.

The invocation for our preceding example is very straightforward:

```
%TRANSP_CAT (DATA = sampsio.assocs, OUT = assoc_tp,
             ID = customer, VAR = Product);
```

14.4.2 Replacing Missing Values

From Table 14.8 we saw that, after transposing, those customer/product combinations that do not appear in the transactional table contain a missing value. Replacing those missing values against a zero value can be done by a simple statement like this:

```
IF bourbon = . THEN bourbon = 0;
```

Or if an indicator variable with 0 and 1 as possible values has to be created:

```
IF bourbon >= 1 THEN bourbon = 1 ELSE bourbon = 0;
```

Note that this can also be processed simply as in the following code. Here we use the property that bourbon NE . evaluates to 1 for all bourbon values other than MISSING, and to 0 for MISSING values.

```
bourbon = (bourbon NE .);
```

For a list of variables this can be done more efficiently by using an ARRAY statement in the SAS DATA step:

```
ARRAY prod {*}  apples artichok avocado baguette bordeaux bourbon
                chicken coke corned_b cracker ham heineken hering
                ice_crea olives peppers sardines soda steak turkey;
 DO i = 1 TO dim(prod);
   IF prod{i} . THEN prod{i}=0;
 END;
 DROP i;
```

As the replacement of missing values of a list of variables occurs very frequently, the preceding code is shown here as a macro version:

```
%MACRO REPLACE_MV(cols,mv=.,rplc=0);
 ARRAY varlist {*}  &cols;
 DO _i = 1 TO dim(varlist);
   IF varlist{_i} = &mv THEN varlist{_i}=&rplc;
 END;
 DROP _i;
%MEND;
```

The explanation of the macro variables is as follows:

COLS

The list of variables for which missing values will be replaced.

MV

The definition of the missing value, default = .

RPCL

The replace value, default = 0.

The invocation of the macro is as follows:

```
DATA assoc_tp;
 SET assoc_tp;
  %REPLACE_MV (apples artichok avocado baguette bordeaux bourbon
               chicken coke corned_b cracker ham heineken hering
               ice_crea olives peppers sardines soda steak turkey);
RUN;
```

An excerpt from the resulting data set is shown in Table 14.11.

Table 14.11: Data set with replaced missing values

	CUSTOMER	bourbon	corned_b	ham	hering	ice_crea	olives	turkey	baguette
1	0	1	1	1	1	1	1	1	0
2	1	0	1	0	1	0	1	0	1
3	2	0	0	1	0	0	0	1	0
4	3	1	0	1	0	1	1	1	0
5	4	0	1	0	1	0	1	1	0
6	5	0	0	1	0	1	0	0	0
7	6	1	0	0	0	1	1	1	0
8	7	1	1	0	0	1	0	0	1
9	8	1	0	0	0	0	1	0	1
10	9	1	1	0	1	0	0	0	0
11	10	0	0	0	0	0	0	1	1
12	11	0	1	0	1	0	0	0	1
13	12	0	1	0	1	0	1	0	0
14	13	0	0	0	0	1	0	0	1

In Appendix C.3 we will introduce a macro for the DATA step version of a transposition of transactional categories. This version also automatically generates the list of variables for the replacment of missing values.

14.5 Creating Key-Value Tables

General
In Chapter 9 we saw an example of a key-value table. In this section we will see how we can create a key-value table from a one-row-per-subject table and how we can re-create a one-row-per-subject table from a key-value table.

Creating a key-value table
In this example we will create a key-value table from the SASHELP.CLASS data set. First we will create an artificial ID variable.

```
DATA class;
 SET sashelp.class;
 ID = _N_;
RUN;
```

Then we will transpose this table by the ID variable and use the NAME variable in the BY statement in order to copy it to the key-value table.

```
PROC TRANSPOSE DATA = class OUT = class_tp;
 BY ID name;
 VAR sex age height weight;
RUN;
```

Then we will compress the leading and trailing blanks and rename the variables KEY and VALUE.

```
DATA Key_Value;
 SET class_tp;
 RENAME _name_ = Key;
 Value = strip(col1);
 DROP col1;
RUN;
```

Finally, we will receive the following table.

Table 14.12: Key-value table

	ID	Name	Key	Value
1	1	Alice	Sex	F
2	1	Alice	Age	13
3	1	Alice	Height	56.5
4	1	Alice	Weight	84
5	2	Barbara	Sex	F
6	2	Barbara	Age	13
7	2	Barbara	Height	65.3
8	2	Barbara	Weight	98
9	3	Carol	Sex	F
10	3	Carol	Age	14
11	3	Carol	Height	62.8
12	3	Carol	Weight	102.5
13	4	Jane	Sex	F
14	4	Jane	Age	12
15	4	Jane	Height	59.8
16	4	Jane	Weight	84.5

14.5.1 Create a One-Row-per-Subject Table from a Key-Value Table

If we receive data in a key-value structure we can create a one-row-per-subject table with the following code.

```
PROC TRANSPOSE DATA = key_value OUT = one_row;
 BY id name;
 VAR value;
 ID key ;
RUN;

DATA one_row;
 SET one_row(RENAME = (Age=Age2 Weight=Weight2 Height=Height2));
 FORMAT Age 8. Weight Height 8.1;
 Age = INPUT(Age2,$8.);
 Weight = INPUT(Weight2,$8.);
```

```
   Height = INPUT(Height2,$8.);
   DROP Age2 Weight2 Height2;
RUN;
```

The DATA step is needed to create numeric variables from textual key-values. Note that we first rename textual variables in the DATA step option in order to be able to give the numerically formatted variables the original name.

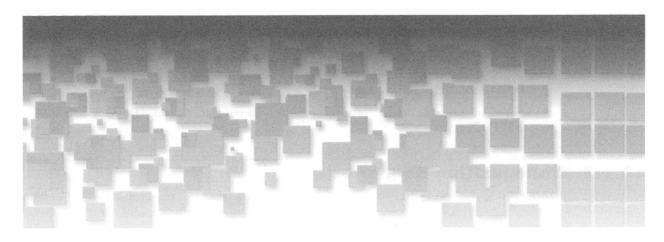

Chapter 15

Transposing Longitudinal Data

15.1 Introduction

In *Chapter 10 – Data Structures for Longitudinal Analysis*, we discussed the different data structures that can be found in the context of longitudinal data.

In the following sections we will show how the data structure can be switched between the three main data mart structures and also between combinations of them. In time series analysis there is a need to prepare the data in different structures in order to be able to specially analyze VALUES and CATEGORIES.

We will use the following abbreviation in our naming:

- the standard time series form (STANDARD)
- the cross-sectional dimensions data structure (CROSS SECTIONAL)
- the interleaved time series data structure (INTERLEAVED)

15.1.1 The Example Data

We will use sales data from a do-it-yourself market for the time period August 15, 2005, to October 14, 2005. For each day the quantity of sold pieces and the sales volume are available. The data are available for three different product categories: ELECTRO, GARDENING, and TOOLS.

Considering our three main entities in longitudinal data structures (see Chapter 10), we have the following:

- TIME: data and a daily aggregation level
- VALUE: quantity and volume
- CATEGORIES: dimension product category with the values ELECTRO, GARDENING, and TOOLS

15.1.2 Transposition Scenarios

By using PROC TRANSPOSE we will now change the data structures for these data in different ways. Starting with standard scenarios where we consider our two value variables or our three product categories only we will change between the following:

- VALUES: standard time series form and interleaved time series
- CATEGORIES: cross-sectional dimensions and standard time series form

Then we will consider our value variables and our product categories in one data set and will change between the following:

- CATEGORIES in cross sections and VALUES in standard form versus CATEGORIES in standard form and VALUES interleaved
- CATEGORIES in cross sections and VALUES in standard form versus CATEGORIES in cross sections and VALUES interleaved
- CATEGORIES in cross sections and VALUES interleaved versus CATEGORIES and VALUES in standard form

Note in the following examples that data in the tables are already sorted that way so that they can be transposed in the TRANSPOSE procedure with the given BY statement. If your data are sorted in a different way, you will need to sort them first.

Data in cross-sectional dimension data structures are usually sorted first by the cross-sectional dimension, and then by date. This sort is omitted in the following examples; otherwise, more rows of the table would be needed to show in the figure in order to visualize the different cross-sectional dimensions. If desired, at the end of each cross-sectional dimension example the data can be sorted with the following statement:

```
PROC SORT DATA = cross_section_result;
 BY <cross section> DATE;
RUN;
```

15.2 Standard Scenarios

General

In this section we will see how we can change between the typical longitudinal data structures:

- standard time series form and interleaved time series (STANDARD and INTERLEAVED)
- cross-sectional dimensions and standard time series form (CROSS SECTIONAL and STANDARD)

15.2.1 From STANDARD to INTERLEAVED and Back

```
PROC TRANSPOSE DATA = diy_standard
               OUT  = diy_intleaved
                      (rename = (Col1 = Value))
               NAME = Type;
  BY date;
  RUN;
```

Date	Quantity	Volume
15/08/05	7321	39079
16/08/05	7926	41357
17/08/05	9507	46972
19/08/05	8607	43731
20/08/05	8034	39402
21/08/05	7775	38277
22/08/05	7723	36687
23/08/05	7413	38813
24/08/05	8229	42677
26/08/05	6914	34210
27/08/05	7419	37048
28/08/05	6730	34626
29/08/05	7228	33361
30/08/05	9444	51607
31/08/05	10830	57812

	Date	Type	Value
1	15/08/05	Quantity	7321
2	15/08/05	Volume	39079
3	16/08/05	Quantity	7926
4	16/08/05	Volume	41357
5	17/08/05	Quantity	9507
6	17/08/05	Volume	46972
7	19/08/05	Quantity	8607
8	19/08/05	Volume	43731
9	20/08/05	Quantity	8034
10	20/08/05	Volume	39402
11	21/08/05	Quantity	7775
12	21/08/05	Volume	38277
13	22/08/05	Quantity	7723
14	22/08/05	Volume	36687
15	23/08/05	Quantity	7413
16	23/08/05	Volume	38813

```
PROC TRANSPOSE DATA = diy_intleaved
               OUT  = diy_standard_back
                      (drop = _name_);
  BY date;
  ID Type;
  VAR value;
  RUN;
```

Note the following:

- PROC TRANSPOSE allows a short and elegant syntax for the change between different longitudinal data structures. The variable names and resulting new categories are not included in the syntax. It can therefore flexibly be used if variables are added.

- Data need to be sorted by DATE because PROC TRANSPOSE uses DATE in the BY statement.

- Transferring data from columns to rows only needs a BY statement and the NAME option in PROC TRANSPOSE.

- Transferring data from rows to columns needs a BY, ID, and VAR statement in PROC TRANSPOSE.

15.2.2 From CROSS SECTIONAL to STANDARD and Back

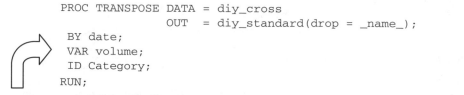

```
PROC TRANSPOSE DATA = diy_cross
                OUT  = diy_standard(drop = _name_);
    BY date;
    VAR volume;
    ID Category;
RUN;
```

	Date	Category	Volume
1	15/08/05	ELECTRO	15725
2	15/08/05	GARDENING	13913
3	15/08/05	TOOLS	9441
4	16/08/05	ELECTRO	15120
5	16/08/05	GARDENING	16315
6	16/08/05	TOOLS	9922
7	17/08/05	ELECTRO	16631
8	17/08/05	GARDENING	18996
9	17/08/05	TOOLS	11345
10	19/08/05	ELECTRO	18080
11	19/08/05	GARDENING	16325
12	19/08/05	TOOLS	9326
13	20/08/05	ELECTRO	15604
14	20/08/05	GARDENING	14690
15	20/08/05	TOOLS	9108
16	21/08/05	ELECTRO	14518
17	21/08/05	GARDENING	14388
18	21/08/05	TOOLS	9371

	Date	ELECTRO	GARDENING	TOOLS
1	15/08/05	15725	13913	9441
2	16/08/05	15120	16315	9922
3	17/08/05	16631	18996	11345
4	19/08/05	18080	16325	9326
5	20/08/05	15604	14690	9108
6	21/08/05	14518	14388	9371
7	22/08/05	13048	15249	8390
8	23/08/05	13857	13974	10982
9	24/08/05	14869	15704	12104
10	26/08/05	12262	13836	8112
11	27/08/05	15011	13438	8599
12	28/08/05	13612	12625	8389
13	29/08/05	11546	13566	8249
14	30/08/05	21352	16918	13337
15	31/08/05	22900	20813	14099
16	02/09/05	15333	15626	8896
17	03/09/05	13156	13306	8082
18	04/09/05	19294	16361	16267
19	05/09/05	15917	15587	15539

```
PROC TRANSPOSE DATA = diy_standard
                OUT  = diy_cross_back(rename = (Col1 = Value))
    NAME = Type;
    BY date;
RUN;
```

As you can see, besides the type of sorting, there is no technical difference between a cross-sectional data structure and an interleaved data structure. Both data structures can be converted into the standard time series form and be retrieved from the standard time series form with the same statements.

15.3 Complex Scenarios

General

When considering our VALUE variables and our product CATEGORIES we have the following scenarios.

- CATEGORIES in cross sections and VALUES in standard form versus CATEGORIES in standard form and VALUES interleaved
- CATEGORIES in cross sections and VALUES in standard form versus CATEGORIES in cross sections and VALUES interleaved
- CATEGORIES in cross sections and VALUES interleaved versus CATEGORIES and VALUES in standard form

15.3.1 CATEGORIES in Cross Sections and VALUES in Standard Form versus CATEGORIES in Standard Form and VALUES Interleaved

In the following example we see the data detailed for both values and categories and show how to simply exchange the columns with rows within one date.

```
PROC TRANSPOSE DATA = diy_cross_standard
               OUT  = diy_intlv_standard
               NAME = Type;
  BY date;
  ID category;
RUN;
```

	Date	Category	Quantity	Volume
1	15/08/05	ELECTRO	2441	15725
2	15/08/05	GARDENIN	3822	13913
3	15/08/05	TOOLS	1058	9441
4	16/08/05	ELECTRO	2429	15120
5	16/08/05	GARDENIN	4365	16315
6	16/08/05	TOOLS	1132	9922
7	17/08/05	ELECTRO	3240	16631
8	17/08/05	GARDENIN	5004	18996
9	17/08/05	TOOLS	1263	11345
10	19/08/05	ELECTRO	2920	18080
11	19/08/05	GARDENIN	4595	16325
12	19/08/05	TOOLS	1092	9326
13	20/08/05	ELECTRO	2559	15604
14	20/08/05	GARDENIN	4215	14690
15	20/08/05	TOOLS	1260	9108
16	21/08/05	ELECTRO	2512	14518
17	21/08/05	GARDENIN	4297	14388
18	21/08/05	TOOLS	966	9371

	Date	Type	ELECTRO	GARDENING	TOOLS
1	15/08/05	Quantity	2441	3822	1058
2	15/08/05	Volume	15725	13913	9441
3	16/08/05	Quantity	2429	4365	1132
4	16/08/05	Volume	15120	16315	9922
5	17/08/05	Quantity	3240	5004	1263
6	17/08/05	Volume	16631	18996	11345
7	19/08/05	Quantity	2920	4595	1092
8	19/08/05	Volume	18080	16325	9326
9	20/08/05	Quantity	2559	4215	1260
10	20/08/05	Volume	15604	14690	9108
11	21/08/05	Quantity	2512	4297	966
12	21/08/05	Volume	14518	14388	9371
13	22/08/05	Quantity	2306	4478	939
14	22/08/05	Volume	13048	15249	8390
15	23/08/05	Quantity	2109	4199	1105
16	23/08/05	Volume	13857	13974	10982
17	24/08/05	Quantity	2524	4538	1167
18	24/08/05	Volume	14869	15704	12104

```
PROC TRANSPOSE DATA = diy_intlv_standard
               OUT  = diy_cross_standard_back
               NAME = Category;
  BY date;
  ID Type;
  ;
RUN;
```

15.3.2 CATEGORIES in Cross Sections and VALUES in Standard Form versus CATEGORIES in Cross Sections and VALUES Interleaved

In this example we show how to create a data structure where both categories and values are represented by a row. Therefore, for each date six rows exist (3 categories x 2 values).

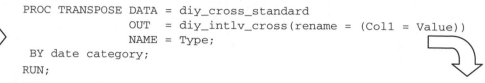

```
PROC TRANSPOSE DATA = diy_cross_standard
               OUT  = diy_intlv_cross(rename = (Col1 = Value))
               NAME = Type;
  BY date category;
RUN;
```

	Date	Category	Quantity	Volume
1	15/08/05	ELECTRO	2441	15725
2	15/08/05	GARDENING	3822	13913
3	15/08/05	TOOLS	1058	9441
4	16/08/05	ELECTRO	2429	15120
5	16/08/05	GARDENING	4365	16315
6	16/08/05	TOOLS	1132	9922
7	17/08/05	ELECTRO	3240	16631
8	17/08/05	GARDENING	5004	18996
9	17/08/05	TOOLS	1263	11345
10	19/08/05	ELECTRO	2920	18080
11	19/08/05	GARDENING	4595	16325
12	19/08/05	TOOLS	1092	9326
13	20/08/05	ELECTRO	2559	15604
14	20/08/05	GARDENING	4215	14690
15	20/08/05	TOOLS	1260	9108
16	21/08/05	ELECTRO	2512	14518
17	21/08/05	GARDENING	4297	14388
18	21/08/05	TOOLS	966	9371

	Date	Category	Type	Value
1	15/08/05	ELECTRO	Quantity	2441
2	15/08/05	ELECTRO	Volume	15725
3	15/08/05	GARDENING	Quantity	3822
4	15/08/05	GARDENING	Volume	13913
5	15/08/05	TOOLS	Quantity	1058
6	15/08/05	TOOLS	Volume	9441
7	16/08/05	ELECTRO	Quantity	2429
8	16/08/05	ELECTRO	Volume	15120
9	16/08/05	GARDENING	Quantity	4365
10	16/08/05	GARDENING	Volume	16315
11	16/08/05	TOOLS	Quantity	1132
12	16/08/05	TOOLS	Volume	9922
13	17/08/05	ELECTRO	Quantity	3240
14	17/08/05	ELECTRO	Volume	16631
15	17/08/05	GARDENING	Quantity	5004
16	17/08/05	GARDENING	Volume	18996
17	17/08/05	TOOLS	Quantity	1263

```
PROC TRANSPOSE DATA = diy_intlv_cross
               OUT  = diy_cross_standard_back2(drop = _name_);
  BY date category;
  VAR value;
  ID type;
RUN;
```

15.3.3 CATEGORIES in Cross Sections and VALUES Interleaved versus CATEGORIES and VALUES in Standard Form

In this example we show how to create a table that has only one row for each date value. We will have all six combinations from CATEGORY and VALUE (3 x 2) in the columns.

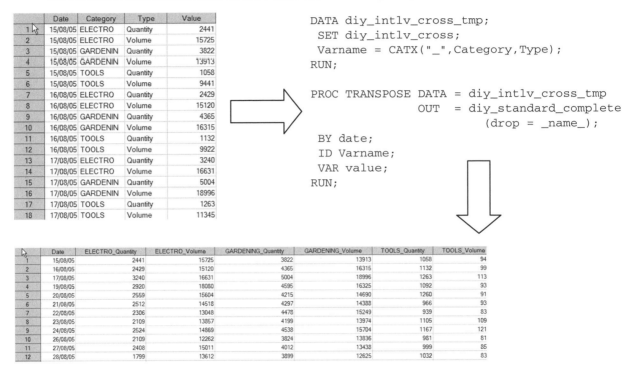

```
DATA diy_intlv_cross_tmp;
  SET diy_intlv_cross;
  Varname = CATX("_",Category,Type);
RUN;

PROC TRANSPOSE DATA = diy_intlv_cross_tmp
               OUT  = diy_standard_complete
                      (drop = _name_);
  BY date;
  ID Varname;
  VAR value;
RUN;
```

Note the following:

- The CATEGORIES in cross sections and VALUES interleaved have some similarity with the key-value table we discussed in Chapter 14.

- In order to create a standard form if more than one cross section and/or interleaved group exists, a preceding DATA step is needed to create a concatenated variable that is used for the transposition.

- The SAS function family CAT* allows concatenation of text strings with different options by automatically truncating leading and trailing blanks.

Chapter 16

Transformations of Interval-Scaled Variables

16.1 Introduction

In this chapter we will deal with transformations of interval variables. With *interval variables* we mean variables such as age, income, and number of children, which are measured on an interval or ratio scale. The most obvious transformation of interval variables is to perform a calculation such as addition or multiplication on them. However, we will also see in this chapter that the transformations are not restricted to simple calculations.

This chapter is divided into six sections, which will deal with the following topics:

Simple derived variables, where we will look at simple calculations between interval variables and the creation of interactions.

Relative derived variables, where we will consider the creation of ratios and proportions by comparing a value to another value.

Time and time intervals, where we will see which calculation methods are available for time intervals and which representations for time variables are possible.

Binning observations into groups, where we will show that an interval value can be grouped with IF_THEN/ELSE clauses or formats.

Transformation of variables, where we will show mathematical or statistical transformations of variables and the filtering of outliers.

Replacement of missing values, where we will deal with replacing missing values with the mean of other variables.

Macros

In this chapter we will introduce the following SAS macros:

- %INTERACT, to create interactions and quadratic terms
- %SHIFT, to shift extreme values toward the center of the distribution
- %FILTER, to filter outliers
- %REPLACE_MV, to replace missing values

16.2 Simple Derived Variables

Overview

The rationale for derived variables is that we create variables that hold information that is more suitable for analysis or for interpretation of the results. Derived variables are created by performing calculations on one or more variables and creating specific measures, sums, ratios, aggregations, or others.

Derived variables can be a simple calculation on one or more variables. For example, we can create the *body mass index* (BMI) with the following expression if WEIGHT is measured in kilograms and HEIGHT is measured in centimeters:

```
BMI = Weight / Height **2;
```

In this section we will deal with interactions, quadratic terms, and derived sums from a list of variables.

16.2.1 Interactions and Quadratic Terms

An *interaction* is a term in a model in which the effect of two or more factors is not simply additive. In the case of two binary input factors, the effect of the co-occurrence of both factors can be higher or lower than the sum of the sole occurrence of each factor. If the business rationale suggests that interactions exist between variables, the interactions will be included in the analysis.

Variables that contain an interaction between two variables can be easily built by a simple multiplication of the variables. The following code computes an interaction variable of AGE and WEIGHT:

```
INT_AGE_WEIGHT = AGE * WEIGHT;
```

Note that by definition the interaction is missing if one of its components is missing, which is true for our expression. In some procedures, e.g., PROC REG, interactions cannot be created in the MODEL statement but need to be present in the input data.

A *quadratic term* and *cubic term* for AGE can be easily created by the following statement:

```
AGE_Q = AGE ** 2;
AGE_C = AGE ** 3;
```

Note that it is up to the programmer to name the AGE variable. AGE2 and AGE3 are also common to express the quadratic and cubic terms. When creating derived variables, the consecutive numbering of variables is very common. Although this "quick and dirty" approach is not the best programming style, we have to consider that the name AGE2 as the quadratic term can be easily confused with an age variable that is created with a different definition.

See also *Chapter 20 – Coding for Predictive Modeling*, for the visualization of quadratic relationships between variables and a discussion of the specification of quadratic terms in predictive modeling.

16.2.2 Coding of Interactions and Quadratic Terms

For a long list of variables, creating all necessary interactions is an exhausting task that can easily be automated with the following macro:

```
%MACRO interact(vars, quadr = 1, prefix = INT);
*** Load the number of items in &VARS into macro variable NVARS;
%LET c=1;
%DO %WHILE(%SCAN(&vars,&c) NE);
  %LET c=%EVAL(&c+1);
%END;
%LET nvars=%EVAL(&c-1);
 %DO i = 1 %TO &nvars;
   %DO j = %EVAL(&i+1-&quadr) %TO &nvars;
      &prefix._%SCAN(&vars,&i)_%SCAN(&vars,&j) =
                    %SCAN(&vars,&i) * %SCAN(&vars,&j);
   %END;
 %END;
%MEND;
```

The explanation of the macro variables is as follows:

VARS

The list of variables for which interactions will be created.

QUADR

Determines whether quadratic terms will also be created (Default = 1) or if only the non-diagonal elements of the variable-matrix will be created (QUADR = 0).

PREFIX

The prefix for the new variable names, Default = INT.

The macro has to be used in a DATA step:

```
DATA class2;
 SET sashelp.class;
 %INTERACT(age weight height,quadr=1);
RUN;
```

Here are two examples of its invocation and the respective code that is created:

```
%INTERACT(age weight height,3,quadr=1);

INT_age_age = age * age;
INT_age_weight = age * weight;
INT_age_height = age * height;
INT_weight_weight = weight * weight;
INT_weight_height = weight * height;
INT_height_height = height * height;
```

and

```
%INTERACT(age weight height,3,quadr=0);

INT_age_weight = age * weight;
INT_age_height = age * height;
INT_weight_height = weight * height;
```

16.2.3 Derived Variables: Sums and Means

In order to calculate simple descriptive measures over variables (columns) it is advisable to use the respective SAS functions such as MEAN, SUM, and others. For details, see SAS Help and Documentation. Instead of the following:

```
SUM_USAGE = USAGE1 + USAGE2 + USAGE3;
MEAN_USAGE = (USAGE1 + USAGE2 + USAGE3)/3;
```

the respective SAS function should be used:

```
SUM_USAGE = SUM(USAGE1 USAGE2 USAGE3);
MEAN_USAGE = MEAN(USAGE1 USAGE2 USAGE3);
```

The SAS functions also have the advantage that they handle missing values correctly. If, for example, USAGE2 is a missing value, the function will not return a missing value but sum the non-missing values. The MEAN function will correctly divide by the number of non-missing values.

16.2.4 Sums of Binary Variables or Expressions

An effective way to reduce the dimensionality in the input data is to create a counter variable that counts the occurrence of certain facts (responses, event…). For example, we have in a survey a set of variables on customer opinion about company A. Each question can be answered with YES/NO/MISSING, where YES is coded with 1, NO with 0, and MISSING with 9.

- Question1   Company A has a better product offer than the competition.
- Question2   Company A has more competitive pricing than the competition.
- Question3   Company A has a better standing in the market.
- Question4   Company A has more qualified employees.
- Question5   Company A supports non-profit and welfare organizations.

We build a variable by summing the YES answers for each subject. This variable will contain information about customer opinion and loyalty about the company.

```
LOYALTY = sum((question1=1), (question2=1), (question3=1),
              (question4=1), (question5=1));
```

We sum the result of the Boolean expressions "QuestionX = 1". Again, we have an example of how an informative variable can be coded simply. This variable ranges from 0 to 5 and is a point score for loyalty. This can be generalized to all situations where we want to count events that fulfill a certain condition. The expression itself can, for example, also be in the form "Value > 200" or "Tariff IN ('START' 'CLASSIC')".

16.3 Derived Relative Variables

General

For some measurement variables it makes sense to set them relative to another measurement variable. The business rationale for derived relative variables is that the absolute number is not meaningful and we have to set the value relative to another value. For example, the relative cost of a hospital per patient day is used for comparison rather than the absolute cost. Basically we can distinguish between two types of derived relative variables, namely proportions and ratios.

We talk about *proportions* if the numerator and the denominator have the same measurement unit. In contrast, *ratios* are calculated by a fraction, where the numerator and denominator have different measurement units.

Proportions can be expressed in percent and are interpreted as "part of the whole." For example, 65% of the employees of company A have used tariff XY in a proportion, whereas the average number of phone calls per day is a ratio, as we divide, for example, the monthly number of phone calls by the number of days per month.

Frequently used denominators for derived relative variables include the following:

- number of employees
- number of days since the start of the customer relationship
- number of days since admission to the hospital or since the start of treatment
- number of months in the observation period
- number of contracts
- total amount of usage (minutes, number of calls, transactions, euro, dollars)

16.3.1 Creating Derived Relative Variables

Coding derived relative variables is basically no big deal. It is just a division of two values. However, the following points can be useful for effective coding:

- Use SAS functions to create sums or means because they can deal with missing values.
  ```
  CALLS_TOTAL = CALLS_FIXED + CALLS_INTERNATIONAL + CALLS_MOBILE;
  ```

  will evaluate to MISSING if one of the variables is missing.

```
CALLS_TOTAL = SUM(CALLS_FIXED, CALLS_INTERNATIONAL, CALLS_MOBILE);
```

will ignore the MISSING value and sum the non-missing values.

- Use an ARRAY statement if more variables have to be divided by the same number:

```
Calls_total = SUM(calls_fixed, calls_international, calls_mobile);
ARRAY abs_vars {*} calls_fixed calls_international calls_mobile;
ARRAY rel_vars {*} calls_fixed_rel calls_international_rel
calls_mobile_rel;
DO i = 1 TO DIM(abs_vars);
  rel_vars{i}._rel = abs_vars{i}/calls_total;
END;
DROP i;
```

16.3.2 Relative Variables Based on Population Means

In addition to the distinction of derived relative variables in ratios and proportions, there is also the distinction of whether the value of the derived variable depends only on values of the subject itself, or whether it is set relative to the values of other subjects (see also *Chapter 11 – Considerations for Data Marts*).

For an explanation consider a simple example. We will use the data from SASHELP.CLASS and only concentrate on the variable WEIGHT. The mean weight of the 19 pupils is 100.03 pounds. We want to use this mean to create derived relative variables that show whether a subject has a value above or below the mean. In the SAS program in the next section we will create these derived variables:

WEIGHT_SHIFT

The mean is subtracted from the values, which results in a shift of the distribution to a zero mean; the shape of the distribution is not changed.

WEIGHT_RATIO

The values are divided by the mean. Assuming a positive mean, all positive values that are smaller than the mean are squeezed into the interval (0,1), whereas positive values larger than the mean are transferred to the interval $(1,\infty)$.

WEIGHT_CENTRATIO

The centered ratio combines the calculations from WEIGHT_SHIFT and WEIGHT_RATIO by shifting the distribution to a zero mean and dividing the values by the mean.

WEIGHT_STD

Subtracting the mean and dividing the values by the standard deviation leads to the standardized values.

WEIGHT_RNK

Ranking the values by their size gives the ordinal-scaled variable WEIGHT_RNK. Ties have been set to the lower rank.

16.3.3 Coding Relative Variables Based on Population Means

```
DATA class;
 FORMAT ID 8.;
 SET sashelp.class(KEEP = weight);
 ID = _N_;
 Weight_Shift = Weight-100.03;
 Weight_Ratio = Weight/100.03;
 Weight_CentRatio = (Weight-100.03)/100.03;
RUN;

PROC STANDARD DATA = class(keep = id weight)  out = class2
              mean=0 std=1;
 VAR weight;
RUN;

DATA class;
 MERGE class class2(RENAME = (weight = Weight_Std));
 BY id;
RUN;

PROC RANK DATA = class OUT = class TIES = low;
 VAR weight;
 RANKS Weight_Rnk;
RUN;

PROC SORT DATA = class; BY weight;RUN;
```

Table 16.1: Different derived relative variables

	ID	Weight	Weight_Shift	Weight_Ratio	Weight_CentRatio	Weight_Std	Weight_Rnk
1	6	50.5	-49.53	0.5048485454	-0.495151455	-2.174693089	1
2	8	77	-23.03	0.7697690693	-0.230230931	-1.011082069	2
3	12	83	-17.03	0.8297510747	-0.170248925	-0.74762297	3
4	1	84	-16.03	0.8397480756	-0.160251924	-0.70371312	4
5	13	84	-16.03	0.8397480756	-0.160251924	-0.70371312	4
6	4	84.5	-15.53	0.844746576	-0.155253424	-0.681758195	6
7	18	85	-15.03	0.8497450765	-0.150254924	-0.65980327	7
8	7	90	-10.03	0.899730081	-0.100269919	-0.440254021	8
9	2	98	-2.03	0.9797060882	-0.020293912	-0.088975222	9
10	14	99.5	-0.53	0.9947015895	-0.00529841	-0.023110447	10
11	3	102.5	2.47	1.0246925922	0.0246925922	0.1086191022	11
12	11	102.5	2.47	1.0246925922	0.0246925922	0.1086191022	11
13	9	112	11.97	1.1196641008	0.1196641008	0.5257626757	13
14	19	112	11.97	1.1196641008	0.1196641008	0.5257626757	13
15	5	112.5	12.47	1.1246626012	0.1246626012	0.5477176006	15
16	10	112.5	12.47	1.1246626012	0.1246626012	0.5477176006	15
17	16	128	27.97	1.2796161152	0.2796161152	1.2283202732	17
18	17	133	32.97	1.3296011197	0.3296011197	1.4478695224	18
19	15	150	49.97	1.499550135	0.499550135	2.1943369697	19

Consider the following points for the possible derived variables:

- The variables WEIGHT_SHIFT, WEIGHT_CENTRATIO, and WEIGHT_STD represent the mean of the original distribution at zero, allowing easy interpretation of the values compared to their mean.

- The variable WEIGHT_RATIO represents the mean of the original distribution at 1 and has values greater than 1 for the original values that are greater than the mean and has values less than 1 for original values that are lower than the mean. For some analyses this makes sense. The drawback with this variable, however, is that for positive values, such as counts, the distribution is not symmetrical.

- From a programming point of view consider the following:

  □ In the preceding code we artificially created an ID variable from the logical _N_ variable in order to have a subject identifier in place.

  □ We needed this subject identifier to merge the results of PROC STANDARD with the original data because PROC STANDARD does not create a new variable but overwrites the existing variables.

  □ We used PROC RANK to create the ordinal variable WEIGHT_RNK and used the TIES=LOW option in order use the lower rank for tied values.

  □ If the variable WEIGHT_RNK does not have the same value for ties, but one observation will randomly get the lower rank and another the higher rank, a variable WEIGHT_RND can be created by adding an artificial random number. The probability that the resulting values of WEIGHT_RND have ties is very low. See the following code:

```
WEIGHT_RND = WEIGHT + UNIFORM(1234) / 10;
```

  □ Here we use the UNIFORM function to create a uniformly distributed number between 0 and 1. We divide these numbers by 10 in order not to overwrite the original digits of the WEIGHT values. In this case, for example, we will get 10.5343 for an age of 10.5. Performing a rank on these values does not change the order of the original values but avoids ties.

16.4 Time Intervals

General

Date and time variables as well as intervals between different points in time are an important group of measurement variables for analysis tables. With *time variables* we mean not only variables that represent time-of-day values, but all variables that represent data that are measured on a time scale.

When calculating *time intervals*, the following points have to be considered:

- definition of the beginning and end of the interval
- definition of how intervals are measured (especially for dates)

16.4.1 Definition of the Beginning and End of the Interval

The definition of the beginning and end of an interval depends on the business questions for which the interval has to be calculated. In many cases the definition of the interval is straightforward.

To show that the calculation of a time interval is not always simple, see the following example with the calculation of age.

In a customer table we have stored the birthdate for a customer, which is obviously the start of the interval when we want to calculate the age.

Here are the possible interval end dates for the calculation:

Calculation	Comment
&SYSDATE	Not recommended! Calculates the age at the start of the SAS session (see the variable &SYSDATE), which does not necessarily need to be the actual date.
Date()	Calculates the age at the current date. Note that this is not necessarily the end date that you need for your analysis. If the data were a snapshot of the last day of the previous month, then age should be calculated for this point in time and not for the current date.
&snapdate (e.g. %LET snapdate = "01MAY2005"d)	Calculates the age for a certain snapshot date. This makes sense if the age will be calculated for a certain point in time and if the value of age will not depend on the date when the data preparation was run. Also, a rerun of the data preparation process at a later time produces consistent results.

Other options for age values at certain events in the subject's lifetime include the following:

- age at start of customer relationship
- age at shift to a certain profitability segment
- age at first insurance claim
- age at contract cancellation
- age at randomization into a clinical trial
- age at first adverse event
- age at recurrence of disease
- age at death

The preceding list should also illustrate that depending on the business questions special definitions of time intervals can make sense. It can make sense to include variables with different definitions for a certain interval in the analysis table, if the data will be analyzed from different points of view or if the importance or predictive power of different variables will be analyzed.

This example also illustrates that in surveys, data collection, data entry systems, and customer databases the date of birth will be collected instead of the age. If we have only the age, we need to know at which point in time it was collected, in order to have the actual age value at a later time.

Having discussed the effect of the definition of the beginning and end of an interval, we will now look at the definition of how intervals are measured.

16.4.2 Definition of How Intervals Are Measured— Dirty Method

Subtracting two date variables from each other results in a number of days between those two dates. In order to calculate the number of years, we need to divide the number of days by 365. Or do we need to divide by 365.25 in order to consider leap years? Or do we need to divide by 365.2422 in order to consider the exact definition of leap years?

When using this method there will always be cases where an age that is not exact is calculated. In the case of a person's age, the age value will be incremented some days before or after the real birthday.

Note that these calculations are only approximations and do not reflect the exact number of years or months between the two dates. Here we are back at the definition of the business problem and the type of analysis we want to perform. If, for example, in data mining, we need a value for a time interval and do not care whether the real interval length is 12.3444 or 12.36 years, and if the definition is consistent for all subjects, we can use this method.

This is the reason why in the creation of data mining marts the situation is very often encountered that derived variables are calculated like this:

```
AGE = (&snapdate - birthdate) / 365.2422;
MonthsSinceFirstContact = (&snapdate - FirstContact) / (365.2422/12);
```

Different from years and months, *weeks* have a constant number of days. Therefore, we can simply divide the number of days between two dates by 7 in order to get the number of weeks.

The SAS function DATDIF calculates the difference in days between two dates. The result equals the subtraction of two dates:

```
NR_DAYS1=DATDIF('16MAY1970'd, '22OCT2006'd,'ACTUAL');
NR_DAYS2='22OCT2006'd - '16MAY1970'd;
```

Both statements return the same results. In the DATDIF function, however, options are available to force a 365-day or a 360-day calculation. See SAS Help and Documentation for details.

16.4.3 Definition of How Intervals Are Measured— Exact Method

Consider the following cases:

- We need the exact length of an interval.
- We need the exact date when a person reaches a certain age.
- We want to consider special calculation definitions such as "number of full months between two dates" and "number of days until the start of the month of a certain date."

In these cases, we have to investigate the functionality of special SAS functions, namely INTCK, INTNX, and YRDIF.

The INTCK function *counts the number of interval boundaries between two dates* or between two datetime values. The function has three parameters—the interval name, which can include YEAR, QTR, MONTH, WEEK; the beginning of the interval; and the end of the interval. For example:

```
INTCK('MONTH','16MAY1970'd,'12MAY1975'd);
```

evaluates to 60, because 60-month borders lie between May 16, 1970, and May 12, 1975.

The INTNX function *increments a date, time, or datetime value by a given number of intervals.* The function has three parameters—the interval name, which can include YEAR, QTR, MONTH, or WEEK; the beginning of the interval; and the number of intervals to increment. For example:

```
INTNX('MONTH','16MAY1970'd,100);
```

evaluates to "01SEP1978," because we have incremented the value of May16, 1970, for 100-month borders.

The age in years can now be calculated with the following formula:

```
AGE = FLOOR((INTCK('MONTH','16MAY1970',&current_date)-(DAY(&current_date)
< DAY(birthdate))) / 12)
```

Note that this formula is taken from a SAS Quick Tip, presented by William Kreuter at http://support.sas.com/sassamples/.

The equivalent to the preceding formula is the SAS function YRDIF. It returns the difference in years between two dates. Here the number of days in leap years is calculated with 366 and in other years with 365. See the following example for a comparison of the two calculation methods:

```
DATA _NULL_;
 years_yrdiff = YRDIF('16MAY1970'd,'22OCT2006'd,'ACTUAL');
 years_divide = ('22OCT2006'd - '16MAY1970'd) / 365.2242;
 output;
 PUT years_yrdiff= years_divide=;
RUN;

years_yrdiff=36.435616438
years_divide=36.437892122
```

The difference between the two values arises from the fact that when dividing by 365.25 we are correct on average. However, for a particular interval the correction for the leap years is not exact. The YRDIF function handles this by exactly calculating the number of days.

16.5 Binning Observations into Groups

General

In some cases it is necessary to bin interval-scaled values into groups. For example, in the case of a variable with hundreds of different values, the calculation of frequencies for each value does not make sense. In these cases the observations are binned into groups and the frequencies are calculated for the groups. Note that the terms *grouping* and *binning* are used interchangeably in this chapter.

In this section we will deal with the possible methods used to group observations. Basically there are three main ways to group observations:

- creating groups of equal number of observations with PROC RANK
- creating groups of equal widths with SAS functions
- creating individual groups with IF-THEN/ELSE statements or SAS formats

16.5.1 Creating Groups of Equal Number of Observations with PROC RANK

In Section 16.3.1 "Creating Derived Relative Variables," we used PROC RANK to create an ordinally ranked variable for an interval variable. Here, we will use PROC RANK to group observations in groups of (almost) equal size. Note that in the case of ties (equal values), equal group sizes are not always possible.

It is possible to achieve equal-sized groups or consecutive rank orders by adding an arbitrary small random number to the variable. With this method ties are avoided. We already showed this in Section 16.3 "Derived Relative Variables." Note, however, that observations with the same value are randomly assigned to different groups.

To bin the observations of the SASHELP.AIR data set into 10 groups we can use PROC RANK as follows:

```
PROC RANK DATA = sashelp.air OUT = air
          GROUPS = 10;
 VAR air;
 RANKS air_grp;
RUN;
```

Note the following:

- The specification of a separate output data set with the OUT= option is in some cases desirable but not necessary.
- One or more variables can be ranked within one PROC RANK invocation.
- The ordering of the ranks (group numbers) can be reversed using the DESCENDING option in the PROC RANK statement.
- The ranks start with 0. If ranks starting from 1 are needed, they have to be incremented in a DATA step.

16.5.2 Creating Groups of Equal Widths with SAS Functions

Groups of equal widths can be easily created with the SAS functions CEIL or FLOOR. See the following example code and the corresponding output for the SASHELP.AIR data set:

```
DATA air;
 SET sashelp.air;
 Air_grp1 = CEIL(air/10);
 Air_grp2 = CEIL(air/10)*10;
 Air_grp3 = CEIL(air/10)*10 - 5;
 Air_grp4 = CEIL(air/10)-10;
RUN;
```

Note the following:

- With the variable AIR_GRP1 we have created a group variable that bins observations greater than 100 up to 110 into group 11, observations greater than 110 up to 120 into group 12, and so on.

- AIR_GRP2, AIR_GRP3, and AIR_GRP4 have the same grouping rule but assign different values to the groups. AIR_GRP2 assigns the maximum value in each group as a group label.

- It is easy to change the code to receive group midpoints just by subtracting 5, as we see in AIR_GRP3.

- AIR_GRP4 gives a consecutive group numbering, which we receive by subtracting the (minimum group number −1).

Table 16.2: Different variables for binning the variable AIR

	DATE	international airline travel (thousands)	Air_grp1	Air_grp2	Air_grp3	Air_grp4
1	JAN49	112	12	120	2	115
2	FEB49	118	12	120	2	115
3	MAR49	132	14	140	4	135
4	APR49	129	13	130	3	125
5	MAY49	121	13	130	3	125
6	JUN49	135	14	140	4	135
7	JUL49	148	15	150	5	145
8	AUG49	148	15	150	5	145
9	SEP49	136	14	140	4	135
10	OCT49	119	12	120	2	115
11	NOV49	104	11	110	1	105
12	DEC49	118	12	120	2	115
13	JAN50	115	12	120	2	115
14	FEB50	126	13	130	3	125
15	MAR50	141	15	150	5	145
16	APR50	135	14	140	4	135
17	MAY50	125	13	130	3	125
18	JUN50	149	15	150	5	145

16.5.3 Creating Individual Groups

The advantage of the preceding two grouping methods is that they can be performed with only a few SAS statements. It is, however, not possible to use them to specify individual group boundaries. For this task, IF-THEN/ELSE statements or PROC FORMAT can be used.

Using IF-THEN/ELSE statements

See the following code for an example:

```
DATA air;
 SET sashelp.air;
 FORMAT air_grp $15.;
 IF      air =  .   THEN air_grp = '00: MISSING';
 ELSE IF air < 220  THEN air_grp = '01: < 220';
 ELSE IF air < 275  THEN air_grp = '02: 220 - 274';
 ELSE                    air_grp = '03: >= 275';
RUN;
```

Note the following:

- Formatting the variable that holds the new group names is advisable. Otherwise, the length of the variable is determined from the first assignment, which can cause truncation of the group names.

- The group names contain a numbering according to their size. This is advisable for a sorted output.

- The vertically aligned coding has the advantage that it is easier to read and to edit.

Using SAS formats

The preceding grouping can also be created by defining a SAS format and assigning the format to the variable AIR during analysis:

```
PROC FORMAT;
 VALUE air
    .  = '00: MISSING'
    LOW -< 220 = '01: < 220'
    220 -< 275 = '02: 220 - 274'
    275 - HIGH = '03: > 275';
RUN;
```

The format can be assigned to the variable AIR by the following statement in every procedure call or DATA step:

```
FORMAT air air.;
```

It is also possible to create a new variable in a DATA step by using the format in a PUT function:

```
DATA air;
 SET sashelp.air;
 Air_grp = PUT(air,air.);
RUN;
```

See the first five observations of the data set AIR in the following table:

Table 16.3: SASHELP.AIR data set with the variable AIR_GRP

	DATE	international airline travel (thousands)	Air_grp
43	JUL52	230	02: 220 - 274
44	AUG52	242	02: 220 - 274
45	SEP52	209	01: < 220
46	OCT52	191	01: < 220
47	NOV52	172	01: < 220
48	DEC52	194	01: < 220
49	JAN53	196	01: < 220
50	FEB53	196	01: < 220
51	MAR53	236	02: 220 - 274
52	APR53	235	02: 220 - 274
53	MAY53	229	02: 220 - 274
54	JUN53	243	02: 220 - 274
55	JUL53	264	02: 220 - 274
56	AUG53	272	02: 220 - 274
57	SEP53	237	02: 220 - 274
58	OCT53	211	01: < 220
59	NOV53	180	01: < 220

16.6 Transformations of Distributions

General

In right-skewed distributions extreme values or outliers are likely to occur. The minimum for many variables is naturally bounded at zero—for example, at all count variables or measurements that logically can start only at zero. Examples of variables that usually have highly skewed distributions and outliers in the upper values are laboratory values, number of events, minutes of mobile phone usage, claim amounts in insurance, or loan amounts in banking.

The presence of skewed distribution or outlier influences has an effect on the analysis because some types of analyses accept only normal (or close to normal) distribution. In order to achieve a close to normal distribution of values you must

- Perform a transformation to maximize the normality of a distribution.
- Filter outliers or replace them with other values.

Note that in some types of analysis a filtering of outliers is not possible, because it is not allowed to leave out certain observations. For example, in medical statistics where the variability of values can be high due to biological variability, observations (patients) must not be excluded from the analysis only because of their extreme values.

In this case it has to be decided whether non-parametric methods of analysis are applied that can deal with extreme values or skewed distributions, or if parametric methods are applied with the knowledge that certain assumptions of these methods are violated.

16.6.1 Transformations to Maximize the Normality

The most frequently used transformation to transform a right-skewed distribution is the log transformation. Note that the logarithm is defined only for positive values. In the case of negative values, a constant has to be added to the data in order to make them all positive. Another transformation that normalizes data is the root transformation, where the quadratic root, but in general the *k*th root, can be taken from the data, where k > 1.

Let's look at the following example where we transform a skewed variable:

```
DATA skewed;
INPUT a @@;
CARDS;
1 0 -1 20 4 60 8 50 2 4 7 4 2 1
;
RUN;
```

Analyzing the distribution can be done as follows.

First we calculate the minimum for each variable in order to see whether we have to add a constant for the logarithm in order to have positive values:

```
PROC MEANS DATA = skewed MIN;
RUN;
```

Second we analyze the distribution with PROC UNIVARIATE and use ODS SELECT to display only the tests for normality:

```
ODS SELECT TestsForNormality  Plots;
 PROC UNIVARIATE DATA = skewed NORMAL PLOT; RUN;
ODS SELECT ALL;
```

We receive the following results:

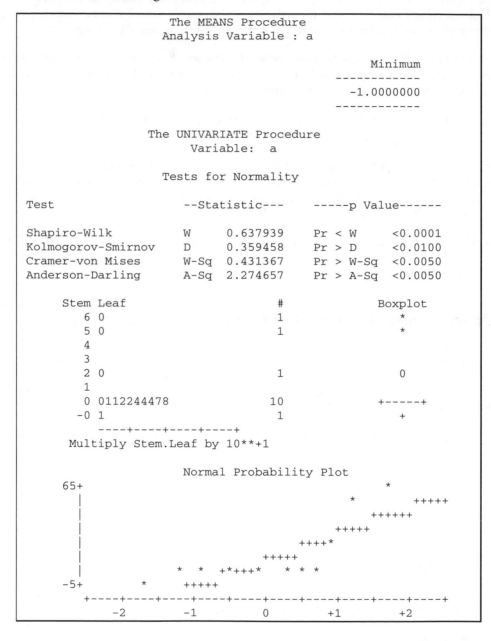

```
                 The MEANS Procedure
                 Analysis Variable : a


                                            Minimum
                                          ------------
                                          -1.0000000
                                          ------------


                The UNIVARIATE Procedure
                     Variable:  a

                  Tests for Normality

Test                     --Statistic---      -----p Value------

Shapiro-Wilk          W     0.637939      Pr < W      <0.0001
Kolmogorov-Smirnov    D     0.359458      Pr > D      <0.0100
Cramer-von Mises      W-Sq  0.431367      Pr > W-Sq   <0.0050
Anderson-Darling      A-Sq  2.274657      Pr > A-Sq   <0.0050

     Stem Leaf                      #             Boxplot
       6 0                          1                *
       5 0                          1                *
       4
       3
       2 0                          1                0
       1
       0 0112244478                 10            +-----+
      -0 1                          1                +
         ----+----+----+----+
      Multiply Stem.Leaf by 10**+1


                  Normal Probability Plot
      65+                                          *
        |                                 *        +++++
        |                                  ++++++
        |                             +++++
        |                        ++++*
        |                   +++++
        |         *   *  +*+++*    *  *  *
      -5+      *      +++++
        +----+----+----+----+----+----+----+----+----+----+
             -2        -1         0        +1        +2
```

We apply a log and a root transformation to the data. We see from this the minimum for variable A is -1; therefore, we add a constant of 2 before applying the log and the root transformation:

```
DATA skewed;
 SET skewed;
 log_a = log(a+2);
 root4_a = (a+2) ** 0.25;
RUN;

ODS SELECT TestsForNormality  Plots;
 PROC UNIVARIATE DATA = skewed NORMAL PLOT; RUN;
ODS SELECT ALL;
```

We receive the following results:

```
                    The UNIVARIATE Procedure
                      Variable:  log_a

                   Tests for Normality

Test                    --Statistic---      -----p Value------

Shapiro-Wilk            W    0.940212    Pr < W       0.4211
Kolmogorov-Smirnov      D    0.182368    Pr > D      >0.1500
Cramer-von Mises        W-Sq 0.069224    Pr > W-Sq   >0.2500
Anderson-Darling        A-Sq 0.410499    Pr > A-Sq   >0.2500


     Stem Leaf                   #              Boxplot
        4 01                     2                 0
        3                                          |
        3 1                      1                 |
        2                                          |
        2 23                     2              +-----+
        1 888                    3              *--+--*
        1 1144                   4              +-----+
        0 7                      1                 |
        0 0                      1                 |
          ----+----+----+----+
```

(*continued*)

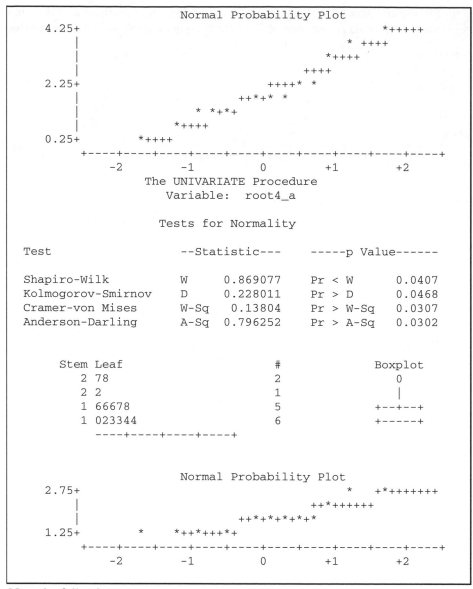

```
                     Normal Probability Plot
    4.25+                                        *+++++
        |                                   *  ++++
        |                                 *++++
        |                              ++++
    2.25+                          ++++*  *
        |                        ++*+*  *
        |                 *  *+*+
        |             *++++
    0.25+      *++++
        +----+----+----+----+----+----+----+----+----+----+
           -2        -1         0        +1        +2
                   The UNIVARIATE Procedure
                   Variable:  root4_a

                     Tests for Normality

Test                   --Statistic---      -----p Value------

Shapiro-Wilk           W    0.869077       Pr < W      0.0407
Kolmogorov-Smirnov     D    0.228011       Pr > D      0.0468
Cramer-von Mises       W-Sq 0.13804        Pr > W-Sq   0.0307
Anderson-Darling       A-Sq 0.796252       Pr > A-Sq   0.0302


       Stem Leaf                   #            Boxplot
          2 78                     2               0
          2 2                      1               |
          1 66678                  5            +--+--+
          1 023344                 6            +-----+
            ----+----+----+----+


                     Normal Probability Plot
    2.75+                                *   +*+++++++
        |                             ++*++++++
        |                        ++*+*+*+*+*
    1.25+            *      *++*+++*+
        +----+----+----+----+----+----+----+----+----+----+
           -2        -1         0        +1        +2
```

Note the following:

- We have used PROC MEANS with the MIN option to calculate the minimum of variable A, even if PROC UNIVARIATE would do this. The advantage of PROC MEANS with many variables that might need transformation is that a tabular output with the minima for all variables is produced, which is very clear.

- We use the options NORMAL and PLOT in the PROC UNIVARIATE statement in order to receive tests for normality and the corresponding plots.

■ We use an ODS SELECT statement to output only those results of PROC UNIVARIATE that are relevant to our task. This is not mandatory, but it is useful for clearness of the output, especially with many variables.

From the Kolmogorov-Smirnov statistic we see, for example, in our case the log transformation performs better than the root transformation.

16.6.2 Handling Extreme Values

With extreme values (outliers) we have the option to either *filter* them or to *replace* them with a certain value. Replacing the value is also called *shifting* and has the advantage of not excluding observations from the analysis, especially those observations that might be important for the analysis.

Shifting the values of a variable that are larger or smaller than a certain value can be done easily with the following statement:

```
IF a > 20 THEN a=20;
IF b < 0 THEN b=0;
```

An alternative is to use the MIN and MAX functions as shown in the following example:

```
a = MIN(a,20);
a = MAX(b,0);
```

The MIN function shifts values that are larger than 20 to the value 20. The MAX function shifts values that are smaller than 0 to the value 0. This is an easy and short way to shift values without using IF-THEN/ELSE clauses.

16.6.3 Macros SHIFT and FILTER

However, if data contain missing values we have to keep in mind that SAS treats a missing value as the smallest possible value. We want to control whether we want to replace missing values to preserve them.

In order to simplify the coding we present the macros SHIFT and FILTER for the respective tasks:

```
%MACRO SHIFT (OPERATION,VAR,VALUE, MISSING=PRESERVE);
 %IF %UPCASE(&missing) = PRESERVE %THEN %DO;
         IF &var NE . THEN &var = &operation(&var,&value);
  %END;
  %ELSE %IF %UPCASE(&missing) = REPLACE %THEN %DO;
         &var = &operation(&var,&value);
  %END;
%MEND;
```

The macro parameters are as follows:

VAR

The name of the variable whose values will be shifted.

VALUE

The value to which the variable values will be shifted.

OPERATION

Possible values are MIN or MAX. MIN shifts values down and MAX shifts values up. Note that the values of this variable are directly used as the MIN or MAX function call. The default value is MIN.

MISSING

Defines whether missing values will be preserved or replaced. Possible values are PRESERVE (= default value) or REPLACE.

Examples:

- Shift extreme values down to 20 and preserve MISSING values:
  ```
  %SHIFT(min,a,20);
  ```

- Replace negative and missing values with 0:
  ```
  %SHIFT(min,a,0,MISSING=REPLACE);
  ```

Note that the advantage of this macro is that it controls missing values. In the case of non-missing values, the direct code is shorter than the macro invocation.

The macro FILTER creates an IF-THEN DELETE statement to filter certain observations:

```
%MACRO FILTER (VAR,OPERATION,VALUE, MISSING=PRESERVE);
 %IF %UPCASE(&missing) = PRESERVE %THEN %DO;
         IF &var NE . AND &var &operation &value THEN DELETE; %END;
 %ELSE %IF %UPCASE(&missing) = DELETE %THEN %DO;
         IF &var &operation &value THEN DELETE; %END;
%MEND;
```

The macro parameters are as follows:

VAR

The name of the variable that will be used in the filter condition.

OPERATION

Specifies the comparison operator. This can be any valid comparison operator in SAS such as <, >, <=, >=, GT, LT, GE, and LE. Note that it has to be specified without quotation marks.

VALUE

The value to which the variable values will be shifted.

MISSING

Defines whether missing values will be preserved or replaced. Possible values are PRESERVE (= default value) or DELETE.

Examples:

Delete all observations that are larger than 20, but do not delete missing values:

```
%FILTER(a,<,20);
```

Delete all observations that are smaller than 0, but do not delete missing values:

```
%FILTER(a,<,0);
```

Delete all observations that are negative or zero, and also delete missing values:

```
%FILTER(a,<=,0,MISSING=DELETE);
```

16.7 Replacing Missing Values

Overview
Replacing missing values can range from a simple replacement with 0 or another predefined constant to a complex calculation of replacement values. In this section we will show a macro that replaces a missing value with a constant. We will also see how PROC STANDARD can be used to replace missing values.

16.7.1 Replacing Missing Values in the DATA Step
Missing values are represented by a period (.) for numerical variables. In Chapter 14, we introduced a macro that replaces missing values with a zero value.

```
%MACRO REPLACE_MV(cols,mv=.,rplc=0);
 ARRAY varlist {*}  &cols;
 DO _i = 1 TO dim(varlist);
    IF varlist{_i} = &mv THEN varlist{_i}=&rplc;
 END;
 DROP _i;
%MEND;
```

In this macro the replacement value can be chosen in order to replace missing values with other values, for example, the mean, that has previously been calculated. The rationale for the macro REPLACE_MV is to replace systematic missing values, e.g., missing values that can be replaced by zeros.

16.7.2 Using PROC STANDARD for Replacement
The following example shows how the REPLACE statement in PROC STANDARD can be used to replace missing values by the mean of the non-missing values.

```
DATA TEST;
INPUT AGE @@;
CARDS;
12 60 . 24 . 50 48 34 .
;
RUN;

PROC STANDARD DATA = TEST REPLACE;
 VAR age;
RUN;
```

In the following results, we see that the missing values have been replaced by their mean 38:

```
Obs      AGE


 1        12
 2        60
 3        38
 4        24
 5        38
 6        50
 7        48
 8        34
 9        38
```

16.7.3 Single and Multiple Imputation

Replacing a missing value by exactly one imputation value is called *single imputation*. In our preceding two cases we performed this kind of missing value replacement. In contrast, *multiple imputation* draws a random sample of the missing values from its distribution. This results not in one imputed data set but in a number of imputed data sets, which hold different values for the imputed values. PROC MI of SAS/STAT performs multiple imputations. PROC MIANALYZE, for example, allows the analysis of these data sets.

16.8 Conclusion

In this chapter we looked at derived variables and transformations. We however skipped those derived variables and transformations that are of particular interest in predictive modeling. *Chapter 20 – Coding for Predictive Modeling* is devoted to this topic.

SAS Enterprise Miner has special data management nodes that allow the binning of observations into groups, the handling of missing values, the transformation interval variables, and others. We will show examples of these functionalities in *Chapter 28 – Case Study 4—Data Preparation in SAS Enterprise Miner*.

We also want to mention that the tasks that we described in the last sections of this chapter go beyond simple data management that can be done with a few statements. This is the point where SAS Enterprise Miner with special data preparation routines for data mining comes into play.

Chapter 17

Transformations of Categorical Variables

17.1 Introduction

In this chapter we will deal with transformations of categorical variables. With *categorical variables* we mean variables such as binary, nominal, or ordinal variables. These variables are not used in calculations—instead they define categories.

In *Chapter 16 – Transformations of Interval-Scaled Variables*, we saw a lot of transformations of interval data. Most of these transformations were based on calculations. With categorical variables calculations are generally not possible. We will therefore investigate other methods that can be used to create derived variables from categorical variables.

In the following sections we will look at these topics:

General considerations for categorical variables such as formats and conversions between interval and categorical variables.

Derived variables, where we will see which derived variables we can create from categorical information.

Dummy coding of categorical variables, where we will show how categorical information can be used in analysis, which allows only interval variables.

Multidimensional categorical variables, where we will show how the information of two or more categorical variables can be combined into one variable.

Lookup tables and *external data,* where we will show how to integrate lookup tables and external data sources.

17.2 General Considerations for Categorical Variables

17.2.1 Numeric or Character Format

Categorical information can be stored in a variable either in the form of a *category name* itself or in the form of a *category code*. Categories can be coded in from number codes, character codes, or abbreviations. In many cases the category code corresponds to a category name that will usually be stored in variables with character formats.

The following table shows possible representations of the variable GENDER:

Table 17.1: Coding of gender

Category Name	Character Code	Numeric Code
MALE	M	1
FEMALE	F	0

In *Chapter 11 – Considerations for Data Marts*, we discussed the properties, advantages, and disadvantages of numeric and character categorical variables.

17.2.2 SAS Formats

SAS formats are an advantageous way to assign a category name depending on the category code. SAS formats are efficient for storage considerations and allow easy maintenance of category names.

The following example shows how formats can be used to assign the category name to the category code. SAS formats for a numeric- and a character-coded gender variable are created. These formats are used in PROC PRINT when the data set GENDER_EXAMPLE is printed.

```
PROC FORMAT;
 VALUE   gender   1='MALE' 0='FEMALE';
 VALUE  $gender_c M='MALE' F='FEMALE';
RUN;

DATA gender_example;
 INPUT Gender Gender2 $;
 DATALINES;
 1  M
 1  M
 0  F
 0  F
 1  M
 ;
RUN;

PROC PRINT DATA = gender_example;
 FORMAT Gender gender. Gender2 $gender_c.;
RUN;
```

Obs	Gender	Gender2
1	MALE	MALE
2	MALE	MALE
3	FEMALE	FEMALE
4	FEMALE	FEMALE
5	MALE	MALE

17.2.3 Converting between Character and Numeric Formats

For some data management tasks it is necessary to convert the format of a variable from character to numeric or vice versa. In many cases this is possible in the DATA step by ignoring the type of variable. In this case, SAS issues warnings in the SAS log:

```
NOTE: Character values have converted to numeric values at the
      places given by: (Line):(Column)
NOTE: Numeric values have converted to character values at the
      places given by: (Line):(Column)
```

These implicit conversions can also cause unexpected results. It is therefore advisable to explicitly convert them by using the functions INPUT or PUT.

The conversion from character to numeric in our preceding gender example is done with the following statement:

```
NEW_VAR = INPUT(gender2, 2.);
```

The conversion from numeric to character in our preceding gender example is done with the following statement:

```
NEW_VAR = PUT(gender, 2.);
```

For a more detailed discussion of type conversions see the SAS Press publication *The Little SAS Book: A Primer*.

17.3 Derived Variables

Overview

Different from numeric variables, not many different types of derived variables are created from categorical data because we cannot perform calculations on categorical data. Therefore, derived variables for categorical data are mostly derived either from the frequency distribution of values or from the extraction and combination of hierarchical codes.

The most common derived variables are:

- Extracting elements of hierarchical codes
- Indicators for the most frequent group
- IN variables in the SET statement to create binary indicator variables

In this section we will see that it is easy to create meaningful derived variables from a categorical variable.

17.3.1 Extracting Elements of Hierarchical Codes

Different digits or characters of numeric or alphanumeric codes can have certain meanings. These codes can be hierarchical or multidimensional.

- In a hierarchical code different elements of the code can define different hierarchies. For example, a product code can contain the PRODUCTMAINGROUP code in the first character.
- In the case of multidimensional codes different characters can contain different sub-classifications. For example a medical disease code contains in the first two digits the disease code and in the third and fourth digits a classification of the location in the body.

By extracting certain digits from a code, derived variables can be created. The following example shows a Product Code that contains the ProductMainGroup in the first digit, and in the second and third digits the hierarchical underlying subproduct group. Extracting the code for the ProductMainGroup can be done as in the following example:

```
DATA codes;
 SET codes;
 ProductMainGroup = SUBSTR(ProductCode,1,1);
RUN;
```

Product Code	ProductMainGroup
216	2
305	3
404	4
233	2
105	1
311	3
290	2

This type of derived variable is frequently used for classifications and aggregations when the original categorical variable has too many different values.

17.3.2 Indicators for Special Categories

With interval data, a lot of transformations such as standardization or differences are possible. With categorical data, the set of possible transformations is not that broad. A possibility to create a derived variable is to compare the value for a subject to the distribution of values for the populations.

Examples of these properties include the following:

- the most frequent group
- the group with the highest average for an interval variable

The creation process of such an indicator variable has two steps:

1. Identify the category. This is done by calculating a simple or advanced descriptive statistics.

2. Create the indicator variable.

If we want to create an indicator variable for the most common ProductMainGroup of the preceding example, we first use PROC FREQ to create a frequency table:

```
PROC FREQ DATA = codes ORDER = FREQ;
 TABLE ProductMainGroup;
RUN;
```

From the output (not printed) we see that ProductMainGroup 2 is the most frequent. The indicator variable can be created with the following statement:

```
IF ProductMainGroup = '2' THEN ProductMainGroupMF = 1;
ELSE ProductMainGroupMF =0;
```

or simply

```
ProductMainGroupMF = (ProductMainGroup = '2');
```

With these derived variables we can create indicators that describe each subject in relation to other subjects. We can therefore determine how a certain subject differs from the population.

17.3.3 Using IN Variables to Create Derived Variables

The logical IN variables in the SET statement of a DATA step easily allow you to create derived variables, which can be used for analysis. These variables are binary variables (indicators) because they hold the information whether a subject has an entry in a certain table or not.

In the following coding example we have data from the call center records and from Web usage. Both tables are already aggregated per customer and have only one row per customer. These two tables are merged with the CUSTOMER_BASE table. In the resulting data set, variables that indicate whether a subject has an entry in the corresponding table or not are created.

```
DATA customer;
 MERGE customer_base (IN=in1)
       Call_center_aggr (IN=in2)
       Web_usage_aggr (IN=in3);
    BY CustomerID;
 HasCallCenterRecord = in2;
 HasWebUsage = in3;
RUN;
```

With this method you add the variables HasCallCenterRecord and HasWebUsage to your data mart. These variables can be used in the following ways:

- As binary input variables holding the property of a subject.
- As subgroup variables for subgroup analysis or exploratory analysis.
- As variables used in context with other derived variables from that table. See also *Chapter 19 – Multiple Categorical Observations per Subject* for a discussion of the case when you aggregate data from a lower hierarchy to the table and you don't have observations for all subjects in the base table. In that context the variables can also be for the creation of interactions.

17.4 Combining Categories

In Section 17.3.1 "Extracting Elements of Hierarchical Codes," we saw a way to reduce the number of categories of a variable. Here we will discuss ways to combine categories in order to reduce the number of different categories. The reduction of the number of different values is important in order to reduce the number of possible subgroups or to reduce the number of degrees of freedom in the analysis.

However, not only the degrees of freedom are reduced. From a business point of view it is also desirable to reduce the number of categories to an interpretable set.

17.4.1 The OTHERS Group

A very frequent way to reduce the number of categories is to combine values into a so-called OTHERS group. In most cases these categories are assigned to the OTHERS group that has a low frequency.

PROC FREQ with the ORDER = FREQ option is an important tool for this task. Using the ProductMainGroup data from the preceding example we create the following output:

```
PROC FREQ DATA = codes ORDER = FREQ;
 TABLE ProductMainGroup;
RUN;
```

```
Product
Main                              Cumulative    Cumulative
Group      Frequency    Percent    Frequency      Percent
----------------------------------------------------------
2              3         42.86         3          42.86
3              2         28.57         5          71.43
1              1         14.29         6          85.71
4              1         14.29         7         100.00
```

We see that the ProductMainGroups 1 and 4 occur only once and we want to assign them to the OTHERS group. This can be done with the following statements:

```
FORMAT ProductMainGroupNEW $6.;
IF ProductMainGroup IN ('1' '4') THEN ProductMainGroupNEW = 'OTHERS';
ELSE ProductMainGroupNEW = ProductMainGroup;
```

Note the following:

- The FORMAT statement can be important here because otherwise the length of the newly created variable corresponds to the length of the first character value that is assigned to it. This might result in truncated category values such as "OTHE".

- The IN operator "IN ('1' '4')" is much more efficient than a long row of OR expressions such as ProductMainGroup = 1 OR ProductMainGroup = 4.

- In the case of many different categories the selection of the relevant groups can be difficult. Here the "matrix selection" of characters out of the Output window can be helpful.

The advantage of "OTHERS groups" is that a high number of potentially low frequent categories are combined into one group. This helps to speed up data preparation and analysis and also makes the interpretation of results easier.

In *Chapter 23 – Scoring and Automation*, we will deal with the problem of new and changing categories that cause problems in the scoring process, which can partly be overcome with OTHERS or UNKNOWN groups.

17.4.2 Coding Tip: Matrix Selections in the SAS Output Window

Matrix selections in the SAS Output window are a powerful possibility to avoid long and annoying code writing.

- In the SAS Output window, the matrix selection can be activated by holding down the ALT key, pressing the left mouse button, and moving the mouse pointer over the desired characters (codes).

- The selected values can be copied to the clipboard and copied to the program editor for the IN clause.

- Note here that during selection in the Output window you must not move after the last line. Otherwise, you are not able to copy the selection to the clipboard.

- The format code from the preceding example might look like the following:

```
FORMAT ProductMainGroupNEW $6.;
IF ProductMainGroup IN (
'1'
'4'
) THEN ProductMainGroupNEW = 'OTHERS';
ELSE ProductMainGroupNEW = ProductMainGroup;
```

Note the following:

- Because we have to add the apostrophe (') manually it might be more efficient to write the list of codes into separate lines.
- The data that are inserted from a matrix selection do not contain line feeds; therefore the appropriate number of empty lines has to be inserted manually before the paste from the clipboard.

Such groupings of categories are not restricted to OTHERS groups, but can also be used to create any composite category.

17.4.3 Coding Tip: Using a Spreadsheet or Long Lists of Categories

Note that the previous way that we described starts to get more and more inefficient with the increasing number of categories. Consider, for example, the case where the list of categories extends over a number of pages. One way to overcome this is the following:

- Store the output of PROC FREQ in a table.
- Export the table to a spreadsheet such as Excel.
- Manually group the categories in the spreadsheet, e.g., by creating a new column.
- Import the table from the spreadsheet in order to merge the new column to the data or create a format on the basis of the old and new categorization.

Here, we show an example of these steps. We assume that the functionality of SAS/ACCESS to PC File Formats can be used, and we create a LIBNAME engine to an Excel spreadsheet. An alternative would be to right-click and select **View in Excel** from the pop-up menu in the SAS Explorer window.

```
LIBNAME lookup 'c:\project1\lookup.xls';

PROC FREQ data = codes ORDER = FREQ;
 TABLE ProductMainGroup / NOPERCENT OUT = lookup.codes1;
RUN;

LIBNAME lookup CLEAR;
```

The table CODES1 in LOOKUP.XLS will then be opened in Excel and a column GROUP_NEW is added. This column is filled with the appropriate category names.

After the new column GROUP_NEW is added to the spreadsheet and the spreadsheet is closed, it can be re-imported. During the re-import a data set is created that can be used directly to create a format with the CNTLIN option on PROC FORMAT.

```
LIBNAME lookup 'c:\project1\lookup.xls';

DATA codes1;
  SET lookup.'codes1'n(RENAME =( ProductMainGroup =start
Group_New=label));
  RETAIN fmtname 'ProductMainGroupNEW' type 'c';
RUN;

PROC FORMAT CNTLIN = codes1;
RUN;
```

The format ProductMainGroupNEW can now be assigned to the variable ProductMainGroup and will contain the new grouping.

```
DATA codes;
  SET codes;
FORMAT ProductMainGroup $ProductMainGroupNEW.;
RUN;
```

17.5 Dummy Coding of Categorical Variables

General

While categorical information in variables such as REGION or GENDER is very straightforward to understand for humans, statistical methods mostly can't deal with it. The simple reason is that statistical methods calculate measures such as estimates, weights, factors, or probabilities, and values such as MALE or NORTH CAROLINA can't be used in calculations.

To make use of this type of information in analytics, so-called *dummy variables* are built. A set of dummy variables represents the content of a categorical variable by creating an indicator variable for (almost) each category of the categorical variable.

17.5.1 Preliminary Example

The variable employment status can take the values EMPLOYED, RETIRED, EDUCATION, UNEMPLOYED. Table 17.2 shows the resulting dummy variables for this variable.

Table 17.2: GLM dummy coding for employment status

Employment_Status	Employed	Retired	Education	Unemployed
Employed	1	0	0	0
Retired	0	1	0	0
Education	0	0	1	0
Unemployed	0	0	0	1

This type of coding of dummy variables is also called GLM coding. Each category is represented by one dummy variable.

17.5.2 GLM Coding (Reference Coding)

In regression analysis, the important point with dummy variables is that if an intercept is estimated, one of the categories can be omitted in the dummy variables. Because the whole set of dummy variables for one subject would sum to 1 and an intercept is present in the design matrix, which is usually also represented by 1, one of *k* categories can be determined from the values of the other *k*–1 variables.

The same is true for binary variables, e.g., gender male/female, where only one binary dummy variable such as MALE (0/1) is needed to represent the information sufficiently.

In regression analysis, for one categorical variable with *k* categories, only *k*–1 of dummy variables is used. The omitted category is referred to as the *reference category* and estimates for the dummy variable are interpreted as differences between the categories and the reference category. A GLM coding, where one category is not represented by a dummy variable but is treated as the reference category, is also referred to as *reference coding*.

In our EMPLOYMENT_STATUS example this means that we provide only three dummy variables (EMPLOYED, UNEMPLOYED, and RETIRED) for the four categories. The coefficients of these three variables are interpreted as the difference from the category EDUCATION.

It is advisable to choose that category as a reference category, which generates interpretable results. For example, for the variable CREDITCARDTYPE with the values NO_CARD, CREDIT_CARD, and GOLD_CARD, the reference category would presumably be NO_CARD in order to have the interpretations of the effects if a certain type of card is present.

17.5.3 Deviation Coding (Effect Coding)

Deviation coding is also referred to as *effect coding*. Here we also have only *k*–1 dummy variables for a categorical variable with *k* categories. Different from the GLM coding where the dummy variable EMPLOYED is 1 for the employed category and 0 for other categories, the dummy variables in deviation coding have 1 for the respective category and –1 for the reference category.

For our employment variables this would look like the following:

Table 17.3: Deviation coding for employment status

EmploymentStatus	Employed	Retired	Education
Employed	1	0	0
Retired	0	1	0
Education	0	0	1
Unemployed	-1	-1	-1

The advantage of this type of coding is that in regression you also get an estimate for unemployment as the negative sum of the estimates of the dummy variables for the other categories. This is very important for the business interpretation of the results, because in many cases the definition and interpretation of a reference category as in GLM coding are not easy.

The estimate for a dummy variable is interpreted as the difference for that category and the average over all categories.

17.5.4 Program Statements

In SAS Enterprise Miner or in PROC LOGISTIC dummy variables are created automatically. SAS Enterprise Miner creates them implicitly, but in PROC LOGISTIC you need to specify them in the CLASS statement followed by the coding method:

```
CLASS employment_status / PARAM = REFERENCE;
```

Dummy codes can also be easily created with IF-THEN/ELSE clauses or SELECT-WHEN clauses. The following example creates dummy variables for the deviation coding method:

```
DATA CUSTOMER;
  SET CUSTOMER;
    SELECT (Employment_Status);
    WHEN ('Employed')   Employed=1;
    WHEN ('Unemployed') Unemployed=1;
    WHEN ('Education')  Education =1;
    OTHERWISE DO;
              Employed  =-1;
              Unemployed=-1;
              Education =-1;
            END;
  END;
RUN;
```

Dummy variables for GLM or reference coding can be created in the same way:

```
DATA CUSTOMER;
  SET CUSTOMER;
    SELECT (Employment_Status);
    WHEN ('Employed')   Employed  =1;
    WHEN ('Unemployed') Unemployed=1;
    WHEN ('Education')  Education =1;
    WHEN ('Retired')    Retired   =1;
RUN;
```

They can, however, also be created by the use of a Boolean expression.

```
DATA CUSTOMER;
  SET CUSTOMER;
    Employed   = (Employment_Status = 'Employed');
    Unemployed = (Employment_Status = 'Unemployed');
    Retired    = (Employment_Status = 'Retired');
    Education  = (Employment_Status = 'Education');
RUN;
```

This type of coding creates short code, because there is no need for IF-THEN/ELSE clauses or SELECT-WHEN clauses. Furthermore the definition of the variables can easily be put into a data mart definitions table (see also *Chapter 24 – Do's and Don'ts When Building Data Marts*). One column contains the variable name and another column contains the definition.

Table 17.4: Definition of dummy variables without IF-THEN/ELSE clauses

Variables	Definition
Employed	(Employment_Status = 'Employed')
Unemployed	(Employment_ Status = 'Unemployed')
Retired	(Employment_ Status = 'Retired')
Education	(Employment_ Status = 'Education')

This would not be possible with IF-THEN/ELSE clauses. A further advantage is that the statements can be used in PROC SQL as well. IF-THEN/ELSE clauses would need to be transferred to CASE/WHEN clauses for SQL.

```
PROC SQL;
CREATE TABLE Customer_Dummy AS
  SELECT *,
  (Employment_Status = 'Employed') AS Employed,
  (Employment_Status = 'Unemployed') AS Unemployed,
  (Employment_Status = 'Retired') AS Retired,
  (Employment_Status = 'Education') AS Education
  FROM customer;
QUIT;
```

17.5.5 Other Types of Dummy Coding

We have introduced only the two most common types of dummy coding; namely, GLM and the deviation coding. GLM coding is very similar to reference coding. Here, one category—the reference category—does not have a separate dummy variable.

Other coding methods exist such as ordinal and polynomial dummy coding, as well as the ordinal and polynomial methods. You can refer to the *SAS/STAT User's Guide* in SAS Online Doc for the LOGISTIC procedure and CLASS statement, or to the SAS Press publication *SAS for Linear Models*.

17.6 Multidimensional Categorical Variables

17.6.1 Rationale

To consider the relationship between two or more categorical variables, cross tables are usually created in statistical analysis. It is, however, also possible to concatenate the values of the respective categorical variables and analyze the univariate distribution of these new variables.

A *multidimensional categorical variable* is a categorical variable whose value is the concatenation of two or more categorical variables. The creation of multidimensional categorical variables can reduce the number of variables in the data set and can also explicitly create relevant information in a variable.

17.6.2 The Principle

The principle of multidimensional categorical variables is technically simple. The concatenation of categorical values is stored in a new categorical variable. Consider the following short example.

Customers can use one or more of three products. Product use is stored in the indicator variables PRODUCTA, PRODUCTB, and PRODUCTC. In order to analyze the usage patterns of the three products a concatenated variable PRODUCTUSAGE is created:

```
ProductString = CAT(ProductA,ProductB,ProductC);
```

In this case, the result is a string variable with three digits and eight possible values (111, 110, 101, 100, 011, 010, 001, 000). The creation of a frequency table of the variable PRODUCTSTRING gives insight about the most frequent product combinations and allows a segmentation of customers. For example, PRODUCTSTRING = 100 can be named the "PRODUCT_A_ONLY customers".

17.6.3 Creating Categorical Interaction Variables

In some cases, business reasons strongly suggest that the combination of values of two or more variables is relevant for analysis—for example, the current tariff of a customer compared to his optimal tariff based on his service usage. Assume that we have the variables ACTUAL_TARIFF and OPTIMAL_TARIFF.

A derived variable HAS_OPTIMAL_TARIFF can easily be created with the following statement:

```
HAS_OPTIMAL_TARIFF = (ACTUAL_TARIFF = OPTIMAL_TARIFF);
```

Note that we assume that the format of the two variables is the same (length, no leading and trailing blanks) so that we compare them on a one-to-one basis.

If we want to create a variable TARIFF_MATRIX that concatenates the actual and the optimal tariffs in one variable we can do this with the following statement:

```
TARIFF_MATRIX = CATX('_',PUT(actual_tariff,$2.),PUT(optimal_tariff,$2.));
```

Note that here we assume that the tariff names (or code) differ in the first two characters. In other cases the PUT function needs to be amended accordingly.

In the following output the results can be seen for selected observations. Note that we have created two meaningful derived variables, one indicating whether the customer has the optimal tariff or not, and another variable describing the deviations between the optimal and actual tariff. Both of these variables can also be important predictor variables for the contract cancellation event of a customer.

Obs	actual_tariff	optimal_tariff	HAS_OPTIMAL_TARIFF	TARIFF_MATRIX
1	STANDARD	STANDARD	1	ST_ST
2	EASY	ADVANCED	0	EA_AD
3	STANDARD	LEISURE	0	ST_LE
4	BUSINESS	ADVANCED	0	BU_AD
5	ADVANED	STANDARD	0	AD_ST

A potential additional important variable can be created by combining the information from HAS_OPTIMAL_TARIFF and TARIFF_MATRIX:

```
IF HAS_OPTIMAL_TARIFF = 1 THEN TARIFF_MATRIX2 = 'OPTIMAL';
ELSE TARIFF_MATRIX2 = TARIFF_MATRIX;
```

This variable allows powerful classification of customers, by segmenting them into an OPTIMAL group and groups of NON-OPTIMAL TARIFFS with the corresponding tariff matrix.

After analyzing the frequencies of the variable TARIFF_MATRIX or TARIFF_MATRIX2, rare classes can be merged together into an OTHERS group. We explained this in the preceding section.

Further examples that are frequently used in this context are changes over time in the categories.

- TARIFF versus PREVIOUS _TARIFF
- PAYMENT_METHOD versus PREVIOUS_PAYMENT_METHOD
- VALUE_SEGMENT versus PREVIOUS_VALUE_SEGMENT
- BEHAVIOR_SEGMENT versus PREVIOUS_BEHAVIOR_SEGMENT

17.6.4 Creating a Concatenated Variable on the Basis of Product Usage Indicators

Assume that an Internet company offers four different product categories via the Web. The purchasing process for each category is handled by a different server domain. Thus we have, in addition to the customer base table, four tables that hold the purchases in the respective product category. For simplicity we will skip the process of creating a one-row-per-subject table for each purchasing history table here and discuss this in Chapter 19. The following DATA step shows how to create a concatenated variable for product usage of the four Products A, B, C, and D.

```
DATA customer;
 MERGE customer_base
       Product_Server_A (IN = Server_A)
       Product_Server_B (IN = Server_B)
       Product_Server_C (IN = Server_C)
       Product_Server_D (IN = Server_D);
 BY customer_id;
 ProdA = Server_A; ProdB = Server_B;
 ProdC = Server_C; ProdD = Server_D;
 ProductUsage = CAT(ProdA,ProdB,ProdC,ProdD);
RUN;
```

We have used the logical variables of the SET statement, indicating whether a customer has an entry in a certain server to create a concatenated variable.

17.6.5 Advanced Concatenated Variables

If we want to enhance this example by also considering the purchasing amount and only indicate those purchase sums higher 1,000, we can do this as in the following example. In our example we assume that each product server tables holds the sum of purchase amounts per customer in the column PURCHASE_SUM.

```
DATA customer;
  MERGE customer_base
        Product_Server_A (IN = Server_A
                          RENAME = (PURCHASE_SUM = SUM_A))
        Product_Server_B (IN = Server_B
                          RENAME = (PURCHASE_SUM = SUM_B))
        Product_Server_C (IN = Server_C
                          RENAME = (PURCHASE_SUM = SUM_C))
        Product_Server_D (IN = Server_D
                          RENAME = (PURCHASE_SUM = SUM_D))
        ;
  BY customer_id;
  ProdA = Server_A; ProdB = Server_B;
  ProdC = Server_C; ProdD = Server_D;
  ProductUsage = CAT(ProdA,ProdB,ProdC,ProdD);
  ProductUsage1000 = CAT((SUM_A > 1000),(SUM_B > 1000),
                         (SUM_C > 1000),(SUM_D > 1000));
  RUN;
```

Again we receive a variable that concatenates 0 and 1. However, 1 is only then inserted if the purchase amount for the respective product exceeds 1,000. We see from this example that we can create meaningful derived variables with only a few lines of code.

Note that the values of a concatenated variable need not be 0 and 1, as we saw in our first example in this section. Missing values for binary variables can be inserted as a period (.) or 9, resulting in the possible values 0, 1, and 9. Also, nominal and ordinal classes as well as counts can be concatenated in order to identify frequent combinations.

17.6.6 Applications of Concatenated Variables

There are many applications of concatenated variables. We have already discussed the interaction and the information reduction into one variable.

Concatenated variables also perform well in the analysis of multiple choice responses in surveys. The multidimensional structure of a lot of yes/no answers is usually complicated to handle. With concatenated variables the most frequent combinations can be identified and rare combinations can be grouped into an OTHERS group.

The associations between several univariate *segmentations* can also be analyzed with concatenated variables. For example, univariate segmentations for age, product usage, and purchase sum can be analyzed for their most frequent combinations. Rare classes can be combined; large classes can be split for another criterion. This type of segmentation is also called *business rule-based segmentation*. In *Chapter 26 – Case Study 2—Deriving Customer Segmentation Measures from Transactional Data*, we will work out a more detailed example on this topic.

Finally, we should mention that the final definition of concatenated variables, in the sense of the number of dimensions or definitions of the OTHERS group, is usually an iterative process. A certain definition is proposed, univariate distributions are analyzed, and the definition is reworked.

17.7 Lookup Tables and External Data

Overview

For a category more detailed data can be available. This information can be a more descriptive name, i.e., a description for a code. In this case we use the term *lookup table*.

It can also be a list of properties that are associated with the category. In this case we are talking about *external data*.

17.7.1 Lookup Tables

Cases where lookup tables are very frequent include the following:

- postal codes
- list of branches or sales outlets
- product categories
- product lists
- tariff plans
- lists of adverse events in clinical research

The dimension tables of a star schema can also be considered lookup tables. If a flat table (one-row-per-subject table) has to be built from a star schema the dimension tables are "merged" with the fact table.

17.7.2 External Data

Examples of external data are geodemographic data. These data can give per region or district additional attributes such as the following:

- age distribution
- gender distribution
- educational status
- income situation

External data are in most cases merged directly with the base table. In some cases a set of formats, one for each attribute in the external data, is created and applied to copies of the region variable.

17.7.3 Creating a SAS Format from a Data Set

Adding lookup tables or external data to a table can be done by a real merge in the sense of combining tables or by using formats to assign the descriptive value to the code.

These formats can be created with PROC FORMAT and a VALUE statement, as we saw in our preceding gender example. For categorical variables with many different values it is more convenient to create formats from SAS data sets. Additionally, the maintenance of those lookup lists is easier and more efficient if the formats can be created from tables.

With a simple example we will show how to create a format from a data set table. We have the following table BRANCHLIST.

Table 17.5: Lookup table branch IDs

	BranchID	BranchName
1	1	Vienna-City
2	2	Vienna-West
3	3	Salzburg
4	4	Wels
5	5	Linz
6	6	Graz

From this table we want to create a SAS format for the variable BRANCHID. This can be done with the following statement:

```
DATA BranchList;
  SET BranchList(RENAME =(BranchID=start BranchName=label));
  RETAIN fmtname 'BranchName' type 'n';
RUN;

PROC FORMAT CNTLIN = BranchList;
RUN;
```

Note that in the case of duplicate rows in the table BRANCHLIST they have to be deleted with the following statement:

```
PROC SQL;
  CREATE TABLE branchlist
  AS SELECT DISTINCT BranchID FROM branchlist;
QUIT;
```

or alternatively

```
PROC SORT DATA = branchlist OUT = branchlist_nodup NODUP;
  BY BranchID;
RUN;
```

Note that the format BRANCHNAME can be assigned to the BRANCHID variable in the base table, which will then display the branch name instead. If we have more details for each branch, such as the square meters of the branch, we can create another format BRANCH_M and assign this format to a copy of the BRANCHID variable.

The alternative to creating a format for each attribute in the lookup or external data table is to merge the whole lookup or external data table to the base table.

In *Chapter 26 – Case Study 2—Deriving Customer Segmentation Measures from Transactional Data*, we will explore a comprehensive example of the creation of a flat table from a star schema.

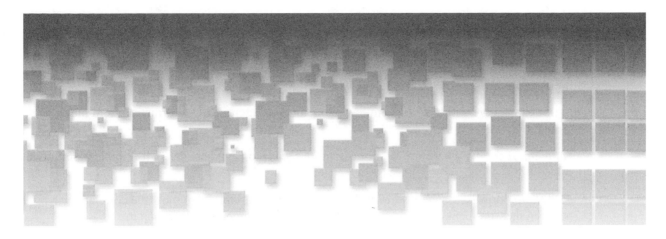

Chapter 18

Multiple Interval-Scaled Observations per Subject

18.1 Introduction

If we have multiple numerical observations per subject, there are two major ways to make them available in a one-row-per-analysis subject. We can either *transpose* each value per subject into a separate column, or we can *aggregate* the multiple observations and condense the inherent information into some descriptive values. The *pure transposition* of values is needed for some analyses such as a repeated measurement analysis of variance.

For many data mining analyses, however, the one-to-one transposition of the values does not make much sense. If we have many observations per subject, we will want to create a set of variables that sufficiently describes the properties of the repeated observation per subject. In contrast to pure transposition, we call this the *aggregation of multiple observations*.

When aggregating multiple observations we can distinguish between two main groups—*static aggregation* and *trend aggregation*. Only in trend aggregation do we consider the timely or sequential ordering of observations and their values and create indicators for trends over time. With static aggregation we aggregate only the values, using descriptive statistics, ignoring their ordering.

In this chapter we will not deal with pure transposition, because we covered this in *Chapter 14 – Transposing One- and Multiple-Rows per Subject Data Structures*, but we will cover the aggregation of multiple observations.

Overview

In this chapter we will look at the following methods of aggregating information from multiple observations.

Static aggregation, where we will discuss various ways to aggregate data with simple descriptive statistics.

Correlation of values, where we will show how derived variables can show the correlation between measurements or the correlation between measurements and a group mean.

Concentration of values, where we will show a special aggregation that provides the information whether the sum per subject concentrates on a few measurements or is distributed over measurements or subhierarchies.

Standardization of values, where we will show the ways to standardize values—for example, by dividing through the subject's mean value and other methods.

Derived variables, where we will deal with ways derived variables can be created that describe the course over time of measurements.

Macros in This Chapter

In this chapter we will introduce the following macro:

- %CONCENTRATE, which calculates the concentration of the total sum of values per subject on the top 50 % of the subhierarchies.

We will also show a number of coding examples.

18.2 Static Aggregation

Overview

In this section we will deal with methods of static aggregation of multiple numeric values per analysis subject. With static aggregations the following topics are of special interest:

- various methods of basic descriptive aggregation measures that condense the information into a set of variables
- correlation of variables
- a special measure that represents the concentration of values on a certain proportion of observations

18.2.1 Example Data

The following example data will be the basis for the examples in this chapter. We have telephone usage data for the last six months for 10 customers. In Chapter 14, we saw that we can easily switch between the LONG and WIDE representation for these data with PROC TRANSPOSE statements.

Table 18.1: One-row-per-subject representation of the example data

CustID	M1	M2	M3	M4	M5	M6
1	52	54	58	47	38	22
2	22	24	30	28	31	30
3	100	120	110	115	100	95
4	43	43	43	.	42	41
5	20	29	35	39	28	44
6	16	24	18	25	30	24
7	80	70	60	50	60	70
8	90	95	80	100	100	90
9	47	47	47	47	47	47
10	50	52	0	50	0	52

Table 18.2: Multiple-rows-per-subject representation of the example data (truncated)

CustID	Usage	Month
1	52	1
1	54	2
1	58	3
1	47	4
1	38	5
1	22	6
2	22	1
2	24	2
2	30	3

If we need the data in a multiple-rows-per-subject structure (WIDE), we can convert them from a one-row-per-subject structure (LONG) with the following statements:

```
PROC TRANSPOSE DATA = WIDE OUT = LONG;
 BY custId;
RUN;

DATA LONG;
 SET LONG;
 FORMAT Month 8.;
 RENAME col1 = Usage;
 Month = compress(_name_,'M');
 DROP _name_;
RUN;
```

Note that the LONG format can be used for creating graphs with PROC GPLOT and for analysis with PROC MEANS.

18.2.2 Aggregating from a One-Row-per-Subject Structure

In order to create aggregations from a table in the WIDE structure (see Table 18.1), we use SAS functions to create the respective aggregations:

```
DATA wide;
 SET wide;
 Usage_Mean = MEAN(of M1-M6);
 Usage_Std  = STD(of M1-M6);
 Usage_Median = MEDIAN(of M1-M6);
RUN;
```

The results in aggregations are shown in Table 18.3:

Table 18.3: Static aggregations with SAS functions

	CustID	M1	M2	M3	M4	M5	M6	Usage_Mean	Usage_Std	Usage_Media
1	1	52	54	58	47	38	22	45.166666667	13.272779161	49.5
2	2	22	24	30	28	31	30	27.5	3.6742346142	29
3	3	100	120	110	115	100	95	106.66666667	9.8319208025	105
4	4	43	43	43	.	42	41	42.4	0.894427191	43
5	5	20	29	35	39	28	44	32.5	8.5965109201	32
6	6	16	24	18	25	30	24	22.833333333	5.0760877324	24
7	7	80	70	60	50	60	70	65	10.488088482	65
8	8	90	95	80	100	100	90	92.5	7.5828754441	92.5
9	9	47	47	47	47	47	47	47	0	47
10	10	50	52	0	50	0	52	34	26.351470547	50

18.2.3 Aggregating from a Multiple-Rows-per-Subject Structure

Starting from data that are in the LONG structure (see Table 18.2), we can use PROC MEANS to create static aggregations of the data on a one-row-per-subject level:

```
PROC MEANS DATA = long NOPRINT NWAY;
 CLASS CustID;
 VAR Usage;
 OUTPUT OUT = aggr_static(DROP = _type_ _freq_)
            MEAN= SUM= N= STD= MIN= MAX=/AUTONAME;
RUN;
```

This results in Table 18.4:

Table 18.4: Static aggregations with PROC MEANS

	CustID	Usage_Mean	Usage_Sum	Usage_N	Usage_StdDev	Usage_Min	Usage_Max
1	1	45.166666667	271	6	13.272779161	22	58
2	2	27.5	165	6	3.6742346142	22	31
3	3	106.66666667	640	6	9.8319208025	95	120
4	4	42.4	212	5	0.894427191	41	43
5	5	32.5	195	6	8.5965109201	20	44
6	6	22.833333333	137	6	5.0760877324	16	30
7	7	65	390	6	10.488088482	50	80
8	8	92.5	555	6	7.5828754441	80	100
9	9	47	282	6	0	47	47
10	10	34	204	6	26.351470547	0	52

Note the following:

- To run the analysis per subject, we use a CLASS statement. A BY statement would do this too, but would require the data be sorted by CustID.

- We use the NWAY option in order to suppress the grand total mean (and possible subtotals if we had more than one class variable), so the output data set contains only rows for the 10 analysis subjects.

- Furthermore, we use the NOPRINT option in order to suppress the printed output from the log, which, in addition to our simple example with 10 observations, can overflow the Output window with thousands of descriptive measures.

- In the OUTPUT statement we specify the statistics that will be calculated. The AUTONAME option creates the new variable name in the form VARIABLENAME_STATISTIC.

- If we have only one statistic specified, we can omit the AUTONAME option and the variables in the aggregated data set will be the same as in the input data set.

- If we want to calculate different statistics for different input variables we can specify the following in the OUTPUT statement:

```
SUM(Usage)  = sum_usage
MEAN(Billing) = mean_billing
```

- In the OUTPUT statement we immediately drop the _TYPE_ and the _FREQ_ variables. We could also keep the _FREQ_ variable and omit the 'N' from the statistic list, as we have done here for didactic purposes.

For a complete discussion of PROC MEANS, see SAS Help and Documentation or the SAS Press publication *Longitudinal Data and SAS*.

18.2.4 Business Rationale

The business rationale for the calculation of static aggregations is that they summarize the properties of data values into a few descriptive statistics. We are in the same situation here as in classic descriptive statistics because we are condensing the information of a number of observations into a few statistics.

The preceding list can easily be longer by adding various descriptive measures such as skewness, kurtosis, median, interquartile ranges, ranges, and so on. If we were in a pure data mining scenario this might have its benefits because we might want to see which aggregations best fit the data. It makes sense however, to calculate only those measures that also have an adequate business interpretation.

- Using the mean is very obvious; it measures the average location of the distribution and allows a differentiation of subjects by their magnitude of values.

- Calculating the standard deviation allows us to distinguish subjects with an erratic behavior of their time series from those with a smooth course over time.

- The number of observations does not measure the values themselves, but gives information about how many periods for which each subject has observations. Here we have to investigate whether the source data contain zeros or missing values for periods without values. In some cases the number of observations with value zeros or the number of missing values can make sense.

There is no rule about which measure to use for which analysis. It is more of a balancing act between a small number of interpretable aggregates and a complete snapshot of the data.

If there are no special business reasons for different statistics measures per variable, it makes sense to decide on one descriptive measure for better interpretability—for example, to use the mean or the median as the location measure, or to use the standard deviation or the interquartile range as the dispersion measure for all variables and not change between them per variable.

18.3 Correlation of Values

18.3.1 Business Rationale

Considering the correlation of variables is an appropriate way to bring the inherent information of multiple observations per subject into a one-row-per-subject table. We perform correlation analysis for each subject in the multiple-rows-per-subject table and add the resulting correlation coefficient to the respective one-row-per-subject data mart. In general we distinguish between two groups of correlation analysis in this context:

- Correlation of different variables

  The business rationale is that the correlation of different product usage variables over time describes the behavior of the customer. For example, we calculate the correlation of telephone usage in different destination zones. The resulting correlation coefficient shows whether a customer increases usage homogenously in different destination zones (= positive correlation) or whether usage from one zone is substituted by usage in another zone (= negative correlation coefficient).

- Correlation of individual values with the group mean or with a reference period

  The business rationale here is that we can see whether a subject's values are homogeneous with the values of the population or whether the subject's individual values or course differs over time.

18.3.2 Correlation of Different Variables

In this case we have in our multiple-rows-per-subject table at least two interval variables whose correlation we want to measure. In our example we have an additional variable BILLING added to our longitudinal data table from the preceding example. We now calculate the correlation of the BILLING and USAGE variables by customer in order to see how the BILLING amount is correlated with the actual usage for different customers.

In practice this can be a good indicator of the correlation of perceived service (= usage) to the customers and the amount they have to pay for it.

Table 18.5: Multiple-rows-per-subject data with two measurement variables

	CustID	Month	Usage	Billing
1	1	1	52	26
2	1	2	54	54
3	1	3	58	58
4	1	4	47	47
5	1	5	38	38
6	1	6	22	22
7	2	1	22	44
8	2	2	24	24
9	2	3	30	44
10	2	4	28	28
11	2	5	31	44
12	2	6	30	44

```
PROC CORR DATA = Corr NOPRINT
           OUTS=CorrSpearman(where = (_type_ = 'CORR') DROP = _name_);
 BY Custid;
 VAR usage;
 WITH billing;
RUN;
```

Note the following:

- We run PROC CORR with a NOPRINT option and save the Spearman correlation coefficients to a table.

- We could also use the Pearson correlation coefficient, which can be specified with an OUTP= option.

- We specify the BY statement to run the analysis per CustID.

We receive one number per customer, holding the correlation of BILLING and USAGE values, as in Table 18.6.

Table 18.6: Correlation coefficients per CustID

	CustID	Usage
1	1	0.8285714286
2	2	0.4458963214
3	3	0.1328422328
4	4	1
5	5	-0.657142857
6	6	1
7	7	0.6363636364
8	8	-0.190692518
9	9	.
10	10	0.8488746876

18.3.3 Correlation of Individual Values with the Group Mean

Similar to the preceding example we calculate the correlation of the individual values per subject, with the mean of values of all subjects. From a business point of view, we measure how homogeneous the course of values of time of a subject is with the overall course of all subjects.

First we calculate the mean usage for each month:

```
PROC MEANS DATA = long NOPRINT NWAY;
 CLASS month;
 VAR usage;
 OUTPUT OUT = M_Mean MEAN=m_mean;
RUN;
```

Then we join the means of all customers per month back to the original data:

```
PROC SQL;
 CREATE TABLE interval_corr
 AS SELECT *
    FROM longitud a, m_mean_tp b
  WHERE a.month = b.month;
QUIT;
```

Note that we use PROC SQL for the join because we want to join the two tables without sorting for the variable MONTH, and then calculate the correlation per customer:

```
PROC CORR DATA = interval_corr
          OUTS = Corr_CustID(WHERE = (_type_ = 'CORR')) NOPRINT;
 BY CustID;
 VAR Usage;
 WITH m_mean;
RUN;
```

Table 18.7: Correlation per customer

	CustID	Usage
1	1	0.2571428571
2	2	-0.81167945
3	3	0.6377481392
4	4	0.4472135955
5	5	0.0857142857
6	6	-0.173931311
7	7	0.2059714602
8	8	0.1765469659
9	9	.
10	10	0.7171371656

We see that customers 3 and 10 have a strong positive correlation with the mean course, and customers 2 and 6 have a negative correlation with the mean course. In our small example however, we have to admit that customers with high values are likely to have a positive correlation because their values contribute much to the overall mean. For customer 9 the correlation is missing because all of that customer's values are the same.

18.4 Concentration of Values

In the case of repeated observations, because of hierarchical relationships we can calculate static aggregations, but we cannot calculate measurements that consider the course over time, because we do not have a course over time.

In this case the concentration of values on a certain proportion of underlying entities is a useful measure to describe the distribution.

Note that this measure is not restricted to repeated measurements because of hierarchical relationships, but it can also be calculated in the case of repeated measurements over time. In this case it measures how the values concentrate on a certain proportion of time periods.

We will look at a simple example where we have usage data per contract. Each customer can have one or more contracts. The data are shown in Table 18.8.

Table 18.8: Usage data per contract and customer

	CustID	ContractID	Usage1
1	1	1	20
2	1	2	40
3	1	3	60
4	1	4	5
5	1	5	2
6	1	6	1
7	2	1	10
8	2	2	10
9	2	3	12
10	2	4	11
11	3	1	40
12	3	2	30
13	3	3	30
14	3	4	10
15	3	5	5
16	4	1	4
17	5	1	1
18	5	2	2
19	5	3	3
20	6	1	1
21	6	2	2
22	6	3	3
23	6	4	4

We see that customer 1 has a high usage concentration on three on his six contracts, whereas customer 2 has a rather equal usage distribution of his contracts.

We will now calculate a measure per customer that shows what percentage of the total usage is done with 50% of the contracts. The business rationale of this measure is that we want to assume different behavior of customers with an equal distribution versus customers with a usage concentration on a few contracts. This variable has proven to be very valuable in event prediction—for example, in marketing analyses for business customers, where several hierarchical levels such as CUSTOMER and CONTRACT exist.

Note that we are not talking here about classic concentration coefficients such as entropy or Gini, but we are defining concentration as the proportion of the sum of the top 50% subhierarchies to the total sum over all subhierarchies per subject.

The following macro calculates the concentration measure per customer:

```
%MACRO Concentrate(data,var,id);

*** Sort by ID and VALUE;
PROC SORT DATA=&data(keep=&var &id) OUT=_conc_;
 BY &id DESCENDING &var;
RUN;

*** Calculation of the SUM per ID;
PROC MEANS DATA=_conc_ NOPRINT;
 VAR &var;
 BY &id;
 OUTPUT out = _conc_sum_(DROP=_type_ _freq_) SUM=&var._sum;
 WHERE &var ge 0;
RUN;

*** Merge the sum to the original sorted data set;
DATA _conc2_;
 MERGE _conc_
       _conc_sum_;
 BY &id;
 IF FIRST.&id THEN &var._cum=&var;
 ELSE &var._cum+&var;
 &var._cum_rel=&var._cum/&var._sum;
 IF LAST.&id AND NOT FIRST.&id THEN skip = 1; ELSE skip=0;
RUN;

*** Calculation of the median per ID;
PROC UNIVARIATE DATA=_conc2_ noprint;
 BY &id;
 VAR &var._cum_rel;
 OUTPUT out=concentrate_&var. MEDIAN=&var._conc;
 WHERE skip =0;
RUN;
%MEND;
```

The macro has three parameters:

DATA

The name of the data set with the original data.

VAR

The name of the variable that holds the value whose aggregation will be measured.

ID

The name of the ID variable of the higher relationship—for example, the customer ID in the case of the "one customer has several contracts" situation.

The macro can be invoked with the following statement for the preceding data:

```
%concentrate(usage,usage1,CustID);
```

And produces the following output table:

Table 18.9: Usage concentration measure per customer

	CustID	usage1_conc
1	1	0.9375
2	2	0.5348837209
3	3	0.7391304348
4	4	1
5	5	0.6666666667
6	6	0.7

The name of the ID variable is copied from the input data set, and the name of the concentration variable is created as <variable name>_CONC. The name of the output data set is CONCENTRATE_<variable name>.

We see from the output that customer 1 has a concentration value of 0.9375, and customer 2 has a value of 0.53488, which is consistent with our input data.

Note that the concentration value can range from 0.5 to 1—0.5 for the equal distribution over subhierarchies and 1 for the concentration of the total sum on the 50% subhierarchies.

In the case of unequal numbers of subentities the macro averages the cumulative concentration at the median entity with the value of the next higher entity. Strictly speaking this is only an approximation but a good compromise between an exact result and a short program.

If concentrations are calculated for more than one variable, we will call the %CONCENTRATE macro for each variable and join the results afterwards.

```
%concentrate(usage,usage1,CustID);
%concentrate(usage,usage2,CustID);

DATA _concentrate_;
 MERGE concentrate_Usage1
       concentrate_Usage2;
 BY CustID;
RUN;
```

18.5 Course over Time: Standardization of Values

General
When we have time series of values available per subject, we can start analyzing their course over time and derive indicators about the patterns over time. When looking at and analyzing patterns over time, it is sometimes relevant to consider relative values instead of absolute values.

Here, the creation of relative values will be called the *standardization of values*. However, we do not mean a statistical standardization in the sense of subtraction of the mean and the division by the standard deviation, but we will consider different ways to make the values relative.

The possibilities include the following:

- Standardizing the values by the individual mean of the subject
- Standardizing the values by the mean over all subjects for the same time period
- Standardizing both—the individual mean of the subject and the mean over all subjects for the same time period

18.5.1 Standardizing the Values by the Individual Mean of the Subject

In this case we see whether a value is above or below its individual average. From a business point of view, this allows us to recognize changes in the subject's individual behavior over time.

To standardize per subject in a one-row-per-subject data mart we use an ARRAY statement to divide each month by the mean of the values:

```
DATA indiv;
 SET wide;
 FORMAT CustID 8. M1-M6 m_avg 8.2;
 ARRAY m {*} M1 - M6;
 ARRAY m_mean {*} M1 - M6;
 m_avg = MEAN(of M1 - M6);
 DO i = 1 TO DIM(m);
  m_mean{i} = ((m{i}-m_avg)/m_avg);
 END;
 DROP i;
RUN;
```

This results in the following table:

Table 18.10: Data standardized by the subject's individual mean

	CustID	M1	M2	M3	M4	M5	M6	m_avg
1	1	0.15	0.20	0.28	0.04	-0.16	-0.51	45.17
2	2	-0.20	-0.13	0.09	0.02	0.13	0.09	27.50
3	3	-0.06	0.12	0.03	0.08	-0.06	-0.11	106.67
4	4	0.01	0.01	0.01	.	-0.01	-0.03	42.40
5	5	-0.38	-0.11	0.08	0.20	-0.14	0.35	32.50
6	6	-0.30	0.05	-0.21	0.09	0.31	0.05	22.83
7	7	0.23	0.08	-0.08	-0.23	-0.08	0.08	65.00
8	8	-0.03	0.03	-0.14	0.08	0.08	-0.03	92.50
9	9	0.00	0.00	0.00	0.00	0.00	0.00	47.00
10	10	0.47	0.53	-1.00	0.47	-1.00	0.53	34.00

For example, we see that customer 1 has a usage drop in months M5 and M6.

18.5.2 Standardizing by the Mean over All Subjects for the Same Time Period

Here we filter the overall trend of the values from the data and show only the individual deviation from the overall trend. If, for example, the values increase in December for all subjects, we will see the absolute values going up in December. After standardization of the monthly means we will see how the individual subject behaves compared to all subjects.

Note that you can use this type of standardization not only to filter seasonal effect, but also to consider the different number of days per month. For example a 10% decrease from January to February or a 3% increase from April to May is probably explained by the number of days in these months rather than a change in the subject's behavior.

For the standardization per month we can calculate the means per month from a one-row-per-subject data mart and output the results to the Output window.

```
PROC MEANS DATA = wide MAXDEC=2;
 VAR M1 - M6;
 OUTPUT OUT = M_Mean MEAN = /AUTONAME;
RUN;
```

From the Output window we copy and paste values to the program editor in order to use the means directly in the DATA step. This is shown in the following example:

```
DATA interval_month;
 SET wide;
 FORMAT M1 - M6 8.2;
M1=M1/ 52.00;
M2=M2/ 55.80;
M3=M3/ 48.10;
M4=M4/ 55.67;
M5=M5/ 47.60;
M6=M6/ 51.50;
RUN;
```

The results are shown in Table 18.11. We see that the resulting data reflect the subject's individual level of values. However, we have corrected the values by the average over all subjects in this month. This allows interpretation, whether a customer really increases or decreases his values.

Table 18.11: Table with standardized values by the mean over all subjects for the same time period

	CustID	M1	M2	M3	M4	M5	M6
1	1	1.00	0.97	1.21	0.84	0.80	0.43
2	2	0.42	0.43	0.62	0.50	0.65	0.58
3	3	1.92	2.15	2.29	2.07	2.10	1.84
4	4	0.83	0.77	0.89	.	0.88	0.80
5	5	0.38	0.52	0.73	0.70	0.59	0.85
6	6	0.31	0.43	0.37	0.45	0.63	0.47
7	7	1.54	1.25	1.25	0.90	1.26	1.36
8	8	1.73	1.70	1.66	1.80	2.10	1.75
9	9	0.90	0.84	0.98	0.84	0.99	0.91
10	10	0.96	0.93	0.00	0.90	0.00	1.01

18.5.3 Standardizing by the Individual Mean of the Subject and the Mean over All Subjects for the Same Time Period

When we look again at the data in Table 18.1, repeated in Table 18.10, we see the following:

- The individual mean per subject corresponds to a standardization of the row mean.
- The mean over all subjects for the same time period corresponds to a standardization of the column mean.

- The last option, standardizing by both, corresponds to a standardization of the row mean and the column mean.

Table 18.12: One-row-per-subject representation of the example data

CustID	M1	M2	M3	M4	M5	M6
1	52	54	58	47	38	22
2	22	24	30	28	31	30
3	100	120	110	115	100	95
4	43	43	43	.	42	41
5	20	29	35	39	28	44
6	16	24	18	25	30	24
7	80	70	60	50	60	70
8	90	95	80	100	100	90
9	47	47	47	47	47	47
10	50	52	0	50	0	52

We see that the result of standardizing by the individual mean of the subject and the mean over all subjects for the same time period equals the calculation of expected values from the margin distributions in a cross table. Comparing to the expected value, we also see how each customer's values deviate from the value that we would expect based on the customer's average usage and the usage in the respective month.

A very elegant way to code the consideration of the individual mean of the subject and the mean over all subjects for the same time period is to use PROC FREQ and an ODS statement. With this method we can also calculate the two components separately.

The first step, which is optional, is to route the output into a file in order to avoid Output window overflow. The reason is that we cannot use a NOPRINT option as in the other example, because we need output to be created for the ODS statement.

```
PROC PRINTTO PRINT='C:\data\somefile.lst';
RUN;
```

We specify that we want to store the output object CrossTabFreqs in the FREQ table.

```
ODS OUTPUT CrossTabFreqs = freq;
```

We run PROC FREQ and specify EXPECTED to calculate the expected values form the margin distributions.

```
PROC FREQ DATA = long;
 TABLE CustID * Month / EXPECTED;
 WEIGHT Usage;
RUN;
```

Finally, we close the ODS statement and route the output back to the Output window.

```
ODS OUTPUT CLOSE;
PROC PRINTTO;
RUN;
```

The resulting output file also contains the margin distributions. We want to keep only the inner cells of the table (TYPE = '11').

```
DATA freq;
 SET freq;
 KEEP CustID Month Expected RowPercent ColPercent;
 WHERE _type_ = '11';
RUN;
```

The resulting table has three additional columns.

- EXPECTED is the expected value after filtering the overall time and the individual customer effect. This value needs to be compared to the actual value and corresponds to the values that we call standardizing by the individual mean of the subject and the mean over all subjects for the same time period.

- ROWPERCENT is the usage proportion of the individual sum of its subject.

- COLPERCENT is the usage proportion of the sum over all subjects for the same time period. In the case of a high number of subjects, these values will get very small; therefore, they might need to be multiplied by a constant.

- Note that the ROWPERCENT and COLPERCENT columns are in absolute terms not the same numbers as we calculated in the preceding examples, but they are proportional to them.

Table 18.13: Table with expected value, row, and column percent

	CustID	Month	Expected	RowPercent	ColPercent
1	1	1	46.188135038	19.188191882	10
2	1	2	49.563421829	19.926199262	9.6774193548
3	1	3	42.72402491	21.402214022	12.058212058
4	1	4	44.500491642	17.343173432	9.381237525
5	1	5	42.279908227	14.022140221	7.9831932773
6	1	6	45.744018355	8.1180811808	4.2718446602
7	2	1	28.121927237	13.333333333	4.2307692308
8	2	2	30.17699115	14.545454545	4.3010752688
9	2	3	26.012782694	18.181818182	6.237006237
10	2	4	27.09439528	16.96969697	5.5888223553
11	2	5	25.742379548	18.787878788	6.512605042
12	2	6	27.85152409	18.181818182	5.8252427184

18.5.4 Discussion of These Methods

To discuss these methods, we have to keep in mind where these standardization methods come from. Their origin is primarily the data mining area, where we try to create derived variables that describe as accurately as possible the subject's behavior over time.

- *Standardizing by the individual mean of the subject* allows us to see whether the subject's values change over time. In event prediction, these variables can be important in order to see whether a subject changes his behavior. For example, if we analyze usage data over data, a drop in the usage data can be a good predictor for an event.

- *Standardizing by the mean over all subjects for the same time period* is a very important measure in describing the subject's behavior. Here we can see whether he behaves differently from other subjects.

> ▪ Standardizing by both, by the individual mean of the subject and the mean over all subjects for the same time period, creates values on a non-interpretable scale. It allows us to filter all effects and to see whether a change took place over time.

The last method, using PROC FREQ and the ODS table CrossTabFreqs, allows the easy calculation of the standardized values on a multiple-rows-per-subject table. The other code example we saw in this section was based on a one-row-per-subject table.

18.6 Course over Time: Derived Variables

We have not yet considered the timely ordering of the observations. When we look at the course of the values over time we can calculate coefficients that describe the patterns of the series of values over time. We will divide these coefficients into three groups:

> ▪ Simple trend measures, where we create only differences and ratios of values over time.

> ▪ Complex trends, where we calculate coefficients of the linear regression for the entire time series or parts of it.

> ▪ Time series analysis, where we calculate measures of time series analysis, including trends but also autoregressive (AR) and moving average (MA) terms.

18.6.1 Simple Trend Coefficients

When looking at the time series data from our preceding example we might be interested in whether a subject drops his usage in the last two months of the observation period. This can be calculated with simple derived variables such as the difference, the ratio, and a standardized difference.

```
DATA SimpleTrend;
 SET wide;
 FORMAT Last2MonthsDiff Last2MonthsRatio Last2MonthsStd 8.2;
 Last2MonthsDiff = MEAN(of m5-m6)-MEAN(of m1-m4);
 Last2MonthsRatio = MEAN(of m5-m6)/MEAN(of m1-m4);
 Last2MonthsStd = (MEAN(of m5-m6)-MEAN(of m1-m4))/MEAN(of m1-m4);
RUN;
```

Note that in the last case we could also divide by the standard deviation of the values. In all cases we use the MEAN function to calculate the mean of the values for the respective months. Note that the SUM function would be appropriate if only we had the same number of observations for each subject.

Table 18.14: Longitudinal data with simple trend indicators

	CustID	M1	M2	M3	M4	M5	M6	Last2MonthsDiff	Last2MonthsRatio	Last2MonthsStd
1	1	52	54	58	47	38	22	-22.75	0.57	-0.43
2	2	22	24	30	28	31	30	4.50	1.17	0.17
3	3	100	120	110	115	100	95	-13.75	0.88	-0.12
4	4	43	43	43	.	42	41	-1.50	0.97	-0.03
5	5	20	29	35	39	28	44	5.25	1.17	0.17
6	6	16	24	18	25	30	24	6.25	1.30	0.30
7	7	80	70	60	50	60	70	0.00	1.00	0.00
8	8	90	95	80	100	100	90	3.75	1.04	0.04
9	9	47	47	47	47	47	47	0.00	1.00	0.00
10	10	50	52	0	50	0	52	-12.00	0.68	-0.32

In Table 18.4 we see the output with the respective columns. From the sign of the difference for the variables LAST2MONTHSDIFF and LAST2MONTHSSTD or the fact the values is above or below 1 for the variable LAST2MONTHSRATIO we can see whether a decrease or increase in usage has occurred.

In our example we are comparing the values of months 1–4 and the values of months 5–6. Other choices such as 1–3 versus 4–6 or 1–5 versus 6 would also be possible. The decision about which months be compared with each other depends on a business consideration, a visual analysis of data courses, or both. Note that in the case of event prediction you might want to create a graph showing the mean course of values for event A versus the mean course of values for event B in order to detect the point in time when values decrease or increase for a certain group.

Note that these variables, unlike static aggregations, do not reflect the location of the values themselves. The purpose of these derived variables is that they tell us about the course over time. They are very suitable to compress the behavior over time into a single variable, which is important for predictive analysis.

The usage of these types of variables has proven to be very predictive in the case of event prediction—for example, in the case of prediction of contract cancellations based on usage drops. Note, however, that the time windows for the time series and the target windows where the event is observed have to be chosen accordingly in order to avoid a sole prediction of those customers who have already canceled their contracts and no marketing actions are effective. See *Chapter 12 – Considerations for Predictive Modeling* for details.

18.6.2 Simple Trend Regression

A trend over time can also be measured by its linear (or higher term) regression coefficients. In this case we fit a regression for each subject. The desired result is a table with an intercept and more important a regression slope per subject. This can be achieved with PROC REG, with the following statements:

```
PROC REG DATA = long NOPRINT
         OUTEST=Est(KEEP = CustID intercept month);
 MODEL usage = month;
 BY CustID;
RUN;
```

Note the following:

- We use a BY statement to perform regression analysis per CustID. The data must be sorted for this variable in the input data set.
- We use the NOPRINT option in order to suppress printed output.
- We use the OUTEST option to create an output data set, which holds the regression coefficients per CustID.
- We specify a simple regression model USAGE = MONTH.
- The data need to be in a multiple-rows-per-subject structure.
- PROC REG is part of the SAS/STAT module.

The output is shown in the following table.

Table 18.15: Regression coefficients per customer

	CustID	Intercept	Month
1	1	66.066666667	-5.971428571
2	2	21.6	1.6857142857
3	3	114.66666667	-2.285714286
4	4	43.744186047	-0.395348837
5	5	20.4	3.4571428571
6	6	16.333333333	1.8571428571
7	7	74	-2.571428571
8	8	89	1
9	9	47	0
10	10	43.6	-2.742857143

For non-centered data the intercept gives the mean value of month 0. The slope coefficient in the variable MONTH shows the average change per month.

The advantage of this approach is that we can get an estimate for the time trend over all observations per subject. If we want to restrict the calculation of the regression slope for a certain time interval we need to add a WHERE statement to the preceding code:

```
WHERE month in (3,4,5,6);
```

This considers only the last four months of the available data.

18.6.3 Combined Trends

In some cases it is desirable to combine the information about the overall trend with the information about a more recent shorter time period. In these cases we create combined indicators by performing two separate regressions.

In the following example, we create linear regressions for the entire time period and a separate regression for months 5 and 6. Thus we can measure the long-term and short-term trend in two separate variables.

We run separate regressions for the respective time periods and store the coefficients in separate data sets with the following code:

```
PROC REG DATA = long NOPRINT
         OUTEST=Est_LongTerm(KEEP = CustID month
                               RENAME = (month=LongTerm));
 MODEL usage = month;
 BY CustID;
RUN;

PROC REG DATA = long NOPRINT
         OUTEST=Est_ShortTerm(KEEP = CustID month
                               RENAME = (month=ShortTerm));
 MODEL usage = month;
 BY CustID;
 WHERE month in (5 6);
RUN;
```

Next we merge the two coefficient data sets together and receive a data set with two derived variables that contain information about the trend:

```
DATA mart;
 MERGE wide
       est_longTerm
       est_shortTerm;
 BY CustID;
RUN;
```

Table 18.16 shows an example:

Table 18.16: Long-term and short-term regression coefficients per customer

	CustID	M1	M2	M3	M4	M5	M6	LongTerm	ShortTerm
1	1	52	54	58	47	38	22	-5.971428571	-16
2	2	22	24	30	28	31	30	1.6857142857	-1
3	3	100	120	110	115	100	95	-2.285714286	-5
4	4	43	43	43	.	42	41	-0.395348837	-1
5	5	20	29	35	39	28	44	3.4571428571	16
6	6	16	24	18	25	30	24	1.8571428571	-6
7	7	80	70	60	50	60	70	-2.571428571	10
8	8	90	95	80	100	100	90	1	-10
9	9	47	47	47	47	47	47	0	0
10	10	50	52	0	50	0	52	-2.742857143	52

The two variables LONGTERM and SHORTTERM show the slopes of the regression. We see that the coefficients in SHORTTERM equal the difference of month 5 (M5) and month 6 (M6) as we had only two periods in the regression.

18.6.4 Creating a Concatenated Variable

In Chapters 16 and 17 we saw ways to group interval observations and to create concatenated derived categorical variables. We can also do this here by classifying the long-term and the short-term trend variable into the groups '+', '=' and '-'.

We create a format for the groups. Note that the limits of −1 and 1 are chosen arbitrarily and can be chosen based on the distribution of coefficients.

```
PROC FORMAT;
 VALUE est LOW -< -1    = '-'
           -1  -  1     = '='
            1   <- HIGH = '+';
RUN;
```

In the next step we create the derived variable LONGSHORTIND. In this variable we concatenate the values of LONGTERM and SHORTTERM and we apply simultaneously the format EST on the data to retrieve values '+', '=' or '-'.

```
DATA mart;
 SET mart;
 LongShortInd = CAT(put(LongTerm,est.),put(ShortTerm,est.));
RUN;
```

We also receive an additional variable showing the group of the long-term trend in the first digit and the group of the short-term trend in the second digit. The concatenated indicator is a powerful classification of the trend over time as it considers both the long-term and short-term trend. For predictive analysis of events this classification provides a powerful segmentation. Again, the choice of the time intervals depends on the business considerations, and a visual analysis of the course over time.

Table 18.17: Long-term and short-term regression coefficients and the concatenated variable per customer

	CustID	M1	M2	M3	M4	M5	M6	LongTerm	ShortTerm	LongShortInd
1	1	52	54	58	47	38	22	-5.971428571	-16	--
2	2	22	24	30	28	31	30	1.6857142857	-1	+-
3	3	100	120	110	115	100	95	-2.285714286	-5	--
4	4	43	43	43	.	42	41	-0.395348837	-1	==
5	5	20	29	35	39	28	44	3.4571428571	16	++
6	6	16	24	18	25	30	24	1.8571428571	-6	+-
7	7	80	70	60	50	60	70	-2.571428571	10	-+
8	8	90	95	80	100	100	90	1	-10	=-
9	9	47	47	47	47	47	47	0	0	==
10	10	50	52	0	50	0	52	-2.742857143	52	-+

18.6.5 Time Series Analysis

The principle of using coefficients of regression analysis to describe the course of the values over time can be generalized to other analysis methods, such as time series analysis. The principle is the same—we perform the analysis on the data and store the coefficients per subject in a table.

There are many options for the parameterization of a time series model. Therefore, there are many possible coefficients that can be created per subject. In this chapter we did not deal with all the possible ways to create and parameterize the time series model, because that would have gone beyond the scope of this book.

We will show a short example of how we can calculate coefficients per subject on the basis of a time series. In our example we will use PROC TIMESERIES, which is part of SAS/ETS. We will use the SASHELP.PRDSAL3 data, which we will first sort by STATE and DATE.

```
PROC SORT DATA = sashelp.prdsal3 OUT = Prdsal3;
 BY state date;
RUN;
```

Next we will run PROC TIMESERIES. Note that we create output data sets for the SEASON, the TREND, the decomposition (DECOMP), and the autocorrelations (CORR).

```
PROC TIMESERIES DATA=prdsal3
                OUTSEASON=season
                OUTTREND=trend
                OUTDECOMP=decomp
                OUTCORR = corr
                MAXERROR=0;
   BY state;
   WHERE product = 'SOFA';
   SEASON SUM / TRANSPOSE = YES;
   TREND MEAN / TRANSPOSE = YES;
   CORR ACOV /  TRANSPOSE = YES;
   DECOMP TCS / LAMBDA = 1600 TRANSPOSE = YES;
   ID date INTERVAL=MONTH ACCUMULATE=TOTAL SETMISSING=MISSING;
   VAR actual ;
RUN;
```

Note the following:

Output data sets for SEASON, TREND, DECOMP, and CORR can be created in one step by using the appropriate statements and output data sets.

The option TRANSPOSE = YES is needed to create an output data set in a one-row-per-subject structure.

The ID statement specifies the time variable and also allows us to perform aggregations in the time hierarchy—for example, by specifying INTERVAL = QTR to perform an analysis per quarter.

Table 18.18 shows the content of the CORR table:

Table 18.18: Table CORR with autocorrelations per subject

	State/Province	Statistic	Lag 0	Lag 1	Lag 2	Lag 3	Lag 4	Lag 5	Lag 6	Lag 7
1	Baja California Norte	ACOV	560691.35556	24972.129572	-14605.64141	31585.102326	-94412.4397	-45551.59922	-30222.94243	-158686.5182
2	British Columbia	ACOV	1009466.3108	-318215.5496	-86825.30208	142706.63292	-147872.8471	139975.88417	-82043.79458	132697.84625
3	California	ACOV	1880815.4266	580008.36858	71596.200849	-66933.41745	-27724.52287	-120409.0442	154836.41621	340805.65284
4	Campeche	ACOV	723516.98873	70615.837586	246645.45298	33983.093921	59995.21715	184044.00371	-154818.4868	111883.99565
5	Colorado	ACOV	2395002.9949	885549.3468	1162890.2763	174213.71273	556999.09054	166151.88356	586637.45637	418385.20508
6	Florida	ACOV	1841954.3297	-181817.3059	-360463.3876	220503.91861	-243447.1074	-117599.02	-113548.3558	-145301.1373
7	Illinois	ACOV	12038013.152	600836.58187	-678896.5898	-2663922.176	-4147209.426	-966878.3897	473004.13144	-1454417.65
8	Michoacan	ACOV	472801.70498	-9193.665851	-37820.75995	46370.512446	12274.431577	36657.710309	175779.59819	15148.715377
9	New York	ACOV	3031486.6608	-497857.8509	852378.26774	-76711.70863	420198.1122	-80055.11688	777719.70849	-322919.8664
10	North Carolina	ACOV	2637225.1066	-146713.006	-288662.2692	366598.9778	437296.70551	-391660.5624	-852292.6885	-245302.7119
11	Nuevo Leon	ACOV	548262.11993	95572.864858	-91028.53622	66457.629575	-26555.59703	151226.46363	159333.42658	-141280.7841
12	Ontario	ACOV	1663501.6725	13250.325798	-95184.59044	394949.84838	-278692.7586	-573027.5682	54478.645881	-336637.1636
13	Quebec	ACOV	871113.09276	-49768.70386	-284381.3479	166253.25222	14739.09763	-141615.4982	29263.905378	57276.437142
14	Saskatchewan	ACOV	1146692.1108	-186271.4055	-160510.9636	-123224.8536	317664.08981	159881.38838	-137395.3964	-180524.3477
15	Texas	ACOV	1844790.9208	248517.76311	-185591.0092	-195311.1771	35048.184546	-269603.5745	192253.296	248007.63756
16	Washington	ACOV	2652133.5979	301523.83984	679617.05724	-334047.4805	387677.35677	353410.0446	-103543.0984	-322059.8778

We see that we get one row per subject, which in this case is the STATE. In our PROC TIMESERIES syntax we specified ACOV as the statistic option, which creates the autocorrelations for various lags. The resulting tables can now be joined per subject ID, which results in a list of variables describing the course of the time series.

If we specify more than one statistic for an output table we get a table with one row per analysis subject and statistic. We can use a WHERE clause to select a certain statistic from this table. Alternatively, we can transpose the table twice and concatenate the variable names LAGx with the STATISTICS value to have one row per analysis subject and all measures in columns. In Chapter 14 we saw an example of transposing a data set twice.

As mentioned earlier, the application of these methods results in a high number of derived variables, which are not always easy to interpret. Therefore, care has to be taken to control the number of derived variables, and time series modeling know-how might be needed to select the appropriate statistic.

Chapter 19

Multiple Categorical Observations per Subject

19.1 Introduction

General

In this chapter we will deal with the case where we have a one-to-many relationship between a subject table and either a table with repeated measurements or a table that holds data from an underlying hierarchy. We will investigate methods of preparing the data for a one-row-per-subject table.

With interval data for certain types of analysis, it can make sense to transpose the data by subject as they are and represent them in columns. We saw this in *Chapter 18 – Multiple Interval-Scaled Observations per Subject*. With categorical data, however, it usually does not make sense to transpose them as they are.

Categorical data are mostly aggregated per subject, and absolute and relative frequencies per categories are created per subject. A *trend* between the categories can be measured in the form of a sequence analysis.

Our general aim in the case of multiple categorical observations per subject is to create a number of derived variables that describe the distribution of the categories per subject. We will see that the number of variables in the one-row-per-subject table can quickly explode with the increasing number of categories and the number of categorical variables.

We will, however, introduce a number of them, because in the data mining paradigm a high number of candidate variables is desirable. We mention again that business considerations play an important role in variable selection.

With categorical variables, in this chapter we mean all kinds of binary, nominal, ordinal, or grouped interval data.

Overview of Derived Variables

In this chapter we will see that we can create meaningful derived variables for categorical data. We will also show that the creation of these variables is very simple and straightforward.

- We will show how we can create variables that hold the absolute and relative frequencies for categories per subject.

- We will show that it makes sense to group categories with rare values to an OTHERS group in order to reduce the number of derived variables.

- In order to reduce the number of derived variables we will show how we can create a concatenated variable that holds the values for each category.

- Simply looking at count statistics such as total and distinct counts also allows the creation of a number of derived variables. We will illustrate this with an example.

- We will show the business interpretation of these derived variables and give an overview of other methods.

19.2 Absolute and Relative Frequencies of Categories

Introduction

In this section we will explore how multiple values of one categorical variable can be aggregated on a one-row-per-subject basis. We will deal only with a static consideration of a categorical variable and will not consider a potential timely order of the observations.

We will deal with example data that are derived from the finance industry.

- We have a CUSTOMER table with variables such as CUST_ID, AGE, GENDER, and START OF CUSTOMER RELATIONSHIP.

- We have a multiple-rows-per-subject ACCOUNTS table with one row per account. Here we have the variables ACCOUNT_ID, ACCOUNT_TYPE, and BALANCE.

In the following examples we will deal with the variable ACCOUNT_TYPE, which can take the values SAVINGS ACCOUNT, CHECKING_ACCOUNT, LOAN_ACCOUNT, CUM SAVINGS ACCOUNT, and SPECIAL ACCOUNT. Example data are shown in Table 19.1.

Table 19.1: Example data for accounts and account types

	Cust_id	Account_id	Account_type
1	1	1	SAVING
2	1	2	CHECKING
3	1	3	SAVING
4	1	4	LOAN
5	2	5	CHECKING
6	2	6	SAVING2
7	3	7	LOAN
8	3	8	MORTGAGE
9	3	9	SAVING
10	3	10	CHECKING
11	4	11	CHECKING
12	5	12	LOAN
13	5	13	SAVING
14	5	14	CHECKING
15	5	15	SAVING2
16	5	16	SPECIAL
17	5	17	SAVING
18	5	18	SAVING

19.2.1 Simple Frequencies

Using PROC FREQ in the following code, we create a table per CUSTOMER with frequencies per account.

```
PROC FREQ DATA = accounts;
 TABLE cust_id * account_type / NOPERCENT NOCOL;
RUN;
```

And we retrieve the following output:

```
Cust_id     Account_type
```

Frequency Row Pct	CHECKING	LOAN	MORTGAGE	SAVINGS	SAVINGS2	SPECIAL	Total
1	1 25.00	1 25.00	0 0.00	2 50.00	0 0.00	0 0.00	4
2	1 50.00	0 0.00	0 0.00	0 0.00	1 50.00	0 0.00	2
3	1 25.00	1 25.00	1 25.00	1 25.00	0 0.00	0 0.00	4
4	1 100.00	0 0.00	0 0.00	0 0.00	0 0.00	0 0.00	1
5	1 14.29	1 14.29	0 0.00	3 42.86	1 14.29	1 14.29	7
Total	5	3	1	6	2	1	18

Note that this table represents the account distribution in a one-row-per-subject structure. We have also intentionally specified the NOPERCENT NOCOL option in order to receive only the COUNTS and the ROW PERCENTS. These two measures are frequently created measures from multiple categorical data per subject.

In the following sections we will show how we can create derived variables on analysis subject levels that hold the absolute and relative frequencies per account type. As a first optional step we will combine categories with low frequency to an OTHERS group, and then we will use PROC FREQ to create the derived variables.

19.2.2 Combining Categories with Rare Frequencies

In some cases it is desirable to combine categories with rare frequencies. First we analyze univariate frequencies, which are created with the following statement:

```
PROC FREQ DATA  = Accounts ORDER = FREQ;
  TABLE Account_type;
RUN;
```

This gives the following output:

Account_type	Frequency	Percent	Cumulative Frequency	Cumulative Percent
SAVINGS	8	44.44	8	44.44
CHECKING	5	27.78	13	72.22
LOAN	3	16.67	16	88.89
MORTGAGE	1	5.56	17	94.44
SPECIAL	1	5.56	18	100.00

We can see that the categories MORTGAGE and SPECIAL have rare frequencies, which we will combine with the following DATA step into a new variable ACCOUNT_TYPE2:

```
DATA Accounts;
  SET Accounts;
  FORMAT Account_Type2 $12.;
  IF UPCASE(Account_Type) IN ('MORTGAGE','SPECIAL') THEN
                                  Account_Type2 = 'OTHERS';
  ELSE Account_Type2 = Account_Type;
RUN;

PROC FREQ DATA  = Accounts ORDER = FREQ;
  TABLE Account_type2;
RUN;
```

This gives the following output with a new group OTHERS.

Account_Type2	Frequency	Percent	Cumulative Frequency	Cumulative Percent
SAVINGS	8	44.44	8	44.44
CHECKING	5	27.78	13	72.22
LOAN	3	16.67	16	88.89
OTHERS	2	11.11	18	100.00

We discussed this method in *Chapter 17 – Transformations of Categorical Variables*, and see here that its application is possible and makes sense in a variety of situations.

19.2.3 Creating a Table with Absolute and Relative Frequencies per Subject

After we have combined the values of ACCOUNT_TYPE we will create absolute and relative frequencies of the categories per subject. This will be done in three steps:

1. Use PROC FREQ to calculate the absolute frequencies.

2. Transpose the data to a one-row-per-subject structure with PROC TRANSPOSE.

3. Calculate the relative frequencies in a DATA step.

First we have to calculate the absolute frequencies.

```
PROC FREQ DATA = Accounts NOPRINT;
 TABLE cust_id * Account_Type2 / OUT = Account_Freqs(DROP = Percent);
RUN;
```

Note that we use the NOPRINT option here in order to avoid an overflow of the Output window. We also drop the PERCENT variable from the output data set because it contains only percentages over all observations, and we will calculate row percents manually in the next step.

```
PROC TRANSPOSE DATA = Account_Freqs
               OUT = Account_Freqs_TP(DROP = _name_ _label_);
 BY Cust_ID;
 VAR Count;
 ID Account_Type2;
RUN;
```

After transposing the data to a one-row-per-customer structure we will perform the following steps:

1. Replace missing values.

2. Calculate the number of accounts per customer as a SUM variable.

3. Calculate the account distribution per customer (row percent).

```
DATA Accounts_Subject;
 SET Account_Freqs_TP ;
 FORMAT Checking_rel loan_rel saving_rel others_rel 8.;
 ARRAY account1 {*} Checking loan saving others;
 ARRAY account2 {*} Checking_rel loan_rel saving_rel others_rel;
 NR_Account = sum(of Checking loan saving others);
 DO i = 1 TO DIM(account1);
   IF account1{i} = . THEN account1{i} = 0;
   account2{i} = account1{i}/Nr_Account * 100;
 END;
 DROP i;
RUN;
```

We used ARRAY statements to replace the missing values and calculate the row percents per customer. The result is shown in Table 19.2.

Table 19.2: Table with absolute and relative frequencies of ACCOUNT_TYPE2 per customer

	Cust_id	CHECKING	LOAN	SAVING	OTHERS	Checking_rel	loan_rel	saving_rel	others_rel
1	1	1	1	2	0	25	25	50	0
2	2	1	0	1	0	50	0	50	0
3	3	1	1	1	1	25	25	25	25
4	4	1	0	0	0	100	0	0	0
5	5	1	1	4	1	14	14	57	14

We created 2x4 derived variables for the account data. The number of derived variables increases with the number of categories. Therefore, it is advisable to group the data to a reasonable number of categories before creating the derived variables for absolute and relative frequencies.

19.3 Concatenating Absolute and Relative Frequencies

Overview

When we look at the data in Table 19.2 we see that we have a number of variables that describe the distribution of ACCOUNT_TYPE2. In Chapter 17, we saw the possibility of concatenating a categorical variable to create a meaningful derived variable. We will now show how that can be used here.

19.3.1 Coding

With the following code we create two variables that hold the absolute and relative frequencies of the four account types in a concatenated variable:

```
DATA Accounts_Subject;
 SET Accounts_Subject;
 Account_Freqs  = CATX('_',Checking, Loan, Savings,Others);
 Account_RowPct = CATX('_',PUT(Checking_rel,3.), PUT (Loan_rel,3.),
                        PUT (Savings_rel,3.), PUT (Others_rel,3.));
RUN;
```

Note that we use the PUT statement in the CATX function in order to truncate the percentage to integer numbers.

The following output shows the most frequent values of ACCOUNT_ROWPCT for a table with 41,916 customers:

Account_RowPct	Frequency	Percent	Cumulative Frequency	Cumulative Percent
0_100_0_0	12832	30.61	12832	30.61
100_0_0_0	9509	22.69	22341	53.30
50_0_0_50	4898	11.69	27239	64.98
33_0_0_67	1772	4.23	29011	69.21
0_0_100_0	1684	4.02	30695	73.23
67 0 0 33	1426	3.40	32121	76.63
0_0_50_50	861	2.05	32982	78.69
50_0_50_0	681	1.62	33663	80.31
25_0_0_75	652	1.56	34315	81.87
33_0_33_33	549	1.31	34864	83.18
75_0_0_25	423	1.01	35287	84.19

We have displayed only those values that have a relative frequency of at least 1%. The other values can be combined into an OTHERS group.

This results in a variable with 12 categories, describing the most frequent combinations of row percentages for the four categories CHECKING, LOAN, SAVINGS, and OTHERS in a multivariate way.

Note the interpretation of this variable:

- 30.6% of the customers have only LOAN accounts at this bank.
- 11.7% of the customers have 50% CHECKING and 50% SAVINGS accounts.

19.3.2 Binning of Relative Values

We can see from the values of the row percentages that we most probably have a low account frequency per subject. The relative frequencies 50, 33, 67 … will mostly likely come from divisions such as 1 out of 2, 1 out of 3, 2 out of 3, 1 out of 4, and so on.

In the case of more distributed values of the row percentages, it makes sense to define a format to group the row percentages into buckets such as decile groups, and use this format in the PUT function in the CATX function.

19.4 Calculating Total and Distinct Counts of the Categories

Overview

Another group of derived variables is the absolute and distinct number of different categories per subject. The following are variables that are very straightforward with multiple categorical variables per subject:

- the number of multiple observations per subject
- the number of distinct categories

Starting from these two variables additional derived variables can be calculated:

- the proportion of distinct categories on all multiple observations per subject
- the indicator that a subject has only distinct categories
- the proportion of distinct categories on all available categories
- the indicator that a subject has all possible categories

This list shows that it is possible to create a number of derived variables just by looking at simple descriptive statistics. From a business point of view these derived variables make sense, because they describe the properties of a subject with respect to the presence of all possible categories.

19.4.1 Coding

In the following code we will use PROC SQL to copy the number of distinct categories overall into a macro variable and calculate the variables listed earlier. Note that PROC SQL has the advantage of being able to calculate DISTINCT counts, which is not possible in PROC FREQ or PROC MEANS.

```
PROC SQL NOPRINT;
 SELECT COUNT(DISTINCT account_type2)
 INTO :NrDistinctAccounts
 FROM accounts;

 CREATE TABLE Account_DistFreq
 AS SELECT Cust_ID,
        COUNT(account_type2) AS Nr_Account,
        COUNT(DISTINCT account_type2) AS Distinct_Count,
        CALCULATED Distinct_Count/CALCULATED Nr_Account*100
                    AS Distinct_Prop FORMAT = 8.1,
        CALCULATED Distinct_Prop = 100 AS OnlyDistinctAccounts,
        CALCULATED Distinct_Count/&NrDistinctAccounts*100
                    AS Possible_Prop FORMAT = 8.1,
        CALCULATED Possible_Prop = 100 AS AllPossibleAccounts
    FROM accounts
    GROUP BY Cust_ID;
QUIT;
```

Note that we use the keyword CALCULATED to reference a derived variable that has just been calculated in the same SELECT statement and is not yet present in the table.

See the resulting output in the following table:

Table 19.3: Derived variables from the number of absolute and different categories

	Cust_id	Nr_Account	Distinct_Count	Distinct_Prop	OnlyDistinctAccounts	Possible_Prop	AllPossibleAccounts
1	1	4	3	75.0	0	75.0	0
2	2	2	2	100.0	1	50.0	0
3	3	4	4	100.0	1	100.0	1
4	4	1	1	100.0	1	25.0	0
5	5	7	4	57.1	0	100.0	1

19.4.2 Interpretation

Note that we have created six derived variables just by looking at the product or service offering. These variables are often valuable to infer certain customer behavior—for example, how a customer chooses form a certain product, service, offering.

Using association or sequence analysis is obviously much more detailed than using the preceding variables. The advantage of the preceding approach is the simple calculation of the derived variables and the straightforward interpretation.

19.5 Using ODS to Create Different Percent Variables

Background

In the preceding sections we used a DATA step to create the row percentages manually. The OUT= option in PROC FREQ does not output row or column percentages to a SAS data set. It is, however, possible to use the Output Delivery System (ODS) to redirect the output of PROC FREQ to a table. This saves the additional need for a DATA step.

19.5.1 Creating Absolute and Relative Frequencies

To create absolute and relative frequencies, we start from the same data as in the preceding example:

```
PROC PRINTTO PRINT = 'c:\data\tmpoutput.out';
RUN;

ODS OUTPUT CrossTabFreqs = Account_Freqs_ODS(WHERE=(_type_ = '11'));
PROC FREQ DATA = accounts;
 TABLE cust_id * account_type2 / EXPECTED MISSING;
RUN;

PROC PRINTTO PRINT= print ;
RUN;
```

We have used CROSSTABFREQS = ACCOUNT_FREQS to copy the output of PROC FREQ to the ACCOUNT_FREQS table. This data set will also contain the expected frequencies because we specified this option in the TABLE statement. CROSSTABFREQS is a keyword of ODS for the output of PROC FREQ.

Note that it is not possible to use the NOPRINT option in this case—the ACCOUNT_FREQS table will not be filled if no output is produced. It is possible, however, to prevent PROC FREQ from filling and possibly overflowing the Output window by redirecting the printed output to a file with PROC PRINTTO.

The ACCOUNT_FREQS table now contains absolute frequencies, relative frequencies, row percentages, column percentages, and an expected number of frequencies. In Table 19.4 we see the output of ODS OUTPUT CROSSTABFREQS =. We see the list of statistics that are calculated.

Table 19.4: Output of ODS OUTPUT CROSSTABFREQS = ;

	Cust_id	Account_Type2	Variables Contributing to Observation	Table Number	Frequency Count	Expected Frequency	Percent of Two-Way Table Frequency	Percent of Row Frequency	Percent of Column Frequency	Frequency Missing
1	1	CHECKING	11	1	1	1.1111111111	5.5555555556	25	20	
2	1	LOAN	11	1	1	0.6666666667	5.5555555556	25	33.333333333	
3	1	OTHERS	11	1	0	0.4444444444	0	0	0	
4	1	SAVING	11	1	2	1.7777777778	11.111111111	50	25	
5	2	CHECKING	11	1	1	0.5555555556	5.5555555556	50	20	
6	2	LOAN	11	1	0	0.3333333333	0	0	0	
7	2	OTHERS	11	1	0	0.2222222222	0	0	0	
8	2	SAVING	11	1	1	0.8888888889	5.5555555556	50	12.5	
9	3	CHECKING	11	1	1	1.1111111111	5.5555555556	25	20	
10	3	LOAN	11	1	1	0.6666666667	5.5555555556	25	33.333333333	
11	3	OTHERS	11	1	1	0.4444444444	5.5555555556	25	50	
12	3	SAVING	11	1	1	1.7777777778	5.5555555556	25	12.5	

19.5.2 Transposing and Merging the Output Tables

Starting from the ACCOUNT_FREQS table we can now use PROC TRANSPOSE to create frequencies and row percentages in a one-row-per-subject structure. Note that PROC TRANSPOSE can transpose only one variable in one step. Therefore, we have to call PROC TRANSPOSE for each variable and then merge the data.

The absolute frequencies are stored in the ACCOUNT_FREQS_TP table:

```
PROC TRANSPOSE DATA = Account_Freqs_ODS(KEEP  = cust_id Account_Type2
Frequency)
               OUT   = Account_Freqs_TP(DROP = _name_ _label_)
               PREFIX = Freq_;
BY Cust_id;
VAR Frequency;
ID Account_Type2;
RUN;
```

The row percentages are stored in the ACCOUNT_ROWPCT_TP table. Note that we do not need to calculate the values as in the preceding section, just transpose the already-calculated values.

```
PROC TRANSPOSE DATA = Account_Freqs_ODS(KEEP  = cust_id Account_Type2
RowPercent _TYPE_)
               OUT   = Account_RowPct_TP(DROP = _name_ _label_)
               PREFIX = RowPct_;
FORMAT RowPercent 8.;
BY Cust_id;
VAR RowPercent;
ID Account_Type2;
RUN;
```

We merge the subresults into one final table:

```
DATA Customer_Accounts;
 MERGE Account_DistFreq
       Account_Freqs_TP
       Account_RowPct_TP;
 BY Cust_ID;
 Account_Freqs  = CATX('_',Freq_Checking, Freq_Loan, Freq_Others,
Freq_Savings);
 Account_RowPct = CATX('_', put(RowPct_Checking,3.),
put(RowPct_Loan,3.), put(RowPct_Others,3.), put(RowPct_Savings,3.));
 RUN;
```

See Table 19.5 for an example of the output data. We see that this is the same output as we generated in Table 19.2.

Table 19.5: Output of CUSTOMER_ACCOUNTS table

	Cust_id	Freq_CHECKING	Freq_LOAN	Freq_OTHERS	Freq_SAVING	RowPct_CHECKING	RowPct_LOAN	RowPct_OTHERS	RowPct_SAVING
1	1	1	1	0	2	25	25	0	50
2	2	1	0	0	1	50	0	0	50
3	3	1	1	1	1	25	25	25	25
4	4	1	0	0	0	100	0	0	0
5	5	1	1	1	4	14	14	14	57

19.5.3 Additional Statistics

We will also mention that we can use the *column percentages* and the *expected frequency* of the output of PROC FREQ to create derived variables.

We did not calculate these values in the preceding section. That would have required an additional calculation of the sums per account type and merge of the values to the frequency table. Again, we benefit from the precalculated values of PROC FREQ.

A table with column percentages per subject can be calculated analogously to the example for the row percentages.

```
PROC TRANSPOSE DATA = Account_Freqs_ODS(KEEP  = cust_id Account_Type2
ColPercent)
               OUT  = Account_ColPct_TP(DROP = _name_ _label_)
               PREFIX = ColPct_;
FORMAT ColPercent 8.2;
BY Cust_id;
VAR ColPercent;
ID Account_Type2;
RUN;
```

Before we transpose the deviations from the expected number of frequencies we have to calculate them in a DATA step. The variable EXPECTED is created in ACCOUNT_FREQS by PROC FREQ:

```
DATA Account_freqs_ODS;
 SET Account_freqs_ODS;
ExpRatio = (Frequency-Expected)/Expected;
 RUN;
```

```
PROC TRANSPOSE DATA = Account_Freqs_ODS(KEEP  = cust_id Account_Type2
ExpRatio)
                 OUT  = Account_ExpRatio_TP(DROP = _name_ _label_)
                 PREFIX = ExpRatio_;
FORMAT ExpRatio 8.2;
BY Cust_id;
VAR ExpRatio;
ID Account_Type2;
RUN;
```

In the next step this table needs to be merged with the customer table. We skip this here because it is analogous to the preceding merge.

19.5.4 Conclusion

The advantage of this method is that the row percentages do not need to be calculated manually in a DATA step and additional statistics such as column percentages or expected frequencies can be calculated. However, each statistic from the output table of PROC FREQ needs to be transposed separately to a one-row-per-subject structure.

19.6 Business Interpretation of Percentage Variables

General

Note that we have calculated 24 derived variables from a single variable ACCOUNT_TYPES.

As mentioned in the first section of this chapter, the creation of derived variables always has to be combined with business consideration and knowledge. Here we introduced the most common derived variables for categories. Not all of them have to be used to make sense in every analysis.

We already discussed the business rationale of the total and distinct counts in the preceding section. The rationale of the absolute frequencies should be obvious. We do not use relative percentages over subjects and categories because their interpretation is not easy.

19.6.1 Percentages

Row percentages represent the relative distribution of categories per subject. With these variables we are able to judge whether a subject concentrates on certain categories compared to his portfolio of categories.

Calculating column percentages allows us to indicate whether a subject's frequency for a category is above or below the average of all subjects for this category.

Calculating the ratio between observed (= frequency) and expected frequencies allows us to quantify whether a subject is dominant or non-dominant in a category compared to his category portfolio and the values of other subjects.

19.6.2 Concatenation

The *concatenated variables* with frequencies or row percentages over categories allow a multivariate analysis of the combinations within one variable. In addition to row percentages a concatenation of absolute values is always important for exploratory purposes.

Note that the values of concatenated row percentages or column percentages or expected ratios are different for each subject. But their separation of the observations into groups is the same. Therefore, we have created only one concatenation of row percentages.

19.6.3 Automation

With a few changes it is possible to include the code in a macro that creates the proposed derived variables just by specifying the variable name. Note, however, that with many categorical variables and possible categories, a *blind* invocation of a macro will explode the number of variables. The *manual* way forces programmers to think of their variables. Furthermore, the possible need to combine rare categories makes the automation more difficult.

19.7 Other Methods

19.7.1 Concentration

In *Chapter 18 – Multiple Interval-Scaled Observations per Subject*, we introduced the *concentration measure,* which measures the concentration of values on a certain proportion of the multiple observations per subject. It is also possible to apply this method to categorical variables. In this case we analyze the proportion of the most common categories that make up a certain proportion of the multiple observations per subject.

In Chapter 19 we introduced a number of derived variables that describe in detail the distribution and characteristics of categorical variables per subject. The applications and importance of concentration for categorical variables are not as important as in the measurement case. That is why we did not cover the creation of these variables here.

19.7.2 Association Analysis and Sequence Analysis

In the preceding methods we have applied only simple descriptive statistics methods to describe the portfolio of categories for a customer. We did not consider a potential sequence ordering of categories.

Association analysis can analyze a list of categories per subject more sophisticated way—for example, by identifying frequent category combinations. *Sequence analysis* additionally considers the time dimension and analyses the sequence of the categories.

Calculating sequences and performing association analysis cannot be performed simply in the SAS language without creating complex macros. Therefore, we will not show a coding example for these methods here. In *Chapter 28 – Case Study 4—Data Preparation in SAS Enterprise Miner*, we will show how SAS Enterprise Miner can help in data preparation. There we will also deal with an example of association and sequence analysis.

Chapter 20

Coding for Predictive Modeling

20.1 Introduction

20.1.1 The Scope of This Chapter

Many data mining analyses deal with predictive modeling tasks. Either the occurrence of an event or a value has to be predicted. Examples of predictive modeling business questions include the prediction of a contract cancellation, the delinquency of a customer's loan, or the claim amount for an insurance customer.

Data mining analysis not only starts with the creation of a predictive model, it also deals with creating and preparing meaningful derived variables. In earlier chapters we saw a number of methods to prepare derived variables and how to insert predictive power into them.

In this chapter we will consider those types of derived variables where we specifically transform variables in order to give them more predictive power. In this chapter we will move toward the world of data mining analyses and reach the limits of this book's scope.

In fact, however, there is no sharp limit between data preparation and analysis. Data preparation and analysis are usually a cyclic process that is repeated over and over. New variables or sets of variables are created and used in analysis.

The task to add predictive power to a variable requires types of transformation that usually go beyond simple transformation. Data mining tools such as SAS Enterprise Miner offer special data mining nodes and data mining functionality for such purposes. In *Chapter 28 – Case Study 4— Data Preparation in SAS Enterprise Miner*, we will see some examples.

In this chapter we will cover the following points of predictive modeling and show code examples and macros in order to prepare data for predictive modeling:

- *Proportions or means of the target variable*, where we show specific ways of how a derived variable can be created that holds the proportion or the mean of the target variable for a certain category.
- We will show how interval variables can be either mathematically transformed or binned into groups.
- We will show a short example of how data can be split into training and validation data.

20.1.2 Macros in This Chapter

In this chapter we will see the following macros:

- %CREATEPROPS, to calculate the proportions or means of a target variable
- %PROPSCORING, to create variables with proportions
- %TARGETCHART, to create a target bar chart for the relationship of input variables to a target variable

20.2 Proportions or Means of the Target Variable

General

The method *proportions or means of the target variable* applies in predictive modeling. Depending on whether the measurement level of the target variable is binary or interval, we speak of proportions or means. The basic idea of this method is that categories of input variables are replaced by interval values. In the case of a binary target variable these values are the proportion of events in this category; in the case of an interval target the values are the mean of the interval variable in this category.

Short example

The following short example explains the principle of a binary target:

We have one categorical variable GENDER with the values MALE and FEMALE, and a binary target variable RESPONSE with the values 0 and 1. Assume that for GENDER = MALE we have the following responses: 1, 0, 0, 1, 0, 1, 0, 0, 0, 1, 0, and for FEMALE we have 0, 1, 1, 1, 0, 1, 0, 1, 1, 0.

In the case of an indicator variable the mean of the variable for a group equals the proportion of event=1 in this group. So we have an event proportion of 36.4% for MALE and of 60% for FEMALE. In the data we would now replace the category MALE with the interval value 36.4 and the category FEMALE with 60.

20.2.1 Properties of This Method

This method has a number of advantages:

- The new variables have very good predictive power.

- Only one degree of freedom is used per variable, because we do not have a categorical variable with two categories, we have an interval variable.

- Interactions can be easily analyzed for categorical data also, without having a large categories interactions matrix.

- This method can also be applied to interval input variables after they have binned into groups The grouping itself can be done on the basis of business considerations or automatically when using PROC RANK with the GROUPS = n option.

- There are no problems with nonlinear relationships between the input variable and the target variable.

Note that this method as described here works only for binary and interval targets.

Also note that the proportions must be calculated only on the training data! Otherwise, we would create a model that generalizes badly, because we would put the information of the target variable directly into the input variables and would dramatically overfit the data.

20.2.2 Step-by-Step Approach

The principle of this method is easy to explain, as we saw earlier. For a small set of input variables it is easy to program step by step.

For each input variable, perform the following steps:

1. Calculate the proportion of the target variable for each category.

```
PROC MEANS DATA = test;
 CLASS gender;
 VAR response;
RUN;
```

2. Create a new variable GENDER2 that holds the proportions of the target variable.

```
IF gender = 'MALE' THEN gender2 = 0.364;
ELSE gender2 = 0.6;
```

For data sets with many variables and many categories, a macro is desirable.

20.2.3 Macro for Proportions or Means of the Target Variable

The following code shows a macro that calculates the proportions or means of a target variable for a set of categorical input variables.

The big advantage of this macro is that there are no manual steps to write IF-THEN/ELSE clauses for each variable. The usage of PUT functions with the respective format is very elegant. The resulting format catalog can be used for scoring new observations.

The macro creates new variables in the data set. For each variable in the VARS= list a corresponding variable with the prefix M is created.

Note that the code of the macro can certainly be enhanced by creating additional macro parameters for individual tasks. It was the intention to create code that is still simple enough to explain the principle. At first the code might appear more complicated than it is because of the many %SCAN() functions calls. To get a clearer picture, look at the SAS log after the macro has run:

```
%MACRO CreateProps(data=,vars=,target=,library=sasuser,
                   out_ds=,mv_baseline=YES,type=c,other_tag="OTHER");
*** Load the number of items in &VARS into macro variable NVARS;
%LET c=1;
%DO %WHILE(%SCAN(&vars,&c) NE);
  %LET c=%EVAL(&c+1);
%END;
%LET nvars=%EVAL(&c-1);

*** Loop over the Variables in Vars;
%DO i = 1 %TO &nvars;
   *** Calculate the MEAN of the target variable for each group;
   PROC MEANS DATA = &data NOPRINT MISSING;
   FORMAT &target 16.3;
    VAR &target;
    CLASS %SCAN(&vars,&i);
    OUTPUT OUT = work.prop_%SCAN(&vars,&i) MEAN=;
   RUN;

   %IF &MV_BASELINE = YES %THEN %DO;
      PROC SORT DATA = work.prop_%SCAN(&vars,&i);
       BY _type_ %SCAN(&vars,&i);
      RUN;
          DATA work.prop_%SCAN(&vars,&i);
        SET work.prop_%SCAN(&vars,&i);
         &target._lag=lag(&target);
        IF _type_ = 1 and %SCAN(&vars,&i) IN ("",".") Then
              &target=&target._lag;
      RUN;
   %END;

   *** Prepare a data set that is used to create a format;
   DATA work.prop_%SCAN(&vars,&i);
       %IF &type = n %THEN %DO;
       SET work.prop_%SCAN(&vars,&i)(rename = (%SCAN(&vars,&i) =
tmp_name));;
            %SCAN(&vars,&i) = PUT(tmp_name,16.);
       %END;
      %ELSE %DO;
            SET work.prop_%SCAN(&vars,&i);;
             IF UPCASE(%SCAN(&vars,&i)) = 'OTHER' THEN
%SCAN(&vars,&i) = '_OTHER';
      %END;
      IF _type_ = 0 THEN DO;
                                %SCAN(&vars,&i)=&other_tag;
                                _type_ = 1;

   END;
     RUN;
```

```
      DATA fmt_tmp;
        SET work.prop_%SCAN(&vars,&i)(RENAME=(%SCAN(&vars,&i) = start
&target=label)) END = last;;
      *WHERE _type_ = 1;
        RETAIN fmtname "%SCAN(&vars,&i)F" type "&type";
      RUN;
      *** Run PROC Format to create the format;
      PROC format library = &library CNTLIN = fmt_tmp;
      RUN;
%end;
*** Use the available Formats to create new variables;
options fmtsearch = (&library work sasuser);
DATA &out_ds;
 SET &data;
 FORMAT %DO i = 1 %TO &nvars;      %SCAN(&vars,&i)_m  %END; 16.3;
 %DO i = 1 %TO &nvars;
    %IF &type = c %THEN IF UPCASE(%SCAN(&vars,&i)) = 'OTHER' THEN
%SCAN(&vars,&i) = '_OTHER';;
    %SCAN(&vars,&i)_m =
INPUT(PUT(%SCAN(&vars,&i),%SCAN(&vars,&i)f.),16.3);
 %END;
RUN;
%mend;
```

The macro itself has the following input parameters:

DATA

The input data set that contains TARGET variables and a list of categorical INPUT variables.

VARS

A list of categorical input variables.

TARGET

The target variable, which can be interval or binary.

LIBRARY

A library where the format catalog will be stored. The default is the SASUSER library.

OUT_DS

The name of the output data set that is created by the macro. This data set will contain additional variables, the variables from the VARS list with the suffix _M.

TYPE

The type of variables in the VARS list—c for character formatted variables, n for numeric formatted variables. Note that the whole macro can run only for character or numeric formatted variables. If both types of variables will be converted to means or proportions, the macro has to be run twice.

MV_BASELINE

Possible values are YES and NO. YES (= default) indicates that for the group of missing values the mean for this group will not be used, but the overall mean (= baseline) will be used.

OTHER_TAG

The name of the category that will be used as the OTHER category. In this category all observations are classified that have values in the respective categorical variable whose value was not found in the training data of this variable.

20.2.4 What Does the Macro Do

Note the following:

- The macro loops over all variables in the VARS list.

- For each category of the variable, the mean of the TARGET variable is calculated and stored in a data set. This data set is then used to create a format.

- A format catalog is created and subsequently used to create new variables in the OUT_DS data set.

- PROC FORMAT can use the OTHER category as a pool for all categories that are not in the training data and therefore not in the list format values. The macro assigns the overall mean (= baseline) to this OTHER category. In the data if the value OTHER for a categorical variable exists, PROC FORMAT would use this value and the mean or proportion for this category for all new categories. Therefore, the macro changes the value OTHER against _OTHER in order to preserve the *other* variable.

- This is a big advantage for the scoring of data with categorical variables whose list of valid values can change frequently. The OTHER group assigns valid values—the overall mean—to new categories.

- The macro derives the format name from the variable name by concatenating an 'F' at the end. This is done in order to avoid an error if a variable ends in a number, because characters format names must not end in a number. Format names can have a length of 32 characters for numeric formats and 31 for character formats. Therefore, the variable names in the VARS list must not exceed 31 for numeric variables and 30 for character variables.

- In many cases it makes sense to use the overall mean for the missing category. In this case no information about the values is replaced by the average.

- The macro assumes a numeric formatted variable for the TARGET variable. In the case of a binary target variable this must be a variable with values 0 and 1. Binary variables with categories such as YES or NO need to be replaced by a 0/1 variable.

Examples

We use the SAMPSIO.HMEQ data. The variable BAD is a binary target variable with values 0 and 1.

```
%CreateProps(DATA=sampsio.hmeq,OUT_DS=work.hmeq,VARS=job,
             TARGET=bad,MV_BASELINE=NO);
```

This results in the creation of a SAS format $JOBF with the following values and labels. Note that this output has been created with the following statement.

```
proc format library = sasuser fmtlib;
run;
```

Table 20.1: Values of SAS format $JOBF with a separate proportion value for missing values

```
         FORMAT NAME: $JOBF    LENGTH:    12    NUMBER OF VALUES:    8
    MIN LENGTH:   1  MAX LENGTH:  40  DEFAULT LENGTH  12  FUZZ:        0
```

START	END	LABEL (VER. V7|V8 09JUN2005:16:31:19)
		0.082437276
Mgr	Mgr	0.2333767927
Office	Office	0.1318565401
ProfExe	ProfExe	0.1661442006
Sales	Sales	0.3486238532
Self	Self	0.3005181347
_OTHER	_OTHER	0.2319932998
OTHER	**OTHER**	0.1994966443

- We see that the category of missing values has its own proportion that is calculated on the basis of the training data.

- The variable REASON originally contains a category OTHER, which has been replaced with _OTHER in order to have the OTHER item free for all categories that are not in the list. This is represented by **OTHER** in the format list.

- The category **OTHER** holds the overall mean of the target variable BAD.

Table 20.2: Original and transformed values of the variable JOB

	JOB	job_m
1	_OTHER	0.232
2	_OTHER	0.232
3	_OTHER	0.232
4		0.082
5	Office	0.132
6	_OTHER	0.232
7	_OTHER	0.232
8	_OTHER	0.232
9	_OTHER	0.232
10	Sales	0.349
11		0.082
12	Office	0.132
13	_OTHER	0.232
14	Mgr	0.233
15	_OTHER	0.232
16	_OTHER	0.232
17	Mgr	0.233
18		0.082
19	_OTHER	0.232
20	Office	0.132

Running the macro %MV_BASELINE=YES causes the MISSING group to be represented by the overall mean.

```
%CreateProps(DATA=sampsio.hmeq,OUT_DS=work.hmeq,VARS=job,
             TARGET=bad,MV_BASELINE=YES);
```

This results in the creation of a SAS format $JOBF with the following values and labels.

Table 20.3: Values of SAS format $JOBF with the overall proportion used for missing values

```
        FORMAT NAME: $JOBF    LENGTH:   12    NUMBER OF VALUES:    8
  MIN LENGTH:   1  MAX LENGTH:  40  DEFAULT LENGTH  12  FUZZ:         0
```

START	END	LABEL (VER. V7|V8 09JUN2005:16:44:12)
		0.1994966443
Mgr	Mgr	0.2333767927
Office	Office	0.1318565401
ProfExe	ProfExe	0.1661442006
Sales	Sales	0.3486238532
Self	Self	0.3005181347
_OTHER	_OTHER	0.2319932998
OTHER	**OTHER**	0.1994966443

A look at the data shows how the values of the variable JOB are replaced by their proportion to the target variable.

Table 20.4: Original and transformed values of the variable JOB with MV_BASELINE = YES

JOB	job_m
Mgr	0.233
ProfExe	0.166
Office	0.132
_OTHER	0.232
	0.199
_OTHER	0.232
ProfExe	0.166
Office	0.132

20.2.5 A Macro for Scoring

With the preceding macro we can create derived variables that can be used in modeling predictive analysis. If we want to use the model in scoring—for example, applying the resulting logic at a later time or to another population—we need to create the variables in the same manner. The preceding macro creates format catalogs that contain the logic of which category will be replaced with which value. The logic in these format catalogs can be used to create variables for scoring.

We will do this in the macro %PROPSCORING. Here we use the proportions or means that have been calculated on the basis of training data to score the validation data set or to score data for other subjects or other time periods.

The macro uses the existing format catalog and creates new variables with values that are replaced on the basis of these formats:

```
%MACRO PropScoring(data=, out_ds=,vars=,library=sasuser,type=c);
options fmtsearch = (&library work sasuser);
%LET c=1;
%DO %WHILE(%SCAN(&vars,&c) NE);
   %LET c=%EVAL(&c+1);
%END;
%LET nvars=%EVAL(&c-1);

DATA &out_ds;
 SET &data;
  FORMAT %DO i = 1 %TO &nvars;     %SCAN(&vars,&i)_m  %END; 16.3;;
  %DO i = 1 %TO &nvars;
    %IF &type = c %THEN IF UPCASE(%SCAN(&vars,&i)) = 'OTHER' THEN
%SCAN(&vars,&i) = '_OTHER';;
    %SCAN(&vars,&i)_m =
INPUT(PUT(%SCAN(&vars,&i),%SCAN(&vars,&i)f.),16.3);
  %END;
 RUN;
%MEND;
```

The macro has the following input parameters, which are similar to the macro %CREATEPROPS:

DATA

The input data set that contains a list of categorical INPUT variables.

VARS

A list of categorical input variables.

LIBRARY

A library where the format catalog will be stored. The default is the SASUSER library.

OUT_DS

The name of the output data set that is created by the macro. This data set will contain additional variables, the variables from the VARS list with the suffix _M.

TYPE

The type of variables in the VARS list—c for character formatted variables, n for numeric formatted variables. Note that the whole macro can run only for either character or numeric formatted variables. If both types of variables will be converted to means or proportions, the macro has to be run twice.

The weight of evidence method

Another possibility to replace the category by an interval-scaled value is to use *Weights of Evidence* (WoE). They are frequently used in the credit-scoring area, where a binary target (good/bad) has to be predicted. They can be calculated only for binary targets.

Similar to the proportions of the target variable, WoE are a measure that holds a quantification of the event rate of the target variable. In order to avoid overfitting, WoE must not be created on the entire data set, but only on the training data.

WoE are defined by the following formula:

WoE(category) = log (p(event,category) / p(non-event/category))

where

- p(event,category) = Number of events in the category / Number of all events

and

- p(non-event,category) = Number of non-events in the category / Number of all non-events

In our example for GENDER in Section 20.2, "Proportions or Means of the Target Variable," we would have the following:

- WoE(MALE) = –0.7367
- WoE(FEMALE) = 0.1206

Compared to the values 0.36 and 0.60 we received with the proportions of the target variable, we see that proportions lower than the baseline proportion result in a negative WoE, and proportions greater than the baseline result in a positive WoE.

20.2.6 Binning of Interval Variables

We have talked only about categorical variables so far. Interval input variables need to be binned into groups before Weights of Evidence values can be calculated or before the %CREATEPROPS or %PROPSCORING macros can be used.

We can bin the values of an interval variable by defining groups of equal size with PROC RANK.

```
PROC RANK DATA = sampsio.hmeq OUT = ranks GROUPS = 10;
 VAR value;
 RANKS value_grp;
RUN;
```

This binning can also be done by manually defining the bins in an IF-THEN/ELSE statement.

```
DATA ranks;
 SET ranks;
      IF Value_grp <= 50000  THEN Value_grp2 = 1;
 ELSE IF Value_grp <= 100000 THEN Value_grp2 = 2;
 ELSE IF Value_grp <= 200000 THEN Value_grp2 = 3;
 ELSE Value_grp2=4;
RUN;
```

Note that these limits are often defined by looking at the distribution of the means or proportions of the target variable over the values of the input variable.

SAS Enterprise Miner offers special functionality to transform an optimal bin input variable for the target variable. Credit Scoring for SAS Enterprise Miner offers the *Interactive Grouping node*, which allows the interactive grouping of input variables and the creation of the respective Weight of Evidence variables.

20.3 Interval Variables and Predictive Modeling

General

The handling of interval variables that have a linear relationship to the target variable is very straightforward. In regression analysis, for example, a coefficient is estimated that describes the relationship of the input variable to the target variable.

If, however, the relationship between the input variable and the target variable is not linear, the interval input variable is either *binned into groups* or *transformed by a mathematical transformation*.

If we have an interval input variable, the relationship to the target variable can be analyzed graphically by a *scatter plot* or by a so-called *target bar chart*. The target variable in this case can be interval or binary.

20.3.1 Mathematical Transformations

Consider a target variable RESPONSE and an input variable AGE that have the following graphical relationship:

Figure 20.1: Scatter plot for RESPONSE and AGE

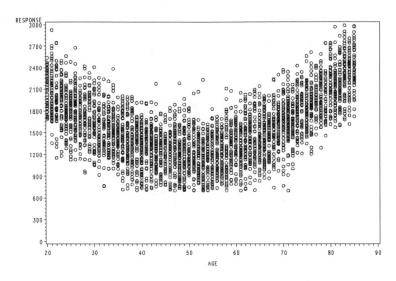

We see a U-shaped relationship between RESPONSE and AGE that requires a quadratic term of AGE.

We now have two options to create the quadratic trend. We can square the values of AGE, or we can first center the values of AGE by the mean and square the centered values.

```
AGE_Qdr = age**2;
AGE_Qdr_Std = (age-50)**2;
```

The second method has the advantage that we explicitly create a variable that has a linear relationship with the target variable, which we can see from the following figure.

Figure 20.2: Relationship between RESPONSE and centered quadratic AGE values

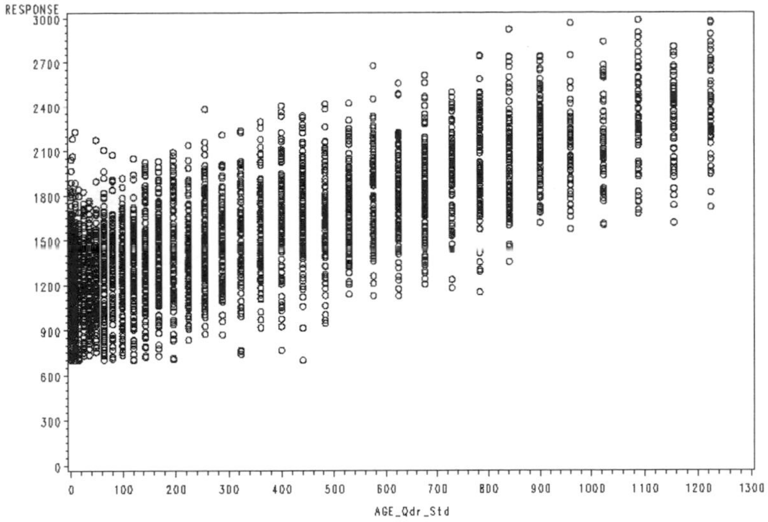

We see that we have transformed that data to an almost linear relationship.

It can be shown that the two alternatives for quadratic variables give the same performance only in regression—for example, by considering the F-value of the linear regression, if we also specify the linear term in the regression equation.

```
MODEL response = AGE AGE_Qdr_Std;
MODEL response = AGE AGE_Qdr;
```

The F-statistic and therefore the predictability of the two preceding models are the same, because the coefficient for the linear trend and the intercept will absorb the difference in the two quadratic methods, because one is centered the other is not centered.

Note, however, that the centered quadratic term has the advantage that it can be specified alone and that the estimation is more stable. It is therefore advisable in the case of quadratic relationships to center them on 0.

Note that types of transformations are not limited to quadratic transformations but can also include more complex mathematical formulas.

20.3.2 Target Bar Charts

An alternative to the scatter plot to visualize relationships for binary-scaled or interval-scaled target variables is the *target bar chart*. The idea of a target bar chart is to create for each group a bar, whose length equals the mean or the proportion of the target variable for this group.

The interval input variables need to be grouped, which can be done either by defining groups of equal size with PROC RANK, which is usually enough for a visual representation of the relationship, or by defining special group limits with IF-THEN/ELSE statements or SAS formats.

To create a target bar chart we will introduce the macro %TARGETCHART:

```
%MACRO TARGETCHART(data=,target=,interval=,class=);
PROC SQL NOPRINT;
 SELECT AVG(&Target) INTO :_Mean FROM &data;
QUIT;
PROC GCHART DATA=&data;
%IF &interval ne %THEN; HBAR &class /TYPE=MEAN FREQ DISCRETE MISSING
SUMVAR=&target SUM MEAN REF=&_mean;
%IF &class ne %THEN; HBAR &interval/type=MEAN FREQ MISSING
SUMVAR=&target SUM MEAN REF=&_mean;
RUN;
QUIT;
%MEND;
```

The macro has the following parameters:

DATA

The name of the data set.

TARGET

The name of the target variable.

INTERVAL

The list of interval input variables.

CLASS

The list of categorical input variables.

The macro can create a target bar chart for nominal-scaled and interval-scaled variables. Note that interval-scaled variables are automatically binned into groups of equal size. The invocation of the macro can be as follows:

```
%TARGETCHART(DATA=age,TARGET=response,interval = age age_qdr_std);
```

The macro will then create two charts, which are shown in Figure 20.3 and Figure 20.4.

Figure 20.3: Target bar chart for AGE

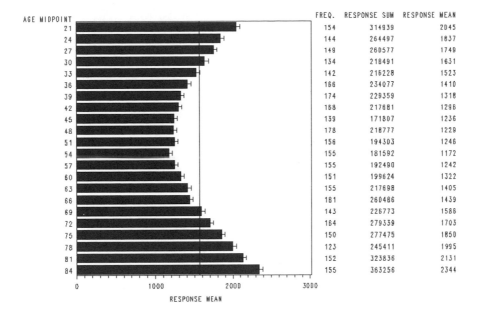

We see the U-shaped relationship between AGE and RESPONSE. The AGE values are binned into groups with a width of 3. For each class a bar is drawn that represents the mean of the target variable RESPONSE with a 95% confidence interval. For each group we see the absolute frequency and the value of the mean of the target variable in this group.

Note that from this type of graphical representation we can see which values of the input variable have which value of the target variable. The information can be derived whether a variable will be transformed—by a quadratic transformation. We can also see which values of the input variable will be grouped together. And the bar chart shows which values of a low absolute frequency might be filtered or transformed to another value.

This type of graphical representation is very helpful in predictive modeling. It is frequently used not only for interval input variables but also for nominal input variables.

Figure 20.4: Target bar chart for AGE_QDR_STD

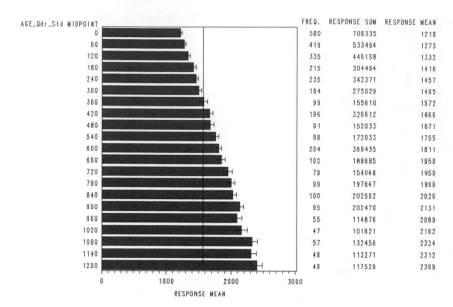

In Figure 20.4 we see the linear relationship between the centered quadratic value to AGE and RESPONSE.

20.3.3 Binning into Groups

As an alternative to mathematical transformation we can bin the values into groups in order to consider nonlinear relationships. In these cases we create a nominal-scaled variable that will be used in predictive modeling.

In our preceding example we grouped the AGE values into categories with the following statement:

```
if age < 30 then Age_GRP = 1;
else if age < 45 then Age_GRP = 2;
else if age < 55 then Age_GRP = 3;
else if age < 75 then Age_GRP = 4;
else Age_GRP = 5;
```

The resulting categorical variables can be analyzed in a target bar chart:

```
%TARGETCHART(DATA=age,TARGET=response,class = age_grp);
```

Figure 20.5: Target bar chart for the variable AGE_GRP

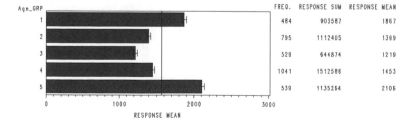

Again, we see the U-shaped relationship between AGE and RESPONSE. However, the predictive model will now consider the input values as nominal categories.

In the case of a U-shaped relationship the advantage of a transformation of the input variable with a quadratic term is very obvious. However, if we have a more complex relationship between the input and target variable that cannot easily be expressed with a mathematical transformation, the creation of nominal categories is very important.

Again we want to mention that the analysis of the relationship of the variable to the target variable, the definition of the mathematical transformation, or the definition of the group limits should be done only on the training data in order to avoid overfitting.

20.4 Validation Methods

In *Chapter 12 – Considerations for Predictive Modeling*, we discussed validation methods to control the generalization error. The *split sample validation* method is very popular and easy to implement. The available data are split into training and validation data. The split is done either by simple random sampling or by stratified sampling.

In our short example we use a random number to decide for each observation whether it will go into the training or validation data set.

```
DATA training_data
     validation_data;
  SET raw_data;
  IF UNIFORM(3) < 0.67 THEN OUTPUT training_data;
  ELSE OUTPUT validation_data;
RUN;
```

We randomly decide for each observation whether it should go into the training or validation data set. The value of 0.67 has the effect that we have a distribution of cases in the ratio 2:1 on training and validation data.

If we want to create a training, validation, and test data set, the code is analogous to the preceding example.

```
DATA training_data
     validation_data
     test_data;
  SET raw_data;
  IF UNIFORM(3) <= 0.50 THEN OUTPUT training_data;
  ELSE IF UNIFORM(4) <= 0.50 THEN OUTPUT validation_data;
  ELSE OUTPUT test_data;
RUN;
```

Note that with this code we create a split of 50:25:25 (2:1:1) to the training, validation, and test data. The statement UNIFORM(4) <= 0.50 applies to only half of the population and therefore splits into 25% (50% * 50%) groups.

20.5 Conclusion

In all the cases mentioned in this chapter, proportions or means of the target variable, Weight of Evidence, optimal binning, and adjusting the distribution of the values of the input variable, it is important to consider that these operations *must be done only on the training data* and not on the complete data set. Otherwise, we would build a model that generalizes badly as we overfit the data.

Chapter 21

Data Preparation for Multiple-Rows-per-Subject and Longitudinal Data Marts

21.1 Introduction

General

In this chapter we will deal with specifics of data preparation for multiple-rows-per-subject and longitudinal data marts. In Chapters 9 and 10 we considered these types of data marts from a data structure and data modeling point of view. This chapter provides information about how these data marts can be filled with data and how derived variables can be created.

There is a lot of information that we can take from Chapters 16 and 17—for example, there is no difference between whether the revenue per number of purchases is calculated per customer only or whether it is calculated per customer and time period. Therefore, ratios, differences, means, event counts, concatenated categorical variables, dummy variables, indicators, and lookup tables can be treated in the same manner in multiple-rows-per-subject data sets.

In this chapter we will deal with those parts of data preparation that are special to multiple-rows-per-subject or longitudinal data sets. We will explore the following:

- We will learn how to prepare data for association and sequence analysis.
- We will learn how we can enhance time series data with information per time interval or with data from higher hierarchical levels.
- We will look at how we can aggregate data on various levels.
- We will see which derived variables can be created using SAS functions.
- We will examine procedures of SAS/ETS for data management, namely PROC EXPAND and PROC TIMESERIES.

References to Other Chapters

- In *Appendix B, The Power of SAS for Analytic Data Preparation*, we will explore special features of the SAS language that are especially important for longitudinal and multiple-rows-per-subject data marts.
- In Chapters 14 and 15, we learned how to transpose from a one-row-per-subject data mart to a multiple-rows-per-subject data mart and vice versa. We also learned how to change between various types of longitudinal data structures.
- In *Chapter 18 – Multiple Interval-Scaled Observations per Subject*, we discussed methods of standardizing numerical values per subject, per measurement periods, or both. These standardization methods apply to multiple-rows-per-subject data marts as well.
- In *Chapter 27 – Case Study 3—Preparing Data for Time Series Analysis*, we will go through a comprehensive example of the preparation of data for time series analysis.

21.2 Data Preparation for Association and Sequence Analysis

General

Data for association or sequence analysis need to be organized in a multiple-rows-per-subject structure. If the data are already in a one-row-per-subject structure, the data need to be transposed to a multiple-rows-per-subject structure.

A well-known type of association analysis is the market basket analysis, where which products are being bought together by customers is analyzed.

For association analysis two variables are needed:

- an ID variable that identifies subjects for which associations will be analyzed
- an item variable that contains the items between which associations might exist

In Table 21.1 we see an example of a data set for association analysis:

Table 21.1: Data set for association analysis

	PurchaseID	PRODUCT
1	0	hering
2	0	corned_b
3	0	olives
4	0	ham
5	0	turkey
6	0	bourbon
7	0	ice_crea
8	1	baguette
9	1	soda
10	1	hering
11	1	cracker
12	1	heineken
13	1	olives
14	1	corned_b
15	2	avocado
16	2	cracker
17	2	artichok
18	2	heineken
19	2	ham
20	2	turkey

This data mart structure can easily be retrieved for transactional systems that store these types of data. In data warehousing environments, these types of data are usually stored in a star schema.

21.2.1 Hierarchies in Association Data Marts

Note that the data structure for association analysis does not always need to be as simple as shown earlier. In many cases we can have a one-to-many relationship between a customer and purchase event. We can have product hierarchies such as product main groups, product groups, and products. The input data set for association analysis can therefore have different candidates for ID values such as the CUSTOMERID or the PURCHASEID. The data set can also have different candidates for the item variables such as product main groups, product groups, or products. Table 21.2 shows an example:

Table 21.2: Data for association analysis with additional hierarchies

	CustomerID	PurchaseID	PRODUCT	ProductGroup
1	1	0	hering	Fish
2	1	0	corned_b	Meat
3	1	0	olives	Fruits
4	1	0	ham	Meat
5	1	0	turkey	Meat
6	1	0	bourbon	Alcoholic Beverages
7	1	0	ice_crea	Others
8	2	1	baguette	Others
9	2	1	soda	Others
10	2	1	hering	Fish
11	2	1	cracker	Others
12	2	1	olives	Fruits
13	2	1	corned_b	Meat
14	1	2	avocado	Fruits
15	1	2	cracker	Others
16	1	2	artichok	Others
17	1	2	heineken	Alcoholic Beverages
18	1	2	ham	Meat
19	1	2	turkey	Meat

We see that customer 1 has made two purchases and that the products are being grouped.

The additional columns CUSTOMERID and PRODUCTGROUP can either be retrieved from lookup tables and merged with this table, or they can be retrieved from SAS formats. We will show an example of this in Chapter 27.

If we want to use only the columns CUSTOMERID and PRODUCTGROUP in the analysis, we see that we will get duplicate rows. For example, customer 1 has bought FISH in both purchases and bought MEAT several times.

Note that it is up to the algorithm of the association or sequence analysis how duplicate items are handled. Duplicate items, for example, can be removed from the preceding table with the following statements:

```
PROC SORT DATA = assoc2(KEEP = CustomerID ProductGroup)
          OUT  = assocs_nodup
          NODUPKEY;
  BY CustomerID ProductGroup;
RUN;
```

The resulting table shows only non-duplicate observations.

Table 21.3: Association data with non-duplicate observations

	CustomerID	ProductGroup
1	1	Alcoholics
2	1	Fish
3	1	Fruits
4	1	Meat
5	1	Others
6	2	Fish
7	2	Fruits
8	2	Meat
9	2	Others

21.2.2 Sequence Analysis

In sequence analysis the TIME or SEQUENCE dimension is added to the association data set. This TIME dimension can be a simple consecutive number that enumerates the items for each ID. For example, in clickstream analysis this situation is found frequently where the user's path through Web pages or areas of the Web site is analyzed.

A sequence variable, however, does not necessarily need to be a consecutive number. The sequence variable can be a timestamp of when a certain event took place such as the purchase of a product or the navigation to a certain page of the Web site.

In the case of timestamps, the sequence variable can also have hierarchies—for example, by defining different time granularities such as days, weeks, or months.

21.2.3 Association Analysis Based on Subject Properties

The application of association analysis is not restricted only to data that are a priori collected in a multiple-rows-per-subject manner as in transactional systems. In some cases it makes sense to use this type of analysis to analyze the association between certain features of subjects. The result of this analysis can be interpreted like a correlation analysis for categorical data and shows which features of a subject are likely to occur together and have a strong relationship to each other.

In Table 14.12 in Chapter 14 we discussed the creation of a key-value structure. The resulting table is again shown in Table 21.4:

Table 21.4: Key-value representation of SASHELP.CLASS

	ID	Name	Key	Value
1	1	Alice	Sex	F
2	1	Alice	Age	13
3	1	Alice	Height	56.5
4	1	Alice	Weight	84
5	2	Barbara	Sex	F
6	2	Barbara	Age	13
7	2	Barbara	Height	65.3
8	2	Barbara	Weight	98
9	3	Carol	Sex	F
10	3	Carol	Age	14
11	3	Carol	Height	62.8
12	3	Carol	Weight	102.5
13	4	Jane	Sex	F
14	4	Jane	Age	12
15	4	Jane	Height	59.8
16	4	Jane	Weight	84.5

In order to be able to use these data in association analysis, we need to concatenate these two columns:

```
Feature = CATX("=",_key_,value);
```

The following table results:

Table 21.5: Concatenation of Key and Value columns

	ID	Name	Key	Value	Feature
1	1	Alice	Sex	F	Sex=F
2	1	Alice	Age	13	Age=13
3	1	Alice	Height	56.5	Height=56.5
4	1	Alice	Weight	84	Weight=84
5	2	Barbara	Sex	F	Sex=F
6	2	Barbara	Age	13	Age=13
7	2	Barbara	Height	65.3	Height=65.3
8	2	Barbara	Weight	98	Weight=98
9	3	Carol	Sex	F	Sex=F
10	3	Carol	Age	14	Age=14
11	3	Carol	Height	62.8	Height=62.8
12	3	Carol	Weight	102.5	Weight=102.5
13	4	Jane	Sex	F	Sex=F
14	4	Jane	Age	12	Age=12
15	4	Jane	Height	59.8	Height=59.8
16	4	Jane	Weight	84.5	Weight=84.5

The ID and Feature columns of this table can now be used in association analysis to analyze which features of a subject are likely to occur together.

Note that for the values of AGE, HEIGHT, and WEIGHT it makes sense to group the interval values—for example, by rounding them—in order to receive classes instead of individual values.

21.3 Enhancing Time Series Data

Overview

In *Chapter 10 – Data Structures for Longitudinal Analysis*, we explored different ways longitudinal data can be structured. We also introduced a general representation of entities such as the generic TIME, VALUE, and CATEGORY entities. In Figure 21.1 we repeat this structure.

Figure 21.1: Generalized representation of entities in longitudinal data structures

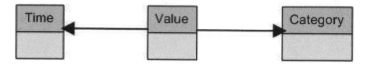

In Table 21.6 we see an example data table for this structure, where the variable MONTH represents the TIME entity, the variable PRODUCT represents the CATEGORY entity, and the variable ACTUAL represents the VALUE entity. We assume that these data cannot be broken down further. We do not have a more granular classification of PRODUCT and we have the data only on a monthly level, which cannot be broken down further.

Table 21.6: Example data for TIME, VALUE, and CATEGORY

	Month	Product	Actual Sales
1	01JAN1993	SOFA	$925.00
2	01FEB1993	SOFA	$999.00
3	01MAR1993	SOFA	$608.00
4	01APR1993	SOFA	$642.00
5	01MAY1993	SOFA	$656.00
6	01JUN1993	SOFA	$948.00
7	01JUL1993	SOFA	$612.00
8	01AUG1993	SOFA	$114.00
9	01SEP1993	SOFA	$685.00
10	01OCT1993	SOFA	$657.00
11	01NOV1993	SOFA	$608.00
12	01DEC1993	SOFA	$353.00
13	01JAN1993	BED	$220.00
14	01FEB1993	BED	$444.00
15	01MAR1993	BED	$178.00
16	01APR1993	BED	$756.00
17	01MAY1993	BED	$329.00
18	01JUN1993	BED	$910.00
19	01JUL1993	BED	$530.00
20	01AUG1993	BED	$101.00
21	01SEP1993	BED	$515.00
22	01OCT1993	BED	$730.00

In the following sections we will show how these data can be enhanced with additional variables. These variables can be used either for aggregation or as possible input variables for time series forecasting.

21.3.1 Adding Aggregation Hierarchies

Assume that in the preceding example we have a lookup table for the product classification of the variable PRODUCT.

Table 21.7: Lookup table for PRODUCT

	PRODUCT	PRODTYPE
1	BED	FURNITURE
2	SOFA	FURNITURE
3	CHAIR	OFFICE
4	DESK	OFFICE
5	TABLE	OFFICE

We can now use this table to join it to Table 21.6 or, if we transfer this table to a SAS format, we can use it as a SAS format. Both methods result in the data as shown in Table 21.8.

Using SQL:

```
PROC SQL;
 CREATE TABLE prdsale_sql
 AS SELECT *
    FROM prdsale AS a,
         lookup  AS b
    WHERE a.product = b.product;
QUIT;
```

Or creating a SAS format:

```
DATA FMT_PG(RENAME =(Product=start ProdType=label));
 SET lookup end=last;
 RETAIN fmtname 'PG' type 'c';
RUN;
PROC FORMAT LIBRARY=work CNTLIN=FMT_PG;
RUN;

DATA prdsale_fmt;
 SET prdsale;
 FORMAT Prodtype $12.;
 Prodtype = PUT(product,PG.);
RUN;
```

Table 21.8: PRDSALE data with a product type variable

	Month	Product	Actual Sales	Prodtype
1	01JAN1993	SOFA	$925.00	FURNITURE
2	01FEB1993	SOFA	$999.00	FURNITURE
3	01MAR1993	SOFA	$608.00	FURNITURE
4	01APR1993	SOFA	$642.00	FURNITURE
5	01MAY1993	SOFA	$656.00	FURNITURE
6	01JUN1993	SOFA	$948.00	FURNITURE
7	01JUL1993	SOFA	$612.00	FURNITURE
8	01AUG1993	SOFA	$114.00	FURNITURE
9	01SEP1993	SOFA	$685.00	FURNITURE
10	01OCT1993	SOFA	$657.00	FURNITURE
11	01NOV1993	SOFA	$608.00	FURNITURE
12	01DEC1993	SOFA	$353.00	FURNITURE
13	01JAN1993	BED	$220.00	FURNITURE
14	01FEB1993	BED	$444.00	FURNITURE
15	01MAR1993	BED	$178.00	FURNITURE
16	01APR1993	BED	$756.00	FURNITURE
17	01MAY1993	BED	$329.00	FURNITURE
18	01JUN1993	BED	$910.00	FURNITURE
19	01JUL1993	BED	$530.00	FURNITURE
20	01AUG1993	BED	$101.00	FURNITURE
21	01SEP1993	BED	$515.00	FURNITURE
22	01OCT1993	BED	$730.00	FURNITURE

The column that we have added can be an aggregation level that allows the aggregation of the data at a higher level. In Section 21.4, "Aggregating at Various Hierarchical Levels," we will show different ways of aggregating these data.

21.3.2 Creating Derived Variables

The data that we add do not necessarily need to be an aggregation level. They can also be the information about whether a certain product was in a promotion at a certain period in time. This would create an additional variable PROMOTION with the values 0 and 1. This variable can be used in time series forecasting to estimate the effect of a promotion.

```
IF product = 'BED' AND '01APR1993'd <= MONTH <= '01AUG1993'd
THEN PROMOTION = 1;
ELSE PROMOTION = 0;
```

The TIME dimension can also provide additional aggregation levels. For example, we can use the monthly values in the variable MONTH to create quarterly or yearly aggregations. This can either be done by explicitly creating a QUARTER or YEAR variable or by using the appropriate format.

21.3.3 Preparing Calendar Data

In addition to using the TIME variable to derive possible aggregation levels, the TIME variable can also be used to join data sets that contain specific information for certain time periods. For example, Table 21.9 contains the number of days in total and the number of Mondays, Tuesdays, and so on, for each month.

Table 21.9: Table with additional attributes per month

	Month	NrDays	Monday	Tuesday	Wednesday	Thursday	Friday	Saturday	Sunday
1	01.2003	31	4	4	5	5	5	4	4
2	02.2003	28	4	4	4	4	4	4	4
3	03.2003	31	5	4	4	4	4	5	5
4	04.2003	30	4	5	5	4	4	4	4
5	05.2003	31	4	4	4	5	5	5	4
6	06.2003	30	5	4	4	4	4	4	5
7	07.2003	31	4	5	5	5	4	4	4
8	08.2003	31	4	4	4	4	5	5	5
9	09.2003	30	5	5	4	4	4	4	4
10	10.2003	31	4	4	5	5	5	4	4
11	11.2003	30	4	4	4	4	4	5	5
12	12.2003	31	5	5	5	4	4	4	4

Table 21.10 contains the number of shops per months. The SHOPS variable can be a good predictor of the number of sold units or the sales amount.

Table 21.10: Table with number of shops per month

	Month	Shops
1	01.2003	12
2	02.2003	12
3	03.2003	14
4	04.2003	14
5	05.2003	14
6	06.2003	14
7	07.2003	14
8	08.2003	17
9	09.2003	17
10	10.2003	17
11	11.2003	17
12	12.2003	17

Tables 21.9 and 21.10 can be joined to the PRDSALE table. This can be done with the following statements. Note that for simplicity we skipped the Monday through Sunday variables:

```
PROC SQL;
 CREATE TABLE prdsale_enh
 AS SELECT a.*, b.NrDays, c.Shops
     FROM prdsale    AS a,
          cal_month  AS b,
          shops      AS c
     WHERE a.month = b.month
       AND a.month = c.month;
QUIT;
```

The result is shown in Table 21.11:

Table 21.11: Data with number of days as number of shops per month

	Month	Product	Actual Sales	NrDays	Shops
1	01.1993	SOFA	$925.00	31	12
2	02.1993	SOFA	$999.00	28	12
3	03.1993	SOFA	$608.00	31	14
4	04.1993	SOFA	$642.00	30	14
5	05.1993	SOFA	$656.00	31	14
6	06.1993	SOFA	$948.00	30	14
7	07.1993	SOFA	$612.00	31	14
8	08.1993	SOFA	$114.00	31	17
9	09.1993	SOFA	$685.00	30	17
10	10.1993	SOFA	$657.00	31	17
11	11.1993	SOFA	$608.00	30	17
12	12.1993	SOFA	$353.00	31	17
13	01.1993	BED	$220.00	31	12
14	02.1993	BED	$444.00	28	12
15	03.1993	BED	$178.00	31	14
16	04.1993	BED	$756.00	30	14
17	05.1993	BED	$329.00	31	14
18	06.1993	BED	$910.00	30	14
19	07.1993	BED	$530.00	31	14
20	08.1993	BED	$101.00	31	17
21	09.1993	BED	$515.00	30	17
22	10.1993	BED	$730.00	31	17

In this section we added variables to the longitudinal data. These variables can either be derived variables that can serve as input variables in time series forecasting or variables that can be used for aggregations, as we will show in the next section.

21.4 Aggregating at Various Hierarchical Levels

General

In the preceding section we created additional hierarchies that can be used for aggregation. Aggregation is important if the analysis will not be performed on the detailed level for which the input data are available, but on a higher level. Reasons for this can be that the business question itself demands the analysis on a certain aggregation level, or that it is not feasible to analyze data on a certain detail level.

In SAS, aggregations can be performed easily with PROC MEANS or PROC SQL. In the last section of this chapter we will introduce PROC EXPAND and PROC TIMESERIES for the aggregation of data. PROC EXPAND and PROC TIMESERIES are part of the SAS/ETS module.

21.4.1 Aggregating with PROC MEANS

Assume we have our data from Table 21.8 and we want to aggregate them on a PRODTYPE level. We can do this with the following statements:

```
PROC MEANS DATA = prdsale_sql NOPRINT NWAY;
 CLASS prodtype month;
 VAR actual;
 OUTPUT OUT = prdtype_aggr(DROP = _type_ _freq_) SUM=;
RUN;
```

The result is shown in Table 21.12:

Table 21.12: Aggregations for PRODTYPE

	Product type	Month	Actual Sales
1	FURNITURE	01JAN1993	$10,887.00
2	FURNITURE	01FEB1993	$12,315.00
3	FURNITURE	01MAR1993	$12,235.00
4	FURNITURE	01APR1993	$13,018.00
5	FURNITURE	01MAY1993	$12,064.00
6	FURNITURE	01JUN1993	$14,196.00
7	FURNITURE	01JUL1993	$15,576.00
8	FURNITURE	01AUG1993	$12,295.00
9	FURNITURE	01SEP1993	$10,737.00
10	FURNITURE	01OCT1993	$11,694.00
11	FURNITURE	01NOV1993	$10,191.00
12	FURNITURE	01DEC1993	$13,687.00
13	OFFICE	01JAN1993	$18,926.00
14	OFFICE	01FEB1993	$17,269.00
15	OFFICE	01MAR1993	$17,638.00
16	OFFICE	01APR1993	$17,563.00
17	OFFICE	01MAY1993	$19,553.00
18	OFFICE	01JUN1993	$19,409.00
19	OFFICE	01JUL1993	$18,002.00
20	OFFICE	01AUG1993	$18,865.00
21	OFFICE	01SEP1993	$17,959.00
22	OFFICE	01OCT1993	$19,661.00

Note that we can simply use a SAS format to get additional aggregations—for example, quarters instead of months. Consider that additional format statement in the following code:

```
PROC MEANS DATA = prdsale_sql NOPRINT NWAY;
 FORMAT month yyq6.;
 CLASS prodtype month;
 VAR actual;
 OUTPUT OUT = prdtype_aggr_qtr(DROP = _type_ _freq_) SUM=;
RUN;
```

Table 21.13: Aggregations for PRODTYPE and QUARTER

	Product type	Month	Actual Sales
1	FURNITURE	1993Q1	$35,437.00
2	FURNITURE	1993Q2	$39,278.00
3	FURNITURE	1993Q3	$38,608.00
4	FURNITURE	1993Q4	$35,572.00
5	OFFICE	1993Q1	$53,833.00
6	OFFICE	1993Q2	$56,525.00
7	OFFICE	1993Q3	$54,826.00
8	OFFICE	1993Q4	$55,398.00

Note that this method is not restricted to date and time formats, but it also works with SAS formats in general. For example, we could use a format for PRODUCT_TYPE that groups them to a higher level.

Here the SAS format is used to overlay the original value, which results in the display and the identification of the aggregation level. However, the value that is effectively stored in the resulting data set does not automatically need to be the beginning of the interval. The effectively stored value of the variable MONTH in Table 21.13 is only the beginning of the interval (e.g., 01JAN1993), if in the source data (see Table 21.12), the data are also aligned at the beginning of the interval.

Date values in the source data can be easily aligned with the INTNX function:

```
Date = INTNX('QUARTER',date,0,'BEGIN')
```

This code aligns the values of DATE to the first date in each quarter.

21.4.2 Aggregating with PROC SQL

We can also use PROC SQL to achieve the same results. The following code shows how to receive an aggregation as shown in Table 21.12:

```
PROC SQL;
 CREATE TABLE prdtype_sql_aggr
 AS SELECT prodtype,
           month,
           SUM(actual) AS Actual
    FROM prdsale_sql
    GROUP BY month, prodtype
    ORDER BY 1,2;
QUIT;
```

Aggregating with PROC SQL has the advantage that a database view can be defined instead of a table. This means that we define the aggregation logic, but we do not physically create the aggregation table. This method is useful if we have a couple of candidate aggregations for the analysis and do not want to redundantly store all aggregation tables. Using the data from a view, however, requires a little more processing time than from a table, because the data have to be aggregated.

The following code shows how to create a view for the aggregation as shown in Table 21.12:

```
PROC SQL;
  CREATE VIEW prdtype_sql_aggr_view
  AS SELECT prodtype,
            month,
            SUM(actual) AS Actual
     FROM prdsale_sql
     GROUP BY month, prodtype
     ORDER BY 1,2;
QUIT;
```

In contrast, PROC SQL has the disadvantage that SAS formats cannot be used in the same direct way as in procedures such as PROC MEANS. Assume you have a SAS format in the preceding SQL statement such as the following:

```
PROC SQL;
  CREATE VIEW prdtype_sql_aggr_view2
  AS SELECT prodtype,
            Month FORMAT = yyq6.,
            SUM(actual) AS Actual
     FROM prdsale_sql
     GROUP BY month, prodtype
     ORDER BY 1,2;
QUIT;
```

This will cause formatted values but not aggregations by the formatted original (underlying) values. In order to circumvent this, the following trick can be applied using a PUT function:

```
PROC SQL;
  CREATE VIEW prdtype_sql_aggr_view3
  AS SELECT prodtype,
            Put(Month,yyq6.) As Quarter,
            SUM(actual) AS Actual
     FROM prdsale_sql
     GROUP BY Quarter, prodtype
     ORDER BY 1,2;
QUIT;
```

Note that we used the PUT statement to produce the year-quarter values and that we gave a new variable name QUARTER which we use in the GROUP BY clause. If we had omitted the creation of the new name and used the variable MONTH, the original values would have been used. In this case, however, the new variable QUARTER will be character formatted and cannot be used for further date calculations. If we want to retrieve a date-formatted variable, we need to use the MDY function in order to create the new date variable explicitly.

21.4.3 Copying Data during Aggregation

In some cases we have variables that will not be aggregated but just copied to the aggregation. Consider the case of Table 21.11 where we have the number of shops in the table. We want to aggregate those data of ACTUAL to the PRODTYPE level, but we just want to copy the values of SHOP per month.

```
PROC SQL;
 CREATE TABLE prdtype_aggr_shops
 AS SELECT product,
           Month,
           SUM(actual) AS Actual,
           AVG(shops) AS Shops
    FROM prdsale_enh
  GROUP BY month, product
    ORDER BY 1,2;
QUIT;
```

In PROC MEANS the code would look like this:

```
PROC MEANS DATA = prdsale_enh NOPRINT NWAY;
 CLASS product month;
 VAR actual;
 ID Shops;
 OUTPUT OUT = prdtype_aggr(DROP = _type_ _freq_) SUM=;
RUN;
```

Or, alternatively:

```
PROC MEANS DATA = prdsale_enh NOPRINT NWAY;
 CLASS product month;
 VAR actual;
 OUTPUT OUT = prdtype_aggr(DROP = _type_ _freq_)
        SUM(Actual) = Actual
        Mean(Shops) = Shops;
RUN;
```

Note that in the first and third cases we calculate the mean of SHOPS, which evaluates for a list of equal values to the value itself. We can also use MEDIAN, MIN, or MAX. In PROC MEANS the ID statement is more elegant. These are possibilities to pass a value through an aggregation. For more variables it would, however, be better to aggregate the data first and then to join fields that contain additional information.

In Chapter 27 we will explore a detailed example of preparing a time series.

21.5 Preparing Time Series Data with SAS Functions

Overview

In this section we will show specific SAS functions that are useful for data preparation with longitudinal data marts. We will cover DATE and TIME functions, and we will specifically show the INTCK and INTNX functions and the definition of INTERVALS.

We will also investigate the power of the LAG and DIF functions.

21.5.1 DATE and TIME Functions

In SAS several variable formats for date and time values exist. A DATE variable can, for example, be 16MAY1970, and a TIME variable can be 05:43:20. There are DATETIME variables that are the concatenation of DATE and TIME variables such as 16MAY1970 05:43:20. In some cases it is usual to refer to DATE, DATETIME, and TIME variables from the preceding definition as *time variables* in order to indicate that they contain information that is measured on the time scale.

In general it is possible to use any interval-scaled time variable(s) to represent time. For example, one variable can contain the YEAR, and another variable can contain the MONTH of the year. It is, however, advisable to use a DATE-formatted variable to represent time and to use a DATETIME-formatted variable to represent time formats. Thus the full power of the SAS functions and formats for time variables can be explored.

In SAS, functions exist that allow the manipulation of time variables. In *Chapter 16 – Transformations of Interval-Scaled Variables*, we introduced the DATDIF, YRDIF, INTNX, and INTCK functions. For additional SAS functions, see SAS Help and Documentation or the SAS Press publication *SAS Functions by Example*.

21.5.2 INTNX and INTCK Functions

We mentioned that the INTNX and INTCK functions use an interval definition such as MONTH, QUARTER, YEAR, or others. For example:

```
INTCK('MONTH','16MAY1970'd,'12MAY1975'd);
```

evaluates to 60, because 60-month borders lie between May 16, 1970, and May 12, 1975. For example

```
INTNX('MONTH','16MAY1970'd,100);
```

evaluates to "01SEP1978," because we have incremented the value of May16, 1970, for 100-month borders.

The *interval* in the INTCK and INTNX functions can be enhanced with a *shift value* and a *multiplier factor*. Possible values for interval include YEAR, SEMIYEAR, QTR, MONTH, WEEK, WEEKDAY, DAY, HOUR, MINUTE, and SECOND. The following gives an overview of intervals with shift and multiplier:

MONTH	Every month
MONTH2	Every two months
MONTH6.2	Every six months, with boundaries in February and August
WEEK	Every week
WEEK4	Every four weeks
WEEK.2	Every week starting with Monday

The following examples show that intervals allow for very flexible definitions of time spans. In combination with the INTCK and INTNX functions, they allow the calculation of a variety of time values.

21.5.3 Aligning Dates

If date values are to be aligned to a certain point in the interval (e.g., the beginning of the interval), but the input data have date values that are distributed over the time interval, the INTNX function can be used:

```
INTNX('MONTH','17OCT1991'D,0, BEG)
```

This returns the date '1OCT1991'D'. Note that for monthly interval data, this could also be achieved with MDY(MONTH('17OCT1991'd),1,YEAR('17OCT1991'd)). However, the INTNX function can take any interval and also shifted intervals.

21.5.4 Computing the Width of an Interval

The width of an interval can be computed using the INTNX function:

```
width = intnx( 'month', date, 1 ) - intnx( 'month', date, 0 );
```

This calculates the number of days for each month in the variable DATE.

21.5.5 Counting Time Intervals

Use the INTCK function to count the number of interval boundaries between two dates.

Note that the INTCK function counts the number of times the beginning of an interval is reached in moving from the first date to the second. It does not count the number of complete intervals between two dates.

For example, the function INTCK('MONTH','1JAN1991'D,'31JAN1991'D) returns 90, because the two dates are within the same month.

The function INTCK('MONTH','31JAN1991'D,'1FEB1991'D) returns 91, because the two dates lie in different months that are one month apart.

When the first date is later than the second date, the INTCK function returns a negative count. For example, the function INTCK('MONTH','1FEB1991'D,'31JAN1991'D) returns –1.

The following example shows how to use the INTCK function to count the number of Sundays, Mondays, Tuesdays, and so forth, in each month. The variables NSUNDAY, NMONDAY, NTUESDAY, and so forth, are added to the data set.

```
data weekdays;
   format date yymmp7.;
   do m = 1 to 12;
   date = mdy(m,1,2005);
   d0 = intnx( 'month', date, 0 ) - 1;
   d1 = intnx( 'month', date, 1 ) - 1;
   nsunday  = intck( 'week.1', d0, d1 );
   nmonday  = intck( 'week.2', d0, d1 );
   ntuesday = intck( 'week.3', d0, d1 );
   nwedday  = intck( 'week.4', d0, d1 );
```

```
            nthurday = intck( 'week.5', d0, d1 );
            nfriday  = intck( 'week.6', d0, d1 );
            nsatday  = intck( 'week.7', d0, d1 );
         output;
       end;
          drop m d0 d1;
      run;
```

Because the INTCK function counts the number of interval beginning dates between two dates, the number of Sundays is computed by counting the number of week boundaries between the last day of the previous month and the last day of the current month. To count Mondays, Tuesdays, and so forth, shifted week intervals are used. The interval type WEEK.2 specifies weekly intervals starting on Mondays, WEEK.3 specifies weeks starting on Tuesdays, and so forth.

Note that this code can also be used to calculate calendar-specific derived variables that we later join to the longitudinal data.

21.5.6 Using Interval Functions for Calendar Calculations

With a little thought, you can come up with a formula involving INTNX and INTCK functions and different interval types to perform almost any calendar calculation.

For example, suppose you want to know the date of the third Wednesday in the month of October 1991. The answer can be computed as

```
      intnx( 'week.4', '1oct91'd - 1, 3 )
```

which returns the SAS date value '16OCT91'D.

Consider this more complex example: How many weekdays are there between 17 October 1991 and the second Friday in November 1991, inclusive? The following formula computes the number of weekdays between the date value contained in the variable DATE and the second Friday of the following month (including the ending dates of this period):

```
      n = intck( 'weekday', date - 1,
          intnx( 'week.6', intnx( 'month', date, 1 ) - 1, 2 ) + 1 );
```

Setting DATE to '17OCT91'D and applying this formula produces the answer, N=17.

Note that the last two examples are taken from SAS Help and Documentation.

21.5.7 LAG and DIF Functions

The LAG and DIF functions are important means to prepare data for time series analysis. In many cases it is necessary to make the value of a previous period available in a column. The following example shows the usage of the LAG and DIF functions for the SASHELP.AIR data set:

```
      DATA air;
       SET sashelp.air;
       lag_air = LAG(air);
       dif_air = DIF(air);
       lag2_air = LAG2(air);
      RUN;
```

Table 21.14: SASHELP.AIR data set with LAGs and DIFs

	DATE	international airline travel (thousands)	lag_air	dif_air	lag2_air
1	JAN49	112	.	.	.
2	FEB49	118	112	6	.
3	MAR49	132	118	14	112
4	APR49	129	132	-3	118
5	MAY49	121	129	-8	132
6	JUN49	135	121	14	129
7	JUL49	148	135	13	121
8	AUG49	148	148	0	135
9	SEP49	136	148	-12	148
10	OCT49	119	136	-17	148
11	NOV49	104	119	-15	136
12	DEC49	118	104	14	119

We see that we created three new columns. LAG_AIR contains the value of the previous period. DIF_AIR contains the difference from the previous period.

We also created a column LAG2_AIR that holds the value of two periods ago. Note that the LAG function is a generic function that can be parameterized by extending its name with a number, for which lags will be produced.

21.5.8 LAG and DIF Functions with Cross-Sectional BY Groups

If we have cross-sectional BY groups as in the SASHELP.PRDSALE data set, we have to make sure that we do not pull down the value of BY group A to BY group B. This can be avoided by using a BY statement in the DATA step. First we create an example data set that gives us the aggregated values for the year 1993 for each product:

```
PROC SQL;
 CREATE TABLE prdsale_sum
 AS SELECT month FORMAT = yymmp6.,
          product,
          SUM(actual) as Actual
    FROM sashelp.prdsale
  WHERE year(month) = 1993
  GROUP BY month, product
    ORDER BY product, month;
QUIT;
```

Then we use a DATA step to calculate the LAG values and use a BY statement:

```
DATA prdsale_sum;
 SET prdsale_sum;
 BY product;
 *** Method 1 - WORKS!;
 lag_actual = LAG(actual);
 IF FIRST.product then lag_actual = .;
 *** Method 2 - Does not work!;
 IF NOT(FIRST.product) THEN lag_actual2 = lag(actual);
RUN;
```

Note that we presented two methods. The second one is shorter, but it does not work, because the LAG function does not work correctly in an IF clause because it "sees" only the rows that satisfy the IF condition. We see the difference of the two methods in the following table:

Table 21.15: PRDSALE_SUM with LAG values

	Month	Product	Actual	lag_actual	lag_actual2
1	93.01	BED	4085	.	.
2	93.02	BED	5025	4085	
3	93.03	BED	4918	5025	5025
4	93.04	BED	6999	4918	4918
5	93.05	BED	5727	6999	6999
6	93.06	BED	7615	5727	5727
7	93.07	BED	8189	7615	7615
8	93.08	BED	5754	8189	8189
9	93.09	BED	4038	5754	5754
10	93.10	BED	5284	4038	4038
11	93.11	BED	4890	5284	5284
12	93.12	BED	6939	4890	4890
13	93.01	CHAIR	5174	.	.
14	93.02	CHAIR	6306	5174	6939
15	93.03	CHAIR	5620	6306	6306
16	93.04	CHAIR	5383	5620	5620
17	93.05	CHAIR	4000	5383	5383
18	93.06	CHAIR	5484	4000	4000

We see that LAG_ACTUAL contains the actual values except for the row when a new BY group starts. LAG_ACTUAL2, however, does not contain the correct value—for example, for row number 14. Instead the correct value of row 12, 6939, is inserted here. This is due to the IF statement that makes the LAG function skip row 13.

In *Appendix A – Data Structures from a SAS Procedure Point of View*, we will see additional SAS functions that make SAS a powerful tool for managing longitudinal data.

21.6 Using SAS/ETS Procedures for Data Preparation

General
In this section we will explore two SAS procedures from the SAS/ETS module, namely PROC EXPAND and PROC TIMESERIES. The SAS/ETS module in general contains procedures for time series analysis and econometric modeling. The two procedures we mention here can be used for data preparation of longitudinal data.

21.6.1 Using PROC EXPAND
With PROC EXPAND you can do the following:

- Collapse time series data from higher frequency intervals to lower frequency intervals or expand data from lower frequency intervals to higher frequency intervals. For example, quarterly estimates can be interpolated from an annual series, or monthly values can be aggregated to produce an annual series. This conversion can be done for any combination of input and output frequencies that can be specified by SAS time interval names.

- ■ Interpolate missing values in time series, either without changing series frequency or in conjunction with expanding or collapsing the series.

- ■ Change the observation characteristics of time series. Time series observations can measure beginning-of-period values, end-of-period values, midpoint values, or period averages or totals. PROC EXPAND can convert between these cases.

The following examples illustrate applications of PROC EXPAND. We use the SASHELP.AIR data set. In order to convert the data from monthly to quarterly values, we can use the following statements. The resulting data set AIR_QTR is shown in Table 21.16.

```
PROC EXPAND DATA = sashelp.air OUT = air_month
            FROM = month TO = qtr;
  CONVERT air / observed = total;
  ID DATE;
RUN;
```

Table 21.16: Quarterly AIR data

	DATE	international airline travel (thousands)
1	1949:1	362
2	1949:2	385
3	1949:3	432
4	1949:4	341
5	1950:1	382
6	1950:2	409
7	1950:3	498
8	1950:4	387
9	1951:1	473
10	1951:2	513
11	1951:3	582
12	1951:4	474
13	1952:1	544
14	1952:2	582
15	1952:3	681
16	1952:4	557

In order to convert the monthly data from SASHELP.AIR to weekly data, the following code can be used. The result is shown in Table 21.17.

```
PROC EXPAND DATA = sashelp.air OUT = air_week
            FROM = month TO = week;
  CONVERT air / OBSERVED = total;
  ID DATE;
RUN;
```

Table 21.17: SASHELP.AIR data converted to weekly data

	DATE	international airline travel (thousands)
1	Sun, 26 Dec 1948	17.913339256
2	Sun, 2 Jan 1949	21.636490213
3	Sun, 9 Jan 1949	24.551013864
4	Sun, 16 Jan 1949	26.671533627
5	Sun, 23 Jan 1949	28.119725557
6	Sun, 30 Jan 1949	29.017265708
7	Sun, 6 Feb 1949	29.485830133
8	Sun, 13 Feb 1949	29.647094885
9	Sun, 20 Feb 1949	29.622736018
10	Sun, 27 Feb 1949	29.534429585
11	Sun, 6 Mar 1949	29.503851641
12	Sun, 13 Mar 1949	29.651760406
13	Sun, 20 Mar 1949	30.02915877
14	Sun, 27 Mar 1949	30.475947973
15	Sun, 3 Apr 1949	30.762273923
16	Sun, 10 Apr 1949	30.657375586
17	Sun, 17 Apr 1949	29.975033264
18	Sun, 24 Apr 1949	28.857344072

For our example we create randomly missing values with the following statements:

```
DATA air_missing;
 SET sashelp.air;
 IF uniform(23) < 0.1 THEN air = .;
RUN;
```

Among others, the values in rows 12 and 15 have been replaced by missing values. We can now use PROC EXPAND to replace the missing values. The result is shown in Table 21.18.

```
PROC EXPAND DATA = air_missing OUT = air_impute;
 CONVERT air / OBSERVED = total;
RUN;
```

Table 21.18: Data set AIR_IMPUTE with missing values replaced

	TIME	international airline travel (thousands)	DATE
1	0	112	JAN49
2	1	118	FEB49
3	2	132	MAR49
4	3	129	APR49
5	4	121	MAY49
6	5	135	JUN49
7	6	148	JUL49
8	7	148	AUG49
9	8	136	SEP49
10	9	119	OCT49
11	10	104	NOV49
12	11	104.59411614	DEC49
13	12	115	JAN50
14	13	126	FEB50
15	14	137.14025632	MAR50
16	15	135	APR50
17	16	125	MAY50

Note that the values are not only replaced by random numbers or means but are replaced by values that correspond to a fitted time series for these values.

21.6.2 Using PROC TIMESERIES

We encountered PROC TIMESERIES in Chapter 18, where we used it to aggregate multiple-rows-per-subject data to one-row-per-subject data. We also saw how we can derive variables that describe trend, season, or autocorrelations.

In this section we will use PROC TIMESERIES to aggregate transactional data. PROC TIMESERIES allows aggregating the data similar as in PROC MEANS or PROC SQL. The advantage of this procedure lies in its possibility to specify options for the aggregations statistic and the handling of missing values.

```
PROC SORT DATA = sashelp.prdsale OUT= prdsale;
 BY product month;
RUN;

PROC TIMESERIES DATA = prdsale
                OUT  = prdsale_aggr3;
 BY product;
 ID month   INTERVAL = qtr
            ACCUMULATE = total
            SETMISS = 0;
 VAR actual;
RUN;
```

PROC TIMESERIES aggregates the data for all variables in the BY statement and for the TIME variable specified in the ID statement.

- The TIME variable in the ID statement is aggregated for the interval specified in the INTERVAL option.

- The ACCUMULATE option allows us to specify how the data will be aggregated. TOTAL causes the creation of sums.

- The SETMISS option specifies how missing values in the aggregated data will be handled. If, as in many cases, a missing value means "No Value," the option SETMISS = 0 is appropriate.

For more details, see the SAS/ETS Help.

P a r t 4

Sampling, Scoring, and Automation

Introduction

In this part of the book we will deal with sampling, scoring, automation, and practical advice when creating data marts.

In *Chapter 22 – Sampling,* we will investigate different methods of sampling such as the simple random sample, the stratified sample, and the clustered sample. We will show a couple of macros and practical programming tips and examples in order to perform sampling in SAS code.

In *Chapter 23 – Scoring and Automation,* we will explain the scoring process and show different ways to perform scoring in SAS such as explicitly calculating the scores or using procedures in SAS/STAT or SAS/ETS. We will also cover pre-checks that are relevant to scoring data in order to allow smooth scoring from a technical and business point of view.

In *Chapter 24 – Do's and Don'ts When Building Data Marts,* we will cover process, data mart handling, and coding do's and don'ts, and present practical experiences in the data preparation process.

Chapter 22

Sampling

22.1 Introduction

General

Sampling means that a number of observations are selected from the available data for analysis. From a table point of view, this means that certain rows are selected. Sampling is the random selection of a subset of observations from the population. The *sampling rate* is the proportion of observations from the original table that will be selected for the sample. In data mining analyses sampling is important, because we usually have a large number of observations. In this case sampling is performed in order to reduce the number of observations to gain performance.

Another reason for sampling is to reduce the number of observations that are used to produce graphics such as scatter plots or single plots. For example, to create a scatter plot with 500,000 observations takes a lot of time and would probably not give a better picture of the relationship than a scatter plot with 5,000 observations.

Sampling is also important in the case of longitudinal data if single plots (one line per subject) are to be produced. Besides the fact that plotting the course of a purchase amount over 18 months for 500,000 customers would take a long time to create, it also has to be considered that 500,000 lines in one graph will not have the visual impression that only 1,000 or 5,000 lines will give.

The rationale behind sampling is that the results can also be obtained by a lower number of observations. There is a lot of theory on sampling. We will not cover these considerations in this book, but in this chapter we will show examples of how different sampling methods can be performed in SAS.

22.1.1 Technical Realization

The selected observations can be copied into a new table, also called the *sample data set*. Selected observations can also be marked by an additional variable in the original table, the sampling indicator, that they will be processed. In the analysis those observations are then filtered, for example, with a WHERE statement, so that they will be included in the analysis.

An alternative to the effective creation of a new data set can also be the definition of a table view or DATA step view. Here the selected observations are not copied to the sample data set but dynamically selected when data are retrieved from that view. This saves disk space, but the analysis takes longer because the sampled observations have to be selected from the data.

22.1.2 Macros in This Chapter

We will introduce the following macros:

- %RESTRICTEDSAMPLE, to draw a simple restricted sample.
- %SAMPLINGVAR, to create a sampling variable that allows flexible oversampling for a binary target variable.
- %CLUS_SAMPLE, to draw an unrestricted clustered sample.
- %CLUS_SAMPLE_RES, to draw a restricted clustered sample.

22.2 Sampling Methods

Overview

There are a number of different sampling methods. We will investigate the three most important for analytic data marts: the simple random sample, the stratified sample, and the clustered sample.

22.2.1 The Simple Random Sample

In the *simple random sample* for each observation it is decided whether the observation is in the sample or not. This decision is made independently of other observations. For each observation a random number between 0 and 1 is created, depending on the sampling rate. All observations that have a random number lower than or equal to the sampling rate are selected.

22.2.2 The Stratified Sample

In the *stratified sample* the distribution of a nominal variable is controlled. In the case of simple stratified sampling the distribution in the original data and in the sampling has to equal. In the case of *oversampling* the sample is drawn in such a way that the proportion of a certain category of a nominal variable in the sample is higher than in the original table.

This is encountered in most cases with binary target variables, where the proportion of events (or responders) will be increased in the sample. If the number of events is rare, all observations that have an event in the target variable are selected for the sample and a certain number of non-events are selected for the sample.

22.2.3 The Clustered Sample

If the data have a one-to-many relationship, a clustered sample is most needed. In the case of a clustered sample, the fact that an observation is in the sample is determined by whether the underlying subject has been selected for the sample. This means that for a subject or cross-sectional group in the sample, all observations are in the sample. In market basket analysis, for example, it does not make sense to sample on basket-item level, but to include the whole basket in the sample. In the case of single plots or time series analysis, for example, it makes sense to include the whole series of measurements for a subject or to exclude the whole series.

We will find this method in the case of multiple-rows-per-subject data marts or cross-sectional or interleaved longitudinal data marts, but not in the case of one-row-per-subject data marts.

22.3 Simple Sampling and Reaching the Exact Sample Count or Proportion

22.3.1 Unrestricted Random Sampling

Considering sampling as a random selection of observations can mean that we select each observation with an appropriate probability in the sample. If we want to draw a 10% sample of the basis population and apply this as shown in the following code, we will see that we reach only approximately 10%, because the *unrestricted random selection* does not guarantee exactly 10% selected cases.

```
DATA SAMPLE;
 SET BASIS;
 IF RANUNI(123) < 0.1 THEN OUTPUT;
RUN;
```

If the sample count is given as an absolute number, say 10,000 observations out of 60,000, we can write the code in the following way:

```
DATA SAMPLE;
 SET BASIS;
 IF RANUNI(123) < 6000/60000 THEN OUTPUT;
RUN;
```

Again, we will not necessarily reach exactly 6,000 observations in the sample.

22.3.2 Restricted Random Sampling

If we want to force the sampling to exactly 6,000 observations or 10%, we have to amend the code as shown in the following example:

```
DATA SAMPLE_FORCE;
 SET BASIS;
 IF smp_count < 6000 THEN DO;
  IF RANUNI(123)*(60000 - _N_) <= (6000 - smp_count) THEN DO;
     OUTPUT;
     Smp_count+1;
  END;
 END;
RUN;
```

In this example the probability for each observation to be in the sample is influenced by the number (proportion) of observations that are in the sample so far. The probability is controlled by the ratio of the actual sample proportion of all previous observations to the desired sample proportion.

In the preceding example the number of observations in the data set is hardcoded. In order to make this more flexible we introduce the macro %RESTRICTEDSAMPLE:

```
%MACRO RestrictedSample(data=,sampledata=,n=);
*** Count the number of observations in the input
    data set, without using PROC SQL or other table scans
    --> Saves Time;
DATA _NULL_;
 CALL SYMPUT('n0',STRIP(PUT(nobs,8.)));
 STOP;
 SET &data nobs=nobs;
RUN;

DATA &sampledata;
 SET &data;
 IF smp_count < &n THEN DO;
  IF RANUNI(123)*(&n0 - _N_) <= (&n - smp_count) THEN DO;
     OUTPUT;
     Smp_count+1;
  END;
 END;
RUN;
%MEND;
```

The macro has three parameters:

DATA

The input data set.

SAMPLEDATA

The sample data set that is created.

N

The sample size as an absolute number.

If we invoke the macro for the SAMPSIO.HMEQ data,

```
%RestrictedSample(data = sampsio.hmeq,sampledata=hmeq_2000,n=2000);
```

we will get a data set HMEQ_2000 with exactly 2,000 observations:

```
NOTE: There were 5960 observations read from the data set
SAMPSIO.HMEQ.
NOTE: The data set WORK.HMEQ_2000 has 2000 observations and 14
variables.
NOTE: DATA statement used (Total process time):
      real time            0.04 seconds
      cpu time             0.04 seconds
```

22.4 Oversampling

General
Oversampling is a common task in data mining if the event rate is small compared to the total number of observations. In this case sampling is performed so that all or a certain proportion of the event cases are selected and a random sample from the non-event is drawn.

For example, if we have an event rate of 1.5% in the population, we might want to oversample the events and create a data set with 15% of events.

22.4.1 Example of Oversampling
A simple oversampling can be performed with the following statements:

```
DATA oversample1;
 SET sampsio.hmeq;
 IF bad = 1 or
    bad = 0 and RANUNI(44) <= 0.5 THEN OUTPUT;
RUN;
```

With these statements we select all event cases and 50% of non-event cases of the observations.

22.4.2 Specifying an Event Rate

If we want to control a certain proportion of event cases, we can specify a more complex condition:

```
%let eventrate = 0.1995;
DATA oversample2;
 SET sampsio.hmeq;
IF bad = 1 or
   bad = 0 and (&eventrate*100)/(RANUNI(34)*(1-&eventrate)
                 +&eventrate) > 25 THEN OUTPUT;
RUN;
```

Note that 0.1995 is the event rate in the SAMPSIO.HMEQ data set. Specifying the value 25 causes the condition to select a number of non-events that we receive approximately a 25% event rate. The distribution in the data set OVERSAMPLE2 is as follows:

```
                    The FREQ Procedure


                               Cumulative    Cumulative
BAD     Frequency    Percent    Frequency      Percent
-------------------------------------------------------
 0         3588       75.11        3588         75.11
 1         1189       24.89        4777        100.00
```

22.4.3 Creating a Flexible Sampling Variable

With the following macro we show an alternative to drawing a sample, in the sense that we create a so-called *flexible sampling variable*. We could certainly amend the preceding macro and the statements so that we do not output the observations to the output data set, but instead create a Boolean sample indicator.

In the case of oversampling, *flexible sampling variable* means a variable that contains values that allow the selection of any oversampling rate. We define this sampling rate with the following statements in a macro:

```
%MACRO SamplingVAR(eventvar=,eventrate=);
FORMAT Sampling 8.1;
IF &eventvar=1 THEN Sampling=101;
ELSE IF &eventvar=0 THEN
   Sampling=(&eventrate*100)/(RANUNI(34)*(1-&eventrate)+&eventrate);
%MEND;
```

The macro is designed to run within a DATA step and has the following parameters:

EVENTVAR

The name of the event variable.

EVENTRATE

The event rate as a number between 0 and 1.

The invocation of the macro in the following example creates a new variable SAMPLING. The columns BAD and Sampling are shown in Table 22.1:

```
DATA hmeq;
  SET sampsio.hmeq;
  %SamplingVar(eventrate=0.1995,eventvar=bad);
RUN;
```

Table 22.1: HMEQ table with the SAMPLING variable

	BAD	Sampling
1670	0	80.5
1671	1	101.0
1672	0	20.1
1673	0	20.7
1674	0	43.5
1675	0	51.7
1676	0	44.8
1677	0	37.9
1678	0	37.0
1679	1	101.0
1680	1	101.0
1681	0	25.0
1682	0	37.9
1683	0	89.9
1684	1	101.0

We see that the SAMPLING variable is greater than 100 for event values and contains appropriate random numbers for non-events. The SAMPLING variable allows specifying in a flexible way which oversampling rate will be chosen for the analysis, without drawing the respective sample again and again. We demonstrate this behavior in the following example with PROC FREQ:

```
PROC FREQ DATA = hmeq;
  TABLE bad;
  WHERE sampling > 30;
RUN;
```

The output shows that we have approximately a 30% sample.

BAD	Frequency	Percent	Cumulative Frequency	Cumulative Percent
0	2811	70.28	2811	70.28
1	1189	29.73	4000	100.00

22.5 Clustered Sampling

Overview

In the case of a multiple-rows-per-subject or longitudinal data mart, simple sampling does not make sense. For one subject or cross-sectional group, we want to have all observations or no observations in the sampled data. Otherwise, we would have the paradoxical situations where in time series analysis, some measurements are missing because they were sampled out, or in market basket analysis, some products per basket are not in the data because they were not sampled.

In this case we want to do the sampling on a subject or BY-group level and then move all observations of the selected subjects or BY groups into the sample. This can be done by a two-step approach where first the subjects or BY groups are sampled. The selected IDs are then merged back with the data in order to select the related observations.

22.5.1 Macro for Clustered Sampling

The following code shows how to perform this in one step in a DATA step. The advantage of this approach is that no merge of data sets, sampled IDs, and the data themselves have to be performed; therefore, the performance is better if the data are already sorted for the ID variable.

The following macro shows the unrestricted sampling approach, where the desired sample proportion or sample count is reached only on average (if a number of samples would be drawn). Note that the macro expects the data to be sorted by the ID variable.

```
%MACRO Clus_Sample (data = , id =, outsmp=, prop = 0.1, n=, seed=12345
);
 /*** Macro for clustered unrestricted sampling
      Gerhard Svolba, Feb 2005
      The macro draws a clustered sample in one DATA step.
      The exact sample count or sample proportion is not
      controlled.
      Macro Parameters:
      DATA  The name of the base data set
            The name of the sample data set will be created
            as DATA_SMP_<sample count (n)>
      ID    The name of the ID Variable, that identifies the
            subject or BY group;
      PROP  Sample Proportion as a number from 0 to 1
      N     Sample count as an absolute number
      SEED  Seed for the random number function;
    Note that PROP and N relate to the distinct ID values;
  ***/
DATA _test_;
  SET &data;
  BY &ID;
  IF first.&id;
RUN;
DATA _NULL_;
  CALL SYMPUT('n0',STRIP(PUT(nobs,8.)));
  STOP;
  SET _test_ nobs=nobs;
RUN;
%IF &n NE %THEN %let prop = %SYSEVALF(&n/&n0);

DATA &outsmp;
  SET &data;
  BY &id;
  RETAIN smp_flag;
  IF FIRST.&id AND RANUNI(&seed) < &prop THEN DO;
    smp_flag=1;
    OUTPUT;
  END;
  ELSE IF smp_flag=1 THEN DO;
    OUTPUT;
    IF LAST.&id THEN smp_flag=0;
  END;
```

```
      DROP smp_flag;
    RUN;
  %MEND;
```

The macro has the following parameters:

DATA

> The name of the base data set.

OUTSMP

> The name of the output data set.

ID

> The name of the ID variable that identifies the subject or BY group.

PROP

> The sample proportion as a number from 0 to 1. Note that the sampling proportion is understood as a proportion of subjects and not as a proportion of observations in underlying hierarchies. PROP is considered only if no value N is specified.

N

> The sample count as an absolute number. Note that the sampling count is understood as a number of subjects and not as the number of observations in underlying hierarchies. If a value for N is specified, the sampling rate is calculated by dividing N by the number of distinct subjects. In this case the PROP parameter is ignored.

SEED

> The seed for the random number function.

The name of the sample data set will be created as DATA_SMP_<sample count (n)>.

From a programming perspective, we avoid the need of a merge of the selected ID data set with the base data set by retaining the SMP_FLAG over all IDs. However, the data need to be sorted first.

22.5.2 Macro for Restricted Clustered Sampling

The following macro does sampling on the basis of the restricted sampling approach where the proportion or sample count is controlled:

```
%MACRO Clus_Sample_Res (data = , id =, outsmp=, prop = 0.1, n=,
seed=12345 );
DATA _test_;
   SET &data;
   BY &ID;
   IF first.&id;
RUN;
DATA _NULL_;
  CALL SYMPUT('n0',STRIP(PUT(nobs,8.)));
  STOP;
  SET _test_ nobs=nobs;
RUN;
```

```
%IF &n EQ %THEN %let n = %SYSEVALF(&prop*&n0);
DATA &outsmp;
 SET &data;
 BY &id;
 RETAIN smp_flag;
 IF smp_count < &n THEN DO;
  IF FIRST.&id THEN DO;
    id_count + 1;
    IF uniform(&seed)*(&n0 - id_count) < (&n - smp_count) THEN DO;
        smp_flag=1;
          OUTPUT;
    END;
   END;
   ELSE IF smp_flag=1 THEN DO;
        OUTPUT;
        IF LAST.&id THEN DO;
            smp_flag=0;
            smp_count + 1;
        END;
   END;
 END;
 DROP smp_flag smp_count;
RUN;
%MEND;
```

The macro has the following parameters:

DATA

The name of the base data set.

OUTSMP

The name of the output data set.

ID

The name of the ID variable that identifies the subject or BY group.

PROP

The sample proportion as a number from 0 to 1. Note that the sampling proportion is understood as a proportion of subjects and not as a proportion of observations in underlying hierarchies. PROP is considered only if no value N is specified.

N

The sample count as an absolute number. Note that the sampling count is understood as a number of subjects and not as the number of observations in underlying hierarchies. If a value for N is specified the sampling rate is calculated by dividing N by the number of distinct subjects. In this case the PROP parameter is ignored.

SEED

The seed for the random number function.

The name of the sample data set will be created as DATA_SMP_<sample count (n)>.

22.5.3 Example of Clustered Sampling

The following short example shows an application of the macro %CLUS_SAMPLE_RES on the data of SASHELP.PRDSAL2. First we need to create a cross-section ID variable, which we want to use to define subjects or cross-sections. We do this by combining STATE and PRODUCT:

```
DATA prdsal2;
  SET sashelp.prdsal2;
  ID = CATX('_',state,product);
RUN;
```

Next we need to sort the data by the ID variable:

```
PROC SORT DATA = prdsal2;
  BY ID;
RUN;
```

Next we apply the macro:

```
%Clus_sample_Restrict ( data = prdsal2,n=25,id=id);
```

We get a resulting data set with 10,656 observations.

```
NOTE: There were 23040 observations read from the data set
WORK.PRDSAL2.
NOTE: The data set WORK.PRDSAL2_SMP_ has 10656 observations and
13 variables
NOTE: DATA statement used (Total process time):
      real time            0.05 seconds
      cpu time             0.06 seconds
```

In order to check whether we really have only 25 ID groups in the sample as specified in the macro, we can apply PROC SQL to count the number of observations:

```
PROC SQL;
  SELECT COUNT(DISTINCT ID)
  FROM prdsal2_SMP_;
QUIT;
```

From the result we see that we get 25 distinct ID values.

22.6 Conclusion

Note that the sampling node in SAS Enterprise Miner offers SIMPLE, STRATIFIED, and CLUSTERED restricted sampling in the sampling node. We, however, wanted to present these methods here, because sampling is in many cases performed before the import of data to the data mining tool—for example, to allow faster data preparation on only a sample of the data or to provide data for visual data exploration where would like to work on samples for better visibility.

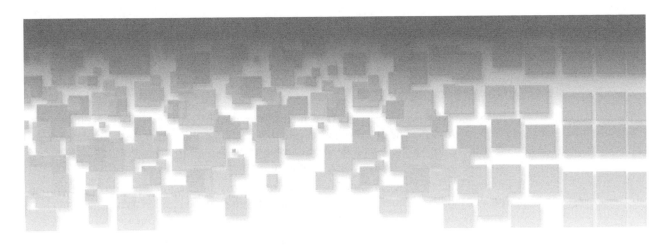

C h a p t e r 23

Scoring and Automation

23.1 Introduction

General

After a data mart is prepared, it can be used for analysis. Some types of analyses, as we saw in Section 2.6, "Scoring Needed: Yes/No," produce a set of rules that can be applied to another data mart in order to calculate a score.

The process of applying rules to a different data mart is called *scoring*. The observations to which the score code is usually applied are either a set of different subjects or the same subject at a later time.

Consider the following examples:

- During analysis the relationship between customer attributes and whether or not a loan has been paid back is modeling using logistic regression. The coefficients for the input variables can be used to predict for other observations the probability that the loan will not be paid back.

- A cluster analysis is performed on seven customer attributes and assigns each customer to one of five clusters. The discrimination rule that calculates the cluster assignments on the basis of the input variables can be used to assign new observations to clusters or to re-assign existing customers after one, two, or more months.

- In medical statistics, survival analysis is performed that models the lifetime of melanoma patients based on a number of risk factors. The results can be applied to new patients in order to predict their lifetime, depending on the values of their risk factors.

- In time series analysis, the number of visitors to a permanent exhibition on a daily basis will be modeled. The course of the time series will be forecasted for future time periods.

In all of these examples, a logic has been derived from analysis. This logic, or these rules, can be used to calculate future unknown values of certain variables. We call these future unknown values *scores* and the logic or rules *scorecard* or *score logic*. The process of applying a scorecard to a data set is called *scoring*. The *score code* is the formulation of the scoring rules or the score logic in program statements.

23.1.1 Example Process

Logistic regression is used to analyze the relationship of various input variables in the event that a customer buys a certain product. The results of this analysis are the coefficients of the logistic regression that describe the relationship between input variables and the event variable in the form of regression coefficients. These coefficients can be used to calculate for a new subject the probability of the purchase event based on the values of the input variables.

23.1.2 Chapter Contents

In this chapter we will look at how to apply the rules that have been retrieved in the analysis to new observations. In our cases a score is a predicted value or a predicted event probability (for a certain period in time).

In this chapter we will deal mainly with three possibilities of scoring:

- explicitly calculating the score values from parameters and input variables
- using the respective SAS analytic procedure for scoring
- scoring with PROC SCORE of SAS/STAT

We will also discuss the following:

- considerations in the scoring process
- pre-checks on the data that are useful before scoring
- the score code that can be produced in SAS Enterprise Miner
- automation of data marts

23.1.3 Macros in This Chapter

We will see the following macros:

- %ALERTNUMERICMISSING, to check the number of missing values for numeric variables.

- %ALERTCHARMISSING, to check the number of missing values for character variables.

- %REMEMBERCATEGORIES, to fill a repository with the list values for categorical values.

- %CHECKCATEGORIES, to compare a list of categories in a data set with the categories in the repository.

- %REMEMBERDISTRIBUTION, to store descriptive measures for interval variables in a repository.

- %SCOREDISTRIBUTION, to compare descriptive measures of data with corresponding values in a repository.

23.2 Scoring Process

23.2.1 The Analysis Data Mart

Analyses are performed on a *modeling data mart*, also called an *analysis data mart*. This data mart is used to create the results and to retrieve the rules that will be used for scoring. In predictive modeling, for example, the analysis data mart contains a target variable, which is being predicted during analysis. The relationship between the input data and the target variable is learned in the analysis.

In large data sets, the modeling data mart can be a sample of available information. In cases of predictive modeling, oversampling can be performed in order to have more events available to allow for better prediction. We saw in *Chapter 12 – Considerations for Predictive Modeling*, that data can be accumulated over several periods in order to have more event data or to have data from different periods in order to allow for a more stable prediction.

If the purpose of an analysis is to create a scorecard, only that information for subjects can be used in the modeling data mart that will also be available in the following periods, when the scoring data marts have to be built. Be aware that ad-hoc data such as surveys are not available for all subjects. We mentioned this in *Chapter 2 – Characteristics of Analytic Business Questions*.

23.2.2 The Scoring Data Mart

According to its name, observations in the scoring data mart are scored using the score code that was created on the basis of the modeling data mart. Different from the modeling data mart, the scoring data mart should not contain a sample of observations but all observations for which a score is needed.

In predictive modeling the scoring data mart usually does not contain the target variable, because it is usually unknown at this time. In the case of a binary or nominal target variable, the score code adds a probability score for the observations that fall in each class. In the case of an interval target the interval value is predicted.

23.2.3 The Scoring Process

Table 23.1 illustrates the scoring process of predictive analysis for a binary event.

Table 23.1: Scoring process

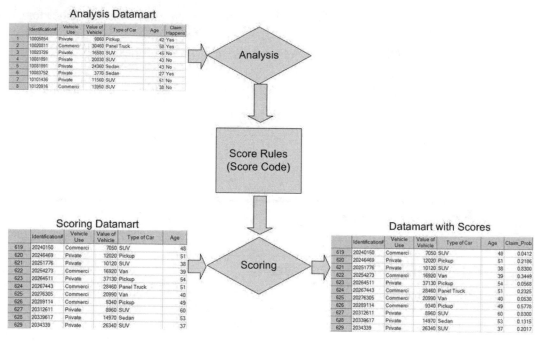

As we can see from Table 23.1 the analysis data mart is used in the analysis to create a model. From this model, score rules or the score code is retrieved. In the scoring process these rules are then applied to the scoring data mart to create a data mart with an additional score column. This column is calculated on the basis of the score rules.

Note that before scoring can start, the data mart has to be prepared in the same manner as the training data mart before the final analysis started. This includes the creation of new variables, the application of the transformation to variables, the replacement of missing values, the creation of dummy variables, and also possibly the filtering of outliers.

The score code that is produced by SAS Enterprise Miner in the Score node contains all data preparation steps that are defined by SAS Enterprise Miner nodes such as the Transform node, the Replacement node, and others. If user-defined code from SAS Code nodes has to be included in the score code, it has to be inserted manually in the Score Code Editor of the Score node.

23.3 Explicitly Calculating the Score Values from Parameters and Input Variables

General

The most direct way to score new observations is to use the parameter estimates and insert them in the formula to calculate predicted values. In linear regression this is done simply by multiplying each variable value with its parameter estimate and summing over all variables plus the intercept. In logistic regression, for example, this can be done in the same way to calculate the linear predictor and then applying the inverse logit transformation to it.

23.3.1 Calculating the Predicted Values

We will use an excerpt of the FITNESS data (see the *SAS/STAT User's Guide*, The REG Procedure, Example 61.1) and apply a linear regression to the oxygen consumption.

```
PROC REG DATA=fitness OUTEST = betas;
 MODEL Oxygen=Age Weight RunTime ;
 RUN;
QUIT;
```

Note that we output the parameter estimates with the OUTEST= option to the BETAS data set. This data set can now be used to merge it with a data set with new observations for which predictions based on the variables will be created. Let's first look at our data set with the new observations NEW_OBS (Table 23.2) and our BETAS data set (Table 23.3).

Table 23.2: Data set with new observations

	AGE	Weight	Runtime
1	35	81	11.6
2	55	91	12.6
3	25	61	9.6

Table 23.3: BETAS data set with regression coefficients

	Label of model	Type of statistics	Dependent variable	Root mean squared error	Intercept	Age	Weight	RunTime	Oxygen
1	MODEL1	PARMS	Oxygen	2.6881812348	93.126150076	-0.173876789	-0.054436515	-3.140386569	-1

By using these two data sets we can create predictions for OXYGEN using the following SAS code:

```
PROC SQL;
CREATE TABLE scores
AS SELECT a.*,
          b.age AS C_age,
          b.weight AS C_weight,
          b.runtime AS C_runtime,
          b.intercept AS Intercept,
          (b.Intercept +
           b.age * a.age +
           b.weight * a.weight +
           b.runtime * a.runtime) AS Score
          FROM NewObs a,
               betas b;
QUIT;
```

As we can see, we merge the data sets together using PROC SQL and calculate the predicted values manually using the respective formula. The estimates for AGE, WEIGHT, and RUNTIME are included only in the SELECT clause for didactic purposes. The resulting data set SCORES is shown in Table 23.4.

Table 23.4: Table with scores

	AGE	Weight	Runtime	C_age	C_weight	C_runtime	Intercept	Score
1	35	81	11.6	-0.173876789	-0.054436515	-3.140386569	93.126150076	46.2026205
2	55	91	12.6	-0.173876789	-0.054436515	-3.140386569	93.126150076	39.040332994
3	25	61	9.6	-0.173876789	-0.054436515	-3.140386569	93.126150076	55.31089184

23.3.2 Creating a Scoring View

For periodic scoring it might be more efficient to create an SQL view that performs the scoring. A scoring view for the preceding example is defined by the following code:

```
PROC SQL;
 CREATE VIEW score_view
 AS SELECT   a.*,
            (b.Intercept         +
             b.age   * a.age    +
             b.weight * a.weight +
             b.runtime * a.runtime) AS Score
       FROM NewObs  a,
            betas      b
 ;
QUIT;
```

The results of the view are the same as shown in Table 23.4. The process of using the view is described in the following points:

- *Scoring new observations*: Just add or replace the observations in the NEWOBS table and query the view SCORE_VIEW to retrieve the score values.

- *Retraining the model with existing variables*: Run PROC REG on new training data again and produce a new BETAS data set that contains the new estimates.

- *Retraining the model with new variables*: Run PROC REG on new training data again and produce a new BETAS data set that contains the new estimates. Then update the view definition in order to consider new variables in the score calculation.

This results in a very flexible scoring environment. Note that the rules (= business knowledge) are stored in the BETAS data set and the scoring algorithm is stored in the view definition.

23.4 Using the Respective SAS/STAT Procedure for Sooring

General

It is possible to use some analytic procedures for model training and scoring in one step. In this case the input data set has to contain the training data and the scoring data.

- The training data must contain the input variables and the target variable.

- The scoring data must contain only input variables—the target variable must be missing.

In this case, SAS/STAT procedures such as PROC REG or PROC LOGISTIC use those observations that have values for both input and target variables for model training and use those observations where the target variable is missing to assign a score to it.

23.4.1 Model Training and Scoring in PROC REG

In our example we use the FITNESS data from the previous section and add the data set NEWOBS to it.

```
DATA fitness2;
 SET fitness NewObs;
RUN;
```

Table 23.5: FITNESS2 data set with training and new observations

	Age	Weight	Oxygen	RunTime
22	54	79.38	46.08	11.17
23	52	76.32	45.441	9.63
24	50	70.87	54.625	8.92
25	51	67.25	45.118	11.08
26	54	91.63	39.203	12.88
27	51	73.71	45.79	10.47
28	57	59.08	50.545	9.93
29	49	76.32	48.673	9.4
30	48	61.24	47.92	11.5
31	52	82.78	47.467	10.5
32	35	81	.	11.6
33	55	91	.	12.6
34	25	61	.	9.6

We see that the last three observations do not have values for the target variable OXYGEN. Running the following PROC REG code, which includes an OUTPUT statement to produce predicted values, will produce a new column that contains the predicted values for all observations:

```
PROC REG DATA=fitness2;
   MODEL Oxygen=Age Weight RunTime ;
   OUTPUT OUT = fitness_out P=predicted;
RUN;
QUIT;
```

The new table FITNESS_OUT is shown in Table 23.6.

Table 23.6: Fitness data with scores

	Age	Weight	Oxygen	RunTime	Predicted Value of
22	54	79.38	46.08	11.17	44.337514887
23	52	76.32	45.441	9.63	49.688039519
24	50	70.87	54.625	8.92	52.562146571
25	51	67.25	45.118	11.08	45.802094978
26	54	91.63	39.203	12.88	38.300606539
27	51	73.71	45.79	10.47	47.366070896
28	57	59.08	50.545	9.93	48.81502513
29	49	76.32	48.673	9.4	50.931958798
30	48	61.24	47.92	11.5	45.331926445
31	52	82.78	47.467	10.5	46.604243314
32	35	81	.	11.6	46.2026205
33	55	91	.	12.6	39.040332994
34	25	61	.	9.6	55.31089184

This method of scoring observations is possible with all SAS/STAT procedures that have an OUTPUT statement, which allows us to create predicted values with the P= option. This scoring method can also be automated with a view, just by changing the concatenation of training and scoring data as in the following:

```
DATA fitness2_view / VIEW = fitness2_view;
 SET fitness NewObs;
RUN;
```

Note that no estimate data set has to be considered, because the estimates are re-created during each invocation of the procedure.

The advantage to this method is that no coding of the scoring formula is needed. The disadvantage is that each scoring run causes the procedure to re-run the model training, which might unnecessarily consume time resources.

23.4.2 Model Training and Scoring in PROC LOGISITC

The same scoring mechanism is available in PROC LOGISITIC. We will show the respective example here for illustration only.

We use the data from Example 42.1 of the PROC LOGISITC documentation in *SAS/STAT User's Guide* and add additional observations without the target variable to them. See Table 23.7.

Table 23.7: REMISSION data set with three additional observations for scoring

	remiss	cell	smear	infil	li	blast	temp	ID
10	0	0.95	0.36	0.34	0.5	0	1.038	10
11	0	0.85	0.39	0.33	0.7	0.279	0.988	11
12	0	0.7	0.76	0.53	1.2	0.146	0.982	12
13	0	0.8	0.46	0.37	0.4	0.38	1.006	13
14	0	0.2	0.39	0.08	0.8	0.114	0.99	14
15	0	1	0.9	0.9	1.1	1.037	0.99	15
16	1	1	0.84	0.84	1.9	2.064	1.02	16
17	0	0.65	0.42	0.27	0.5	0.114	1.014	17
18	0	1	0.75	0.75	1	1.322	1.004	18
19	0	0.5	0.44	0.22	0.6	0.114	0.99	19
20	1	1	0.63	0.63	1.1	1.072	0.986	20
21	0	1	0.33	0.33	0.4	0.176	1.01	21
22	0	0.9	0.93	0.84	0.6	1.591	1.02	22
23	1	1	0.58	0.58	1	0.531	1.002	23
24	0	0.95	0.32	0.3	1.6	0.886	0.988	24
25	1	1	0.6	0.6	1.7	0.964	0.99	25
26	1	1	0.69	0.69	0.9	0.398	0.986	26
27	0	1	0.73	0.73	0.7	0.398	0.986	27
28	.	1	0.5	0.8	2.7	0.464	0.89	28
29	.	0.5	0.59	0.89	1.9	0.498	0.886	29
30	.	1	0.53	0.83	1.7	0.498	0.886	30

Again we run PROC LOGISTIC and use an OUTPUT statement to produce predicted values:

```
PROC LOGISTIC DATA=Remission OUTEST=betas;
   MODEL remiss(event='1')=cell smear infil li blast temp;
   OUTPUT OUT = remission_out P=p;
RUN;
QUIT;
```

Finally, we receive a new column in the table with the predicted probability for the event REMISS, as shown in Table 23.8.

Table 23.8: REMISSION table with scores, created with PROC LOGISTIC

	remiss	cell	smear	infil	li	blast	temp	ID	_LEVEL_	p
19	0	0.5	0.44	0.22	0.6	0.114	0.99	19	1	0.0000638741
20	1	1	0.63	0.63	1.1	1.072	0.986	20	1	0.6778453292
21	0	1	0.33	0.33	0.4	0.176	1.01	21	1	0.015910081
22	0	0.9	0.93	0.84	0.6	1.591	1.02	22	1	0.0074448461
23	1	1	0.58	0.58	1	0.531	1.002	23	1	0.2476837478
24	0	0.95	0.32	0.3	1.6	0.886	0.988	24	1	0.8510967326
25	1	1	0.6	0.6	1.7	0.964	0.99	25	1	0.9384648502
26	1	1	0.69	0.69	0.9	0.398	0.986	26	1	0.4611781908
27	0	1	0.73	0.73	0.7	0.398	0.986	27	1	0.2794687948
28	.	1	0.5	0.8	2.7	0.464	0.89	28	1	0.9999204383
29	.	0.5	0.59	0.89	1.9	0.498	0.886	29	1	0.0033975134
30	.	1	0.53	0.83	1.7	0.498	0.886	30	1	0.9972367669

23.4.3 Scoring Observations Using PROC FASTCLUS

PROC FASTCLUS can be used to perform a k-means clustering for observations. The observations in the input data set are assigned to clusters on the basis of the parameterization of the procedure and of their variable values. Adding new observations to the input data set will cause the procedure to assign them to clusters as well. However, in this case the clustering of all observations might change because the cluster assignment rules depend on all, including new, observations.

Scoring new observations without changing the cluster assignment rules can be achieved by using a SEED data set in PROC FASTCLUS. The following example illustrates this process, based on a clustering of the FITNESS data.

We perform a clustering based on four variables with the following code:

```
PROC FASTCLUS DATA = fitness
               MAXCLUSTERS=5
               OUTSEED = ClusterSeeds;
   VAR age weight oxygen runtime;
RUN;
```

Note that we used the MAXCLUSTERS= option to request five clusters, and we used the OUTSEED= option to save the seeds of the k-means clustering in a data set.

A look at the CLUSTERSEEDS table in Table 23.9 shows us that the cluster midpoints are stored for each cluster.

Table 23.9: CLUSTERSEEDS table, output by the OUTSEED= option in PROC FASTCLUS

	CRIT	CLUST	_FREQ_	_RMSSTD_	_RADIUS_	_NEAR_	_GAP_	Age	Weight	Oxygen	RunTime
1	3.4203647374	1	5	3.6757127098	9.8371446442	5	10.462681222	47	88.362	41.4832	12.322
2	3.4203647374	2	10	3.6528656897	10.41895845	5	9.8003207606	48.3	72.86	44.7539	10.961
3	3.4203647374	3	2	3.4463953089	4.8739389871	2	14.108450445	52.5	60.16	49.2325	10.715
4	3.4203647374	4	4	4.5060018401	10.753097472	5	11.810896955	43.5	73.48	56.198	8.9625
5	3.4203647374	5	10	3.4694135495	12.574612918	2	9.8003207606	48.1	81.613	49.0438	9.967

In the NEWCLUSOBS data set we have three new observations that will be assigned to clusters, as shown in Table 23.10.

Table 23.10: Data set NEWCLUSOBS with three new observations

	AGE	Weight	Oxygen	Runtime
1	35	81	39.3	11.6
2	55	91	59.3	12.6
3	25	61	69.3	9.6

We can now use PROC FASTCLUS to assign the new observations to clusters:

```
PROC FASTCLUS DATA = NewClusObs
              OUT  = NewClusObs_scored
              SEED = ClusterSeeds
              MAXCLUSTERS=5
              MAXITER = 0;
   VAR age weight oxygen runtime;
RUN;
```

Note that we specify an OUT data set that contains the scored data. With SEED= we specify our existing SEED data set that contains the cluster assignment rules. It is important to specify MAXITER=0 in order to perform cluster assignments only and to prevent the procedure from starting new iterations from the existing seeds. The resulting data set is shown in Table 23.11.

Table 23.11: Data set with cluster assignments

	AGE	Weight	Oxygen	Runtime	Cluster	Distance to Cluster Seed
1	35	81	39.3	11.6	1	14.264876103
2	55	91	59.3	12.6	5	15.743192066
3	25	61	69.3	9.6	4	25.885695089

23.5 Scoring with PROC SCORE of SAS/STAT

General

The SCORE procedure is available in SAS/STAT. This procedure multiplies values from two data sets, one containing raw data values and one containing coefficients. The result of this multiplication is a data set that contains linear combinations of the coefficients and the raw data values.

Note that PROC SCORE performs an element-wise multiplication of the respective variables in the raw and coefficients data set and finally sums the elements. No additional transformations such as log or logit can be performed within PROC SCORE. Therefore, PROC SCORE can be used only for scoring of a linear regression directly. For more details, see SAS Help and Documentation for the SCORE procedure.

Example

We again use the fitness data and run PROC REG for a prediction of oxygen and store the coefficients in the data set BETAS:

```
PROC REG DATA=fitness OUTEST = betas;
 MODEL Oxygen=Age Weight RunTime ;
RUN;
QUIT;
```

The data set NEWOBS contains three observations that will be scored with PROC SCORE using the BETAS data set:

```
PROC SCORE DATA  = NewObs
           SCORE = betas
           OUT   = NewObs_scored(RENAME = (Model1 = Oxygen_Predict))
           TYPE  = PARMS;
 VAR Age Weight RunTime;
RUN;
```

Note that we specify three data sets with PROC SCORE—the raw data (DATA=), the coefficients (SCORE=), and the scored data (OUT=).

23.6 Using the Respective SAS/ETS Procedure for Scoring

General

Most of the SAS/ETS procedures deal with time series forecasting. Therefore, scoring in this context means the writing forward of the forecasted values of the time series.

23.6.1 Scoring in PROC AUTOREG

Scoring in PROC AUTOREG is performed in the same manner as we showed in the PROC REG and PROC LOGISTIC sections. In the input data set, observations are added. These observations have missing values for the target variable and respective values for the input variables. If time was used as an independent variable, forecasts over time can be produced by adding observations with the respective values for time.

In the code an OUTPUT statement is added, which can have the following form:

```
OUTPUT OUT = autoreg_forecast P=predicted_value PM = structural_part;
```

Note that two predicted values can be created. P= outputs the predicted value, formed from both the structural and autoregressive part of the model. PM= outputs the structural part of the model only.

23.6.2 Scoring in PROC FORECAST

PROC FORECAST allows us to create forecasts by specifying the LEAD= option in the PROCEDURE statement. Note that PROC FORECAST produces forecasts only on the time variable itself and no additional input variables are possible. Therefore, different from PROC AUTOREG, no input variables for future periods are available.

23.6.3 Scoring in PROC ARIMA

In PROC ARIMA, forecasts are produced with the FORECAST statement. In this statement LEAD= defines the number of periods to be forecasted.

PROC ARIMA can produce forecasts either on the time variable only or on additional input variables. See the CROSSCORR= option in the IDENTIFY statement. The future values of additional input variables can be forecasted as well and can be used in the forecasting for future periods automatically.

Values of input variables that have been used during model definition with the CROSSCORR= option in the IDENTIFY statement can also be provided in the input data set. In this case the input data set contains values for those variables in future periods, as well as missing values for the dependent variable and for those input variables that will be forecasted in the procedure itself.

The following code shows an example of time series forecasting on the basis of the SASHELP.CITIMON data set. An ARIMA model for the total retail sales (the variable RTRR) will be created, and the values of the variable EEGP, the gasoline retail price, will be used. The following statement shows how PROC ARIMA can be used to forecast the EEGP values and then use them and their forecasts for a forecast of RTRR:

```
PROC ARIMA DATA = sashelp.citimon;
 IDENTIFY VAR =eegp(12);
 ESTIMATE P=2;
 IDENTIFY VAR = RTRR CROSSCORR = (eegp(12));
 ESTIMATE P=1 Q=1 INPUT=(eegp);
 FORECAST LEAD=12 INTERVAL = month ID = DATE OUT = results;
RUN;
QUIT;
```

Note that the emphasis of this example is not on creating the best possible ARIMA model for the underlying questions but on illustrating how scoring can be used in PROC ARIMA.

23.7 The Score Code That Can Be Produced in SAS Enterprise Miner

SAS Enterprise Miner produces score code automatically. This score code is, by default, DATA step code that reflects all data management that has been performed on the input data set and the scoring rules. This approach differs from the methods we described earlier, because all scoring rules are translated into DATA step code. For example, in the case of logistic regression the linear predictor is sequentially increased for the coefficients for each input variable. In the case of a decision tree the score code contains IF-THEN/ELSE clauses.

Besides SAS DATA step code, SAS Enterprise Miner can also produce C score code, JAVA score code, and PMML code. For more information, see Chapter 28, Section 28.5.

23.8 The Pre-checks on the Data That Are Useful before Scoring

General

In the case of periodic scoring, data marts need to be re-created or updated each period. Over time it can happen that the content of a data mart does not correspond any longer to the content of the training data mart on which the basis of the model was built. The most common cases of non-correspondence are the following:

- variables that have a large number of missing values
- categorical variables that have new categories
- the distribution of interval variable changes

In this section we will look at methods of checking and alerting these cases before the scoring starts.

23.8.1 Variables That Have a Large Number of Missing Values

Most analytical methods, except decision trees, do not accept missing values. In this case the whole observation is excluded from the analysis.

In some analyses, missing values are replaced by some values during missing value imputation. However, these replacement methods assume that in most cases, the number of missing values do not exceed a certain percentage.

Missing values of a variable for all observations in a data set are likely to appear over time if the definition of data sources, variables, and so forth, changes in the source systems. Therefore, it makes sense that the number of missing values is monitored and reported before each scoring run.

The number of missing values for numeric variables can be checked with the following macro:

```
%MACRO AlertNumericMissing (data=,vars=_NUMERIC_,alert=0.2);
PROC MEANS DATA = &data NMISS NOPRINT;
 VAR &vars;
 OUTPUT OUT = miss_value NMISS=;
RUN;
```

```
PROC TRANSPOSE DATA  = miss_Value(DROP = _TYPE_)
               OUT   = miss_value_tp;
RUN;
DATA miss_value_tp;
 SET miss_value_tp;
 FORMAT Proportion_Missing 8.2;
 RETAIN N;
 IF _N_ = 1 THEN N = COL1;
 Proportion_Missing = COL1/N;
 Alert = (Proportion_Missing >=  &alert);
 RENAME _name_ = Variable
        Col1 = NumberMissing;
 IF _name_ = '_FREQ_' THEN DELETE;
RUN;
TITLE Alertlist for Numeric Missing Values;
TITLE2 Data = &data -- Alertlimit >= &alert;
PROC PRINT DATA = miss_value_tp;
RUN;
TITLE;TITLE2;
%MEND;
```

The macro uses **PROC MEANS** to calculate the number of missing values for the numeric variables and produces a list of all numeric variables with the number of missing values, the percentage of missing values, and an alert, based on a predefined alert percentage.

The macro has three parameters:

DATA

The data set to be analyzed.

VAR

The list of variables that will be checked. Note that the default is all numeric variables. Also note that here we use the very practical logical variable _NUMERIC_.

ALERT

The value between 0 and 1. Variables that have a proportion of missing values greater than "Alert" will be flagged with alert = 1.

The macro can be called as in the following example:

```
%AlertNumericMissing(data = sampsio.hmeq,alert=0.1);
```

And creates the following result:

```
                    Alertlist for Numeric Missing Values                78
                   Data = sampsio.hmeq -- Alertlimit >= 0.1
                                      13:51 Wednesday, November 2, 2005


                          Number     Proportion_
         Obs    Variable   Missing     Missing       N      Alert

          1     BAD          0          0.00        5960      0
          2     LOAN         0          0.00        5960      0
          3     MORTDUE     518         0.09        5960      0
          4     VALUE       112         0.02        5960      0
          5     YOJ         515         0.09        5960      0
          6     DEROG       708         0.12        5960      1
          7     DELINQ      580         0.10        5960      0
          8     CLAGE       308         0.05        5960      0
          9     NINQ        510         0.09        5960      0
         10     CLNO        222         0.04        5960      0
         11     DEBTINC    1267         0.21        5960      1
```

For the number of missing values of character variables the following macro counts the number of missing values:

```
%MACRO ALERTCHARMISSING(data=,vars=,alert=0.2);
*** LOAD THE NUMBER OF ITEMS IN &VARS INTO MACRO VARIABLE NVARS;
%LET C=1;
%DO %WHILE(%SCAN(&vars,&c) NE);
  %LET C=%EVAL(&c+1);
%END;
%LET NVARS=%EVAL(&C-1);
*** CALCULATE THE NUMBER OF OBSERVATIONS IN THE DATA SET;
DATA _NULL_;
  CALL SYMPUT('N0',STRIP(PUT(nobs,8.)));
  STOP;
  SET &data NOBS=NOBS;
RUN;
PROC DELETE DATA = work._CharMissing_;RUN;
%DO I = 1 %TO &NVARS;
 PROC FREQ DATA = &data(KEEP =%SCAN(&VARS,&I))  NOPRINT;
    TABLE %SCAN(&vars,&I) / MISSING OUT = DATA_%SCAN(&vars,&I)(WHERE
=(%SCAN(&vars,&I) IS MISSING));
 RUN;
 DATA DATA_%SCAN(&vars,&i);
 FORMAT VAR $32.;
  SET data_%SCAN(&vars,&i);
  VAR = "%SCAN(&vars,&i)";
  DROP %SCAN(&vars,&i) PERCENT;
 RUN;
 PROC APPEND BASE = work._CharMissing_ DATA = DATA_%SCAN(&vars,&i)
FORCE;
 RUN;
%END;
PROC PRINT DATA = work._CharMissing_;
RUN;
```

```
DATA _CharMissing_;
 SET _CharMissing_;
 FORMAT Proportion_Missing 8.2;
 N=&N0;
 Proportion_Missing = Count/N;
 Alert = (Proportion_Missing > &alert);
 RENAME var = Variable
        Count = NumberMissing;
 *IF _NAME_ = '_FREQ_' THEN DELETE;
RUN;
TITLE ALERTLIST FOR CATEGORICAL MISSING VALUES;
TITLE2 DATA = &DATA -- ALERTLIMIT >= &ALERT;
PROC PRINT DATA = _CharMissing_;
RUN;
TITLE;TITLE2;

%MEND;
```

The macro performs a call of PROC FREQ for each variable and concatenates the results for the frequencies of the missing values. Note that the macro deletes a potential existing data set WORK._CHARMISSING_.

The macro has three parameters:

DATA

The data set to be analyzed.

VARS

The list of variables that will be checked.

ALERT

The value between 0 and 1. Variables that have a proportion of missing values greater than "Alert" will be flagged with alert = 1.

It can be called, as in the following example:

```
%ALERTCHARMISSING(DATA=sampsio.hmeq,VARS=job reason);
```

And creates the following output:

```
              Alertlist for Categorical Missing Values              82
                Data = sampsio.hmeq -- Alertlimit >= 0.2
                              13:51 Wednesday, November 2, 2005

                         Number    Proportion_
  Obs    Variable        Missing      Missing        N      Alert

   1     job               279        0.05          5960       0
   2     reason            252        0.04          5960       0
```

23.8.2 Categorical Variables with New Categories

With categorical variables it happens very frequently that new categories appear in the data over time. For example, if a new price plan, tariff, or product is launched, the code or name for it might appear after some time in the scoring data mart. If this category was not present at the time of model creation, problems will occur in the scoring process.

- In the case of dummy variables for categorical variables, usually no error will be issued, but the observation with a new category will typically have zeros for all dummy variables. In this case, a wrong score will be calculated for these observations.

- In case of a SAS Enterprise Miner score code, a new category will cause a warning in the scoring process. This will lead to an average score for this observation—for example, in the form of the baseline probability in event prediction.

- The macro shown in *Chapter 20 – Coding for Predictive Modeling*, for the calculation of proportions for categories, introduced an option that provides the average event rate or average mean for a new category, which leads to a correct scoring.

Here we will introduce a macro that checks whether new categories in categorical variables exist in the scoring data mart that were not present in the training data mart.

The macro %REMEMBERCATEGORIES has to be applied on the training data mart. It creates and stores a list for each variable in the VARS data set. These data sets have the prefix CAT. This list of categories will be used by macro %CHECKCATEGORIES to compare whether new categories in the scoring data mart exist. The list of categories will be stored in data sets with the prefix SCORE. The result of the comparison is stored in data sets with the prefix NEW. The list of new categories is printed in the SAS Output window.

```
%MACRO RememberCategories(data =, vars=,lib=work);
*** Load the number of itenms in &VARS into macro variable NVARS;
%LET c=1;
%DO %WHILE(%SCAN(&vars,&c) NE);
   %LET c=%EVAL(&c+1);
%END;
%LET nvars=%EVAL(&c-1);

%DO i = 1 %TO &nvars;
PROC FREQ DATA = &data NOPRINT;
 TABLE %SCAN(&vars,&i) / MISSING OUT = &lib..cat_%SCAN(&vars,&i)(DROP
= COUNT PERCENT);
RUN;
%END;

%MEND;
```

The macro %REMEMBERCATEGORIES has the following parameters:

DATA

The data set that contains the training data.

VARS

The list of variables that will be checked. Note that here we cannot use the logical variable for all categorical variables, _CHAR_, because we need to scan through the macro list of values.

LIB

The SAS library where the list of categories will be stored.

The macro %CHECKCATEGORIES is defined as follows:

```
%MACRO CheckCategories(scoreds=, vars=,lib=work);
*** Load the number of items in &VARS into macro variable NVARS;
%LET c=1;
%DO %WHILE(%SCAN(&vars,&c) NE);
  %LET c=%EVAL(&c+1);
%END;
%LET nvars=%EVAL(&c-1);

%DO i = 1 %TO &nvars;
PROC FREQ DATA = &scoreds NOPRINT;
 TABLE %SCAN(&vars,&i) / MISSING OUT =
&lib..score_%SCAN(&vars,&i)(DROP = COUNT PERCENT);
RUN;


PROC SQL;
    CREATE TABLE &lib..NEW_%SCAN(&vars,&i)
  AS
    SELECT %SCAN(&vars,&i)
    FROM &lib..score_%SCAN(&vars,&i)
  EXCEPT
    SELECT %SCAN(&vars,&i)
    FROM &lib..cat_%SCAN(&vars,&i)
  ;
QUIT;
TITLE New Categories found for variable %SCAN(&vars,&i);
PROC PRINT DATA = &lib..NEW_%SCAN(&vars,&i);
RUN;
TITLE;
%END;
%MEND;
```

The macro %CHECKCATEGORIES has the following parameters:

SCOREDS

The data set that contains the scoring data.

VARS

The list of variables that will be checked. Note that here we cannot use the logical variable for all categorical variables, _CHAR_, because we need to SCAN through the macro list of values.

LIB

The SAS library where the list of categories will be stored.

The basic idea of this set of macros is to create a repository of the list of values at the time of model training. At each scoring the list of values of the scoring data mart is compared with the list of values in the training data mart. This has the advantage that the training data mart does not need to be available each time a scoring is done, and it saves processing time because the list for the training data mart is generated only once. Note, however, that the data sets that are created by the macro %REMEMBERCATEGORIES must not be deleted and must be present when the macro %CHECKCATEGORIES is invoked.

The macros can be invoked, as in the following example:

```
%RememberCategories(data=sampsio.hmeq,vars=job reason);
```

This creates the data sets WORK.CAT_JOB and WORK.CAT_REASON, which contain a list of values that are available in the training data.

Consider the new data in Table 23.12.

Table 23.12: Scoring data with new categories

	job	reason
1	Office	HomeImp
2	Sales	New
3	SAS-Consultant	DebtInc

Now we invoke the macro %CHECKCATEGORIES to compare the values in the data set with our repository:

```
%CheckCategories(scoreds = hmeq_score,vars=job reason);
```

And we receive the output:

```
New Categories found for variable reason              92
        13:51 Wednesday, November 2, 2005
                Obs     reason
                 1      DebtInc
                 2      New
New Categories found for variable job                 91
        13:51 Wednesday, November 2, 2005


                Obs     job
                 1      SAS-Consultant
```

23.8.3 The Distribution of Interval Variables

When the distribution of interval variables changes we usually don't have the problem that the scoring does not run or that it creates missing values. It is, however, important to consider changes or shifts in the distribution of interval variables because it can affect the validity of the model for this set of observations.

We introduce here a set of macros to compare the distribution of numeric variables in the training data mart with the distribution of numeric variables in the scoring data mart. Again, we create a repository of descriptive statistics for the training data mart and compare these statistics with the values of the scoring data mart.

```
%MACRO
RememberDistribution(data=,vars=_NUMERIC_,lib=work,stat=median);
 PROC MEANS DATA = &data NOPRINT;
  VAR &vars;
  OUTPUT OUT = &lib..train_dist_&stat &stat=;
 RUN;

PROC TRANSPOSE DATA  = &lib..train_dist_&stat(DROP = _TYPE_ _FREQ_)
               OUT   = &lib..train_dist_&stat._tp(RENAME = (_NAME_ =
Variable

                                               Col1    =
Train_&stat));
RUN;
%MEND;
```

The macro %REMEMBERDISTRIBUTION has the following parameters:

DATA

The data set that contains the training data.

VARS

The list of variables that will be checked. Note that the default is all numeric variables. Also note that here we use the very practical logical variable _NUMERIC_.

LIB

The SAS library where the statistics describing the distributions will be stored.

STAT

The descriptive statistic that will be used for comparison. The default is MEDIAN. Valid values are those statistics PROC MEANS can calculate (see SAS Help and Documentation for details).

The macro creates data sets in the &lib library with the name TRAIN_DIST_&stat_TP.

The macro %SCOREDISTRIBUTION calculates the same statistics for the score data sets and joins these statistics to those of the training data set in order to perform a comparison.

```
%MACRO
ScoreDistribution(data=,vars=_NUMERIC_,lib=work,stat=median,alert=0.1)
;
 PROC MEANS DATA = &data NOPRINT;
  VAR &vars;
  OUTPUT OUT = &lib..score_dist_&stat &stat=;
 RUN;

PROC TRANSPOSE DATA  = &lib..score_dist_&stat(DROP = _TYPE_ _FREQ_)
               OUT   = &lib..score_dist_&stat._tp(RENAME = (_NAME_ =
Variable

                                               Col1    =
Score_&stat));
RUN;

PROC SORT DATA = &lib..train_dist_&stat._tp;
 BY variable;
RUN;
```

```
PROC SORT DATA = &lib..score_dist_&stat._tp;
 BY variable;
RUN;


DATA &lib..compare_&stat;
 MERGE &lib..train_dist_&stat._tp &lib..score_dist_&stat._tp;
 BY variable;
 DIFF = (Score_&stat - Train_&stat);
 IF Train_&stat NOT IN (.,0) THEN
      DIFF_REL = (Score_&stat - Train_&stat)/Train_&stat;
 Alert = (ABS(DIFF_REL) > &alert);
RUN;

TITLE Alertlist for Distribution Change;
TITLE2 Data = &data -- Alertlimit >= &alert;
PROC PRINT DATA = &lib..compare_&stat;
RUN;
TITLE;TITLE2;

%MEND;
```

The macro %SCOREDISTRIBUTION has the following parameters:

DATA

The data set that contains the score data.

VARS

The list of variables that will be checked. Note that the default is all numeric variables. Also note that here we use the very practical logical variable _NUMERIC_.

LIB

The SAS library where the statistics describing the distributions will be stored.

STAT

The descriptive statistic that will be used for comparison. The default is MEDIAN. Valid values are those statistics PROC MEANS can calculate (see SAS Help and Documentation for details).

ALERT

The alert level as the absolute value of the relative difference between training and scoring data.

The results for the following calls are shown:

```
%RememberDistribution(data=sampsio.dmthmeq,stat=median);
%ScoreDistribution(data=sampsio.dmvhmeq,stat=median);
```

```
            Alertlist for Distribution Change    09:51 Friday, July 8, 2005    5
                       Data = sampsio.dmvhmeq -- Alertlimit >= 0.1

                     Train_        Score_
  Obs    Variable     mean          mean          DIFF      DIFF_REL     Alert

    1    bad          0.20          0.20          0.00      0.00000        0
    2    clage      182.41        176.92         -5.50     -0.03013        0
    3    clno        21.16         20.80         -0.36     -0.01709        0
    4    debtinc     34.06         33.35         -0.71     -0.02096        0
    5    delinq       0.50          0.44         -0.06     -0.11886        1
    6    derog        0.23          0.26          0.03      0.14998        1
    7    loan     19444.55      18006.21      -1438.34     -0.07397        0
    8    mortdue  74413.93      72348.39      -2065.54     -0.02776        0
    9    ninq         1.13          1.24          0.10      0.09150        0
   10    value   104675.37     100985.21      -3690.17     -0.03525        0
   11    yoj          8.98          9.13          0.15      0.01618        0
```

23.8.4 Summary

We see that with a simple macro we can easily monitor and alert changes in distributions. It is up to the automatization process in general how to act from these alerts—for example, whether a conditional processing will follow. We will discuss this in the next section.

23.9 Automation of Data Mart Creation in General

General

In this section we will briefly touch on the possible automation of data mart creation. With this topic, however, we are reaching the limits of this book's scope. It is up to a company's IT strategists to determine how the creation of data marts will be automated.

It is possible to automate data mart creation with pure SAS code, as we will show in the following sections.

Automatizing a SAS program is very closely related to the SAS macro language and also closely related to efficient SAS programming in general. It would go beyond the scope of this book to go into too much detail about this. If you are interested, see SAS Help and Documentation or SAS Press books on this topic.

Example 1: Using Macro Parameters for the Control of SAS Programs

In this example we show how macro variables can easily be used to parameterize SAS programs. In this case data set names are created with the postfix of the period in the format YYYYMM. This makes sense for both the differentiation of the data sets that are created and the selection of the appropriate table from the source systems.

```
****************************************************************
***   Example 1 -- Simple Example;
**************************************************************;

%LET Period = 200507;
%LET Varlist = CustID Age Gender Region;

DATA customer_&period;
 SET source.customer_&period;
 -- some other statements --;
 KEEP &Varlist;
RUN;
```

Example 2: Filling Macro Parameters with Values from the Current Date

We see a simple example that allows the automatic creation of a macro variable that contains the actual year and month in a YYYYMM format.

```
****************************************************************
***   Example 2 -- Calculate YYYYMM Variable from Current Date;
**************************************************************;

DATA _NULL_;
 CALL
symput('MONTH_YEAR',TRIM(YEAR(TODAY()))||PUT(MONTH(TODAY()),z2.));
RUN;
%PUT &MONTH_YEAR;
```

Example 3: Conditional Processing

In this example we show how conditional processing can be performed within a macro and checking the &SYSERR variable. We see that we can use SAS statements, a macro call, or a %INCLUDE of an external file for processing. We check whether the error code is less than or equal to a certain limit. With the default LIMIT=0, the macro will stop executing when a warning is issued.

```
*****************************************************************
***   Example 3 -- Conditional Processing;
****************************************************************;

%MACRO Master(limit=0);
 DATA _NULL_;RUN; *** Start with a valid statement to initialize
SYSERR;
 %IF &syserr <= &limit %THEN %DO; PROC PRINT DATA = sashelp.class;
RUN; %END;
 %IF &syserr <= &limit %THEN %CREATE_DATA SETS;;
 %IF &syserr <= &limit %THEN %INCLUDE 'c:\programm\step03.sas';
%MEND;
%MASTER;
```

Note that not only the &SYSERR variable can be used to control the flow of a program. Generic checks on data sets such as the number of observations can also be used to make a conditional next step in the program. The macros we introduced in the previous sections are well suited to decide whether a next step—for example, the scoring—will be performed or not.

Using SAS Data Integration Studio

In many cases, however, it is highly advisable to use a specialized tool such as SAS Data Integration Studio for this purpose. SAS Data Integration Studio provides a graphical process flow tool to create and document data management process flows. It also facilitates the creation of conditional processing. SAS Data Integration Studio is part of the SAS Data Integration Server package and is very well suited to the automation and documentation of data preparation processes.

Chapter 24

Do's and Don'ts When Building Data Marts

24.1 Introduction

In this chapter we will look at what should be done and what should be avoided when building data marts. The approach of this chapter is not to teach but to share experiences from practice.

The do's and don'ts as they are called in this chapter are divided into three sections:

- process do's and don'ts
- data mart handling do's and don'ts
- coding do's and don'ts

24.2 Process Do's and Don'ts

24.2.1 Use Check Sums, Cross-Checks, or Peer Review

When writing a program that prepares an analysis data mart, the programmer or analyst is often seduced into giving full attention to the program or its clever coding and forgetting to think of the plausibility of the created numbers in the data set. It is, therefore, highly advisable to include (at a couple of stages and before the final created data mart) a *plausibility review of the data*.

At those stages it is, in most cases, sufficient to create simple descriptive statistics such as means or quantiles for interval variables or frequencies for categorical variables and to discuss and review these results from a business point of view. In many cases it makes sense to include business experts for such a review. Business experts are those people who are involved in the processes where data are collected and created, and who know whether data are plausible or not.

At these reviews, you should challenge the data mart content with questions like these:

- Is it possible that patients have on average 79 days per year in the hospital?

- May customers with a certain tariff have a contract with our company that is less than 12 months?

- Is it possible that bank customers have cash withdrawals on average of 3.493 per month?

In these cases a cross-check with data from the source systems or operative systems might be needed in order to have the proof that the aggregated data are correct.

24.2.2 Do Not Start the Analysis Too Early

For the analyst, it is in many cases a hard decision to stop the fluid process of data mart programming and perform a data mart review. This is often strengthened by the problem that data mart creation takes more time than expected—for example, due to delayed delivery dates of the source data. If the data mart is finally finished, analysts tend to immediately continue the process of analysis either because the project plan is already delayed and they want to deliver results, or because the boring and exhausting data preparation phase is finally finished and they now want to start with the interesting analysis phase.

If, however, we start the analysis too early, before we are confident that our data are correct, we might have to start again with data correction. The following list shows a few of the steps we might need to run twice:

- define analysis metadata (what is the analysis variable, what is the ID variable, and so forth)

- prepare statistical data such as replacing missing values, filtering extreme values, transforming variables, grouping values, and optimal binning with relationship to the targeting

- explore graphical data

These problems get severe if preliminary results have already been discussed in workshops with interested parties and potentially promising analysis results have already been communicated throughout different departments.

After data preparation, however, it makes sense to step back and check the content of the data as described earlier, or to check the process of data creation in the sense of, where does the data for a certain column in the analysis data mart come from. Note that this break in the process does not necessarily last days or weeks; it can be a short break for coffee or tea and a short review of the data processing steps that have been performed so far. This does not take much time, but it can save a lot of time.

24.3 Data Mart Handling Do's and Don'ts

24.3.1 Do Not Drop the Subject ID Variable from the Data Mart

At some point in the analysis the analyst probably wants to get rid of all unused variables. In this case it often happens that all variables that are not directly used in the analysis as dependent or independent variables are dropped from the table.

Also in this case it often happens that the subject ID variable is dropped from the data mart. This is not necessarily a problem in the current analysis; but, if at a later stage additional data will be joined to the table, or if scores for different models or different points in time will be compared, the subject ID variable will be needed.

24.3.2 Attend to Even One "Disappeared" Observation

The number of observations in a table can change for various reasons—for example, for filtering. In this case the number of observations changes intentionally. The number of observations in a table is, however, most likely to change after a merge has been done with another table. In this case it is mandatory after the merge to observe the notes in the SAS log that relate to the number of observations in the respective tables.

When working with data sets that have many observations, we tend to downplay the importance of a few observations in the merge. The fact, however, that observations "disappear" during data merges shows that we probably have problems with our definition of the merges or with the consistency of the merging keys.

It makes sense to control the merge by using logical IN variables as data set options. The following example shows the code:

```
DATA CustomerMart;
 MERGE Customer (IN = InCustomer)
       UsageMart (IN = InUsageMart);
 BY CustID;
 IF InCustomer;
RUN;
```

The preceding code ensures that all observations from the customer table are output to CUSTOMERMART and only the matching observations from USAGEMART are output. If we omit the restriction for the logical variable, we would have all observations, including the non-matching cases, from the CUSTOMER and USAGEMART tables.

Also note that the following warning in the SAS log might be a cause for lost or unexpected added observations.

```
WARNING: Multiple lengths were specified for the BY variable by
input data sets. This may cause unexpected results.
```

It makes sense to have equal lengths and formats for the merging key in the respective tables. For more details, see SUGI paper 98-28, which is available at http://www2.sas.com/proceedings/sugi28/098-28.pdf.

24.3.3 Use Meaningful Units for Interval Variables

In many cases, interval variables are technically stored in different units because they are used in the business language. There is no rule about which unit best fits which variable. In general it is best to use units with the following characteristics:

- understandable and interpretable from a business point of view
- technically efficient, in the sense that they do not have too many trailing zeros or too many zeros after the comma

For example, in the case of time values, the appropriate unit can easily be achieved by multiplying or dividing by 60 or 3,600.

In the telecommunications industry the number of call minutes per month is a very central variable and is commonly used in the unit "minutes". In profiling customers, however, when we look at monthly or yearly sums, it might make sense to convert the minutes to hours for better comparability and interpretability.

24.3.4 Show the Unit in the Name of the Variable

In the case of weights, durations, lengths, and so on, it is necessary to include the unit in the variable name or at least in the label of the variable.

```
LABEL weight = "Weight in kg";
LABEL weight = "Weight (kg)";
```

The duration of phone calls per month is measured in minutes of use. MINUTESOFUSE is in this case a variable name that tells you what it means.

```
RENAME usage = MinutesOfUse;
```

24.3.5 Check the Distribution, the List of Values, and the Number of Missing Values

We saw in *Chapter 23 – Scoring and Automation*, which methods can be used to compare training and scoring data. The number of missing values, the list of values for categorical variables, and the description of the distribution of interval variables can also be used to check the distribution of a variable over time or among systems.

24.3.6 Do Not Loop through Unnecessary Static Attributes through Aggregation Steps

When creating a one-row-per-subject data mart and multiple observations per subject exist for some entities, the base table should be clearly defined. The definition of the base table usually depends on the business question. The base table also holds the static attributes. The aggregates per subject of the multiple-rows-per-subject data sets are joined to this table.

Even if some SAS procedures can loop through static attributes with a COPY statement during the aggregation process, this should be avoided especially for large data sets for performance reasons, but also for small data sets, in order to improve the structure of the data mart creation process.

24.3.7 Do Not Refuse to Use a Spreadsheet as a Central Definition of Your Variables

If a data mart creation process reaches the complexity that metadata for each variable will be maintained, it is advisable to use tools such as SAS Data Integration Studio. If, however, such a tool is not in place and metadata on variables will be stored, using a spreadsheet as a central definition of your variables should be considered.

This spreadsheet has one line per variable in the data mart. The following is a list of possible relevant columns:

- Variable name
- Variable order number
- Variable number in the data set
- Type (categorical or interval)
- Format
- Upper and lower limit for interval variables
- Expression for a derived variable
- Label
- Variable category
- Source system
- Number of missing values
- Measurement type: interval, ordinal, nominal, binary
- Variable usage: target, input
- Profiling status: Is the variable a profiling variable?
- Variable importance: Here a grading for the variables can be created in order to sort them based on their suggested importance. For example, 0 can mean the variable will be dropped from the analysis data mart.

Note that these spreadsheet lists have the following advantages:

- Variable properties are centrally documented.
- The spreadsheet can be used to sort variables based on certain criteria.
- The values of certain columns can be used to create the respective SAS code—for example, the variable for a KEEP statement, based on the sorting by variable importance or on the number of missing values.

This table can also hold the formulas for the definition of derived variables. Examples for this include the following:

```
AGE **2
(CURRENT_TARIFF = LAST_TARIFF)
```

The spreadsheet can then be used to create the formula by using the columns Variable_Name and Formula. In the second example, an indicator variable is created without using an IF-THEN/ELSE clause just by comparing the values.

The starting point for such a table can be the routing of the output of PROC CONTENTS to SAS table with the following statements:

```
PROC CONTENTS DATA = sashelp.citimon
             OUT  = VarList(KEEP = name type length
                                   varnum label format);
RUN;
```

The resulting table can then be exported to Microsoft Office Excel, simply by right-clicking it in the SAS Explorer window and selecting **View in Excel** from the menu. The table can then be edited in Excel by sorting and creating new columns as described earlier.

24.4 Coding Do's and Don'ts

24.4.1 Use Data Set Options If Possible

SAS statements such as WHERE or KEEP can also be used as DATA step options. This has the advantage that performance is increased because only those observations and columns that are needed are read into the procedure or DATA step.

See the following code:

```
DATA class;
 SET sashelp.prdsal;
WHERE product = 'SOFA';
KEEP actual year quarter country;
RUN;
```

This can be enhanced by the following coding:

```
DATA class;
 SET sashelp.prdsale(WHERE = (product = 'SOFA')
                     KEEP = actual year quarter country);
RUN;
```

Only a subset of observations and columns is important to the DATA step.

24.4.2 Do Not Ignore the SAS Log

To speak from a correct data processing perspective, it is not sufficient that the resulting data set has a plausible number of observations and the columns are filled with plausible values. Additionally the SAS log has to be checked. We mentioned these points earlier:

- warning because of multiple lengths of the BY variable
- monitoring the change of the number of observations

Additional important points include the following:

- Errors or warnings in general have to be observed.
- The fact that uninitialized variables or missing values were used in an operation should be observed.
- Uninitialized variables often result from typos. See the following short example:

The following code with a typo for variable WEIGHT gives the log shown below:

```
DATA test;
 SET sashelp.class;
 Age_Weight = Age * Weihgt;
RUN;
```

```
NOTE: Variable Weihgt is uninitialized.
NOTE: Missing values were generated as a result of performing
an operation on missing values.
      Each place is given by: (Number of times) at
(Line):(Column).
      19 at 25:19
```

Structure your code

When writing SAS programs in the SAS Program Editor the efficient use of line breaks, blanks and tab stops can easily make a program more readable. This is shown in the following example:

```
data class; set sashelp.class;
if age <= 13 then age_class = 1;
else if age <= 14 then age_class =2;
else age_class = 3;
label age="Age (years)" height="Height (inch)" weight="Weight
(pound)";
rename sex=gender name = firstname;
run;
```

Simply structuring the program leads to the following:

```
data class;
 set sashelp.class;

      if age <= 13 then age_class = 1;
 else if age <= 14 then age_class = 2;
 else                  age_class = 3;
 label age    = "Age (years)"
       height = "Height (inch)"
       weight = "Weight (pound)"
 ;
 rename sex    = gender
        name   = firstname
 ;
run;
```

Also note that the additional time and effort in coding a program that is well structured visually is usually between 10 and 15% of the total time effort. The benefit is a more readable and transparent program.

24.4.3 Use Formats and Labels

In addition to the benefits of the LABEL or FORMAT statement to define the properties of a variable, the ATTRIB statement can be used as a central point of variable property definition.

The following example shows how the ATTRIB statement can be used to define properties of two new variables in a data set:

```
DATA class;
 SET sashelp.class;
 ATTRIB NrChildren  LABEL  = "Number of children in family"
                    FORMAT = 8.;
 ATTRIB district    LABEL = "Regional district"
                    FORMAT = $10.;
RUN;
```

24.4.4 Use Shortcuts for Frequently Used Code Segments

The SAS windowing environment allows you to define keyboard macros that create code in the SAS Program Editor. This can be used to automate the creation of frequently used code segments.

It can also be used to automate the creation of standardized comment lines.

In many cases the programmer is too "busy" to insert a comment line that would make the code more readable. Keyboard macros can help here. For example, the following code can be automated with one key:

```
*****************************************************************
*** Author: Gerhard Svolba
*** Date:
*** Project:
*** Subroutine:
***
*** Change History:
***
*****************************************************************;
```

24.4.5 Do Not Forget to Document and Comment Your Code

Writing about the rationale for documenting and commenting code and the reasons for not doing so could be a book on its own. We, therefore, want to make only a few notes here on documenting and commenting.

Documenting code is fairly simple, and we have seen that we define macros to insert templates. The importance of well-documented code increases with the existence of one or more of the following facts:

- More than one person is currently programming the data preparation.

- Data are not prepared for a one-shot analysis, but the code will be used over and over.

- The lifetime of the project is more than three months. This includes the chance that after three months, knowledge of the data preparation steps is required. Note that three months seem to be very short, but usually after this time, programmers tend to forget the details of their code.

- The code elements should be reusable for a different project.

Documentation is also very important if we use only SAS programs for data preparation, and we do not make use of tools such as SAS Data Integration Studio. In this case the result of data preparation tends to be a "spaghetti code" with %INCLUDE and SAS macro calls. If in this case documentation is not available, it is very hard to reuse or interpret the code at a later time.

In these cases it often happens that a program is re-programmed from scratch rather than re-using parts of the program.

24.4.6 Coding Tip: Using a Spreadsheet for Program Code Generation

A spreadsheet can be used to generate SAS programming statements, especially in the case of a list of IF-THEN/ELSE statements or statements that include lists of variable names.

For example, if we have the variable name in a spreadsheet and create a new column where we include the variable label, we can easily add a column with the =" values and a column with the " values that are needed for the LABEL statement.

	A	B	C	D
1	Variable Name		Variable Label	
2	Age	="	Age (Years)	"
3	Height	="	Height (cm)	"
4	Name	="	Pre Name	"
5	Sex	="	Gender	"
6	Weight	="	Weight (kg)	"

The contents of the cells A2:D6 only have to be copied to the SAS Program Editor between a column with LABEL and a column with a ";".

Another example is the preparation of code for categories. If for many categories the " values have to be added, it might be helpful to do this in a spreadsheet such as Excel. In this case, we export the list of values to Excel. In our case this is column B and copy down the " in columns A and C.

	A	B	C
1	"	2	"
2	"	3	"
3	"	1	"
4	"	4	"

We can also use the spreadsheet to copy down programming statements. In this case we export the codes that are represented in column B and manually edit the other columns A, C, D, and E. Then we copy the content of the cells to the SAS Program Editor to include the code in a program:

	A	B	C	D	E
1	IF ProductMainGroup="	1	" THEN ProductMainGroupNEW = "	100	";
2	IF ProductMainGroup="	2	" THEN ProductMainGroupNEW = "	100	";
3	IF ProductMainGroup="	3	" THEN ProductMainGroupNEW = "	200	";
4	IF ProductMainGroup="	4	" THEN ProductMainGroupNEW = "	200	";

```
IF ProductMainGroup="1" THEN ProductMainGroupNEW = "100";
IF ProductMainGroup="2" THEN ProductMainGroupNEW = "100";
IF ProductMainGroup="3" THEN ProductMainGroupNEW = "200";
IF ProductMainGroup="4" THEN ProductMainGroupNEW = "200";
```

We see how we make use of the automatic copy in editing and preparing SAS code.

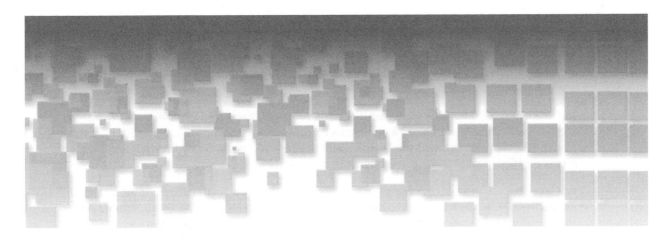

Part 5

Case Studies

Introduction

In this part we will cover four case studies for data preparation. These case studies refer to the content we presented in earlier chapters and put together in the context of a concrete question. In the case studies we will show example data and complete SAS code to create from the input data the respective output data mart. The following case studies will be examined:

- building a customer data mart
- deriving customer segmentation measures from transactional data
- preparing data for time series analysis
- preparing data in SAS Enterprise Miner

In *Case Study 1—Building a Customer Data Mart,* we will create a one-row-per-subject data mart from various data sources. We will show how data from multiple-rows-per-subject tables have to be aggregated to a one-row-per-subject structure and how relevant derived variables can be created. Finally, we will create a table that can be used directly for analyses on a CUSTOMER level such as predictive analysis or segmentation analysis.

In *Case Study 2—Deriving Customer Segmentation Measures from Transactional Data,* we will see how we can create a one-row-per-subject customer data mart from tables with hierarchies from a star schema. We will see how the data can be effectively joined together and how we create derived variables for marketing analysis purposes.

In *Case Study 3—Preparing Data for Time Series Analysis,* we will deal again with data with hierarchies. Here we have to prepare data marts for time series analysis by combining tables for shops, sales organizations, products, product groups, and orders. We will also deal with performance considerations and the selection of different aggregation levels.

In *Case Study 4—Preparing Data in SAS Enterprise Miner,* we will explore how SAS Enterprise Miner can assist the data preparation process. We will show example process flows where we deal with the following SAS Enterprise Miner nodes: Input Data Source node, Sample node, Data Partition node, Time Series node, Association node, Transform Variables node, Filter node, Impute node, Score node, Metadata node, SAS Code node, and Merge node.

Chapter 25

Case Study 1—Building a Customer Data Mart

25.1 Introduction

This chapter provides an example of how a one-row-per-subject table can be built using a number of data sources. In these data sources we have in the CUSTOMER table only one row per subject; in the other tables, we have multiple rows per subject.

The task in this example is to combine these data sources to create a table that contains one row per customer. Therefore, we have to aggregate the multiple observations per subject to one observation per subject.

Tables such as this are frequently needed in data mining analyses such as prediction or segmentation. In our case we can assume that we create data for the prediction of the cancellation event of a customer.

This example mainly refers to the content that was introduced in Chapters 8, 12, 17, and 18.

25.2 The Business Context

General

The business question is to create an analytical model that predicts customers that have a high probability to leave the company. We want to create a data mart that reflects the properties and behaviors of customers. This data mart will also contain a variable that represents the fact that a customer has left the company at a certain point in time. This data mart will then be used to perform predictive modeling.

25.2.1 Data Sources

To create this data mart, we use data from the following data sources:

- *customer data*: demographic and customer baseline data
- *account data*: information customer accounts
- *leasing data*: data on leasing information
- *call center data*: data on call center contacts
- *score data*: data of value segment scores

From these data sources we want to build one table with one row per customer. We have therefore the classic case of the one-row-per-subject paradigm that we need to bring the repeated information from—for example, the account data to several columns.

25.2.2 Considerations for Predictive Modeling

In our data, the fact that a customer has left can be derived from value segment. The SCORE table contains the VALUESEGMENT for each customer and month. If a customer cancels his contract, his value segment is set to value '8. LOST' in this month. Otherwise, the value segment has values such as '1. GOLD' or '2. SILVER' in order to represent the customer value grading.

We will define our snapshot date as June 30, 2003. This means that in our analysis we are allowed to use all information that is known at or before this date. We use the month July 2003 as the *offset window* and the month August 2003 as the *target window*. This means we want to predict with our data the cancellation events that took place in August 2003. Therefore, we have to consider the VALUESEGMENT in the SCOREDATA table per August 2003.

The offset window is used to train the model to predict an event that does not take place immediately after the snapshot date but some time after. This makes sense, because retention actions cannot start immediately after the snapshot date, but some time later when data are loaded into the data warehouse and scored, then the retention actions are defined. For details, see Figure 25.1. Customers that cancelled in the offset month, July 2003, are excluded from the training data. Keeping them as non-event customers would (in the data) reduce the signal of customers that cancel, as we have cancelling customers as non-events in the offset window and as events in the target window.

Figure 25.1: Observation, offset, and target windows

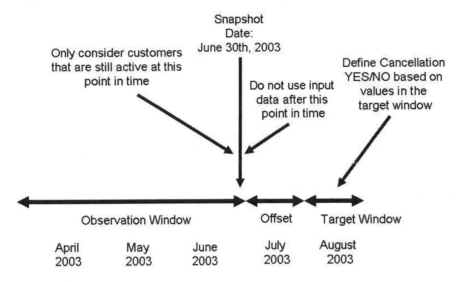

25.2.3 Reasons for Multiple Observations per Customer

We have different reasons for multiple observations per customer in the different tables:

- In the score data, we have multiple observations because of scores for different points in time.

- In the account data, we aggregated data over time, but we have multiple observations per customer because of different contract types.

- In the call center data, we have a classic form of a transactional data structure with one-row-per-call-center contact.

25.3 The Data

We have five tables available:

- CUSTOMER
- ACCOUNTS
- LEASING
- CALLCENTER
- SCORES

These tables can be merged by using the unique CUSTID variable. Note that, for example, the account table already contains data that are aggregated from a transactional account table. We already have one row per account type that contains aggregated data over time.

The data model and the list of available variables can be seen in Figure 25.2.

Figure 25.2: Data model for customer data

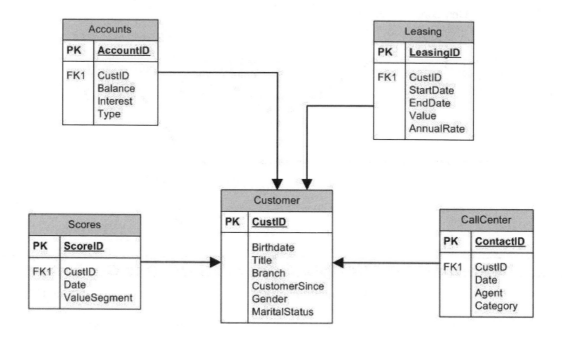

Table 25.1 through Table 25.5 give a snapshot of the contents of the tables:

Table 25.1: CUSTOMER table

	CustID	Birthdate	Title	Branch	CustomerSince	Gender	MaritalStatus
1	1000002	26DEC1958		Fil1	01JAN2000	Male	Married
2	1000005	25JUN1947	Ing.	Fil4	01APR1999	Male	Single
3	1000006	10DEC1945		Fil4	01SEP1996	Female	Married
4	1000007	02JUN1934		Fil1	01SEP1997	Male	Married
5	1000008	15DEC1957	Dr.	Fil3	01JAN1996	Male	Single
6	1000009	11MAR1959		Fil2	01JUL2001	Male	Single
7	1000014	23AUG1952		Fil4	01MAY1996	Male	Single
8	1000015	12MAY1959		Fil2	01FEB1999	Male	Single
9	1000016	11FEB1967		Fil2	01FEB2001	Male	Married
10	1000018	.		Fil2	01JUL2003	Male	Single
11	1000019	30DEC1936	Ing.	Fil2	01JAN2002	Male	Single
12	1000021	12AUG1959		Fil2	01OCT2003	Male	Divorce
13	1000022	10JUL1961		Fil1	01NOV1993	Male	Married
14	1000026	20JAN1972		Fil1	01DEC1995	Male	Married

Table 25.2: ACCOUNTS table

	CustID	AccountID	Balance	Interest	Type
1	1000002	16853	1550.42	5.00	Savings Account
2	1000002	16853	1550.42	5.00	Loan
3	1000005	13296	3775.31	6.00	Savings Account
4	1000006	21592	2376.43	2.00	Funds
5	1000007	16789	1812.72	5.00	Funds
6	1000007	16789	1812.72	5.00	Savings Account
7	1000008	15567	3350.65	2.00	Funds
8	1000009	11907	1191.82	4.00	Savings Account
9	1000009	11907	1191.82	4.00	Funds
10	1000009	11907	1191.82	4.00	Loan
11	1000014	8800	1500.46	4.00	Savings Account
12	1000014	13513	1500.46	5.00	Savings Account
13	1000015	16853	2801.09	5.00	Savings Account
14	1000016	166	1662.83	1.00	Savings Account
15	1000016	166	1662.83	1.00	Savings Account
16	1000018	20703	0.00	2.00	Funds

Table 25.3: LEASING table

	Leasing ID	CustID	StartDate	EndDate	Value	AnnualRate
1	1001	1000002	13SEP1999:00:00:00	02APR2004:00:00:00	521763	254.69454187
2	1002	1000005	18FEB2002:00:00:00	21JAN2009:00:00:00	855215	232.51593453
3	1003	1000006	26FEB2002:00:00:00	17APR2017:00:00:00	560362	167.36627547
4	1004	1000007	23FEB2002:00:00:00	30APR2010:00:00:00	785661	84.376219214
5	1005	1000007	07OCT1999:00:00:00	07OCT2007:00:00:00	950047	84.376219214
6	1006	1000008	20OCT1999:00:00:00	16APR2009:00:00:00	5276	109.1474312
7	1007	1000009	16OCT1998:00:00:00	20MAR2005:00:00:00	243731	85.069300694
8	1008	1000009	03OCT2001:00:00:00	18JUL2017:00:00:00	348232	85.069300694
9	1009	1000014	14MAR2000:00:00:00	02APR2003:00:00:00	564728	92.513475696
10	1010	1000015	01JUL1997:00:00:00	19OCT2001:00:00:00	393984	189.54488281
11	1011	1000018	27AUG1995:00:00:00	28MAR2010:00:00:00	651408	179.8417421
12	1012	1000021	07DEC1998:00:00:00	21APR2013:00:00:00	461405	179.8417421

Table 25.4: CALLCENTER table

	CustID	ContactID	Date	Agent	Category
1	1000008	1	19JUL2003:00:00:00	58	Telebanking
2	1000014	2	08APR2003:00:00:00	94	Complaint
3	1000014	3	02MAR2003:00:00:00	56	Complaint
4	1000018	4	12JUN2003:00:00:00	28	Telebanking
5	1000028	5	23FEB2003:00:00:00	36	Telebanking
6	1000034	6	20MAR2003:00:00:00	24	Telebanking
7	1000035	7	24MAY2003:00:00:00	21	Telebanking
8	1000035	8	25JUN2003:00:00:00	81	Telebanking
9	1000037	9	06JAN2003:00:00:00	32	Complaint
10	1000039	10	26JUN2003:00:00:00	70	Complaint
11	1000040	11	28APR2003:00:00:00	31	Complaint
12	1000040	12	19MAY2003:00:00:00	68	Complaint
13	1000041	13	18JUL2003:00:00:00	12	Telebanking
14	1000050	14	04JUL2003:00:00:00	99	Telebanking

Table 25.5: SCORES table

	CustID	ScoreID	Date	ValueSegment
1	1000002	1000001	01JAN2003	3. BRONZE
2	1000002	1000002	01FEB2003	2. SILBER
3	1000002	1000003	01MAR2003	1. GOLD
4	1000002	1000004	01APR2003	3. BRONZE
5	1000002	1000005	01MAY2003	2. SILBER
6	1000002	1000006	01JUN2003	2. SILBER
7	1000002	1000006	01JUL2003	2. SILBER
8	1000002	1000006	01AUG2003	8. LOST
9	1000005	1000010	01APR2003	1. GOLD
10	1000005	1000011	01MAY2003	3. BRONZE
11	1000005	1000012	01JUN2003	3. BRONZE
12	1000005	1000012	01JUL2003	3. BRONZE
13	1000005	1000012	01AUG2003	3. BRONZE
14	1000006	1000013	01JAN2003	2. SILBER
15	1000006	1000014	01FEB2003	1. GOLD
16	1000006	1000015	01MAR2003	3. BRONZE
17	1000006	1000016	01APR2003	1. GOLD
18	1000006	1000017	01MAY2003	3. BRONZE
19	1000006	1000018	01JUN2003	3. BRONZE
20	1000006	1000018	01JUL2003	3. BRONZE
21	1000006	1000018	01AUG2003	3. BRONZE

Transactional Data or Snapshot

As we can see from the data examples, only the SCORES and CALLCENTER tables are real transactional tables. When processing data from these tables, we need to make sure that no observations after our snapshot data of 30JUN2003 are used.

The other tables that we consider to be snapshots of 30JUN2003 are the ones in which we do not have to worry about the correct time span any more.

25.4 The Programs

1. We will create aggregations on a subject level of the ACCOUNTS, LEASING, CALLCENTERS, and SCORES tables.

2. We will join the resulting tables together and create a CUSTOMER DATA MART table.

25.4.1 Transformation for ACCOUNTS

With the ACCOUNTS table, we want to perform two aggregations:

- derive sums per customer
- calculate an account type split based on the balance

We use PROC MEANS to create an output data set ACCOUNTMTP that contains both aggregations per customer and aggregations per customer and account type. Note that we have intentionally omitted the NWAY option in order to have the full set of _TYPE_ categories in the output data set.

```
PROC MEANS DATA = accounts NOPRINT;
 CLASS CustID Type;
 VAR Balance Interest;
 OUTPUT OUT = AccountTmp(RENAME = (_FREQ_ = NrAccounts))
             SUM(Balance) = BalanceSum
             MEAN(Interest) = InterestMean;
 RUN;
```

In the next step we use a DATA step with multiple output data sets to create the following:

- ACCOUNTSUM, which contains the aggregations on the CUSTOMER level
- ACCOUNTTYPETMP, which is our basis for the account type split that we need to transpose later on

Note that the creation of two output data sets is very elegant and we can rename the respective variables with a data set option. Based on the value of the _TYPE_ variable, we decide into which data set we output the respective observations.

```
DATA AccountSum(KEEP = CustID NrAccounts BalanceSum InterestMean)
     AccountTypeTmp(KEEP = CustID Type BalanceSum);
 SET AccountTmp;
 Type = COMPRESS(Type);
 IF _TYPE_ = 2 THEN OUTPUT AccountSum;
 ELSE IF _TYPE_ = 3 THEN OUTPUT AccountTypeTmp;
 RUN;
```

Finally, we need to transpose the data set with ACCOUNTTYPES to have a one-row-per-customer structure:

```
PROC TRANSPOSE DATA = AccountTypeTmp
               OUT = AccountTypes(DROP = _NAME_ _LABEL_);
 BY CustID;
 VAR BalanceSum;
 ID Type;
 RUN;
```

25.4.2 Transformation for LEASING

For the leasing data we simply summarize the value and the annual rate per customer with the following statement:

```
PROC MEANS DATA = leasing NWAY NOPRINT;
 CLASS CustID;
 VAR Value AnnualRate;
 OUTPUT OUT = LeasingSum(DROP = _TYPE_
                        RENAME = (_FREQ_ = NrLeasing))
                   SUM(Value) = LeasingValue
                   SUM(AnnualRate) = LeasingAnnualRate;
 RUN;
```

25.4.3 Transformation for CALLCENTER

We first define the snapshot date for the data in a macro variable:

```
%let snapdate = "30JUN2003"d;
```

Then we aggregate the call center data over all customers using the FREQ procedure:

```
PROC FREQ DATA = callcenter NOPRINT;
 TABLE CustID / OUT = CallCenterContacts(DROP = Percent RENAME =
(Count = Calls));
WHERE datepart(date) < &snapdate;
RUN;
```

And we calculate the frequencies for the category COMPLAINT:

```
PROC FREQ DATA = callcenter NOPRINT;
 TABLE CustID / OUT = CallCenterComplaints(DROP = Percent RENAME =
(Count = Complaints));
 WHERE Category = 'Complaint' and datepart(date) < &snapdate;
RUN;
```

25.4.4 Transformation for SCORES

We create three output data sets—one with the value segment in the actual (snapshot date) month, one with the value segment of one month ago, and one with the value segment in a "future" month in order to calculate the target variable.

```
DATA ScoreFuture(RENAME = (ValueSegment = FutureValueSegment))
     ScoreActual
     ScoreLastMonth(RENAME = (ValueSegment = LastValueSegment));
 SET Scores;
 DATE = INTNX('MONTH',Date,0,'END');
 IF Date = &snapdate THEN OUTPUT ScoreActual;
 ELSE IF Date = INTNX('MONTH',&snapdate,-1,'END') THEN OUTPUT
ScoreLastMonth;
 ELSE IF Date = INTNX('MONTH',&snapdate,2,'END') THEN OUTPUT
ScoreFuture;
 DROP Date;
RUN;
```

Note that we have used the INTNX function, which allows the relative definition of points in time in a very elegant way, for various purposes:

- We align the values of DATE in the SCORES table to the END of each month. In this case we actually move the date values that come with the first of the month to the end of each month in order to be able to compare them with our snapshot date, which is set to the end of the month.

- We calculate how many months a certain date is before or after our specified snapshot date. In our case the first INTNX function will evaluate to 01MAY2003, and the second will evaluate to 01AUG2003.

25.4.5 Creation of the Customer Data Mart

Finally, we create the customer data mart by joining all the intermediate result tables to the CUSTOMER table.

First we define a macro variable for the snapshot date in order to have a reference point for the calculation of age. (Note that we repeated this macro variable assignment for didactic purposes here).

```
%let snapdate = "30JUN2003"d;
```

Next we start defining our CUSTOMERMART. We use ATTRIB statements in order to have a detailed definition of the data set variables, by specifying their order, format, and label. Note that we aligned the FORMAT and LABEL statements for a clearer picture.

```
DATA CustomerMart;

ATTRIB /* Customer Baseline */
CustID          FORMAT = 8.    LABEL = "Customer ID"
Birthdate       FORMAT = DATE9. LABEL = "Date of Birth"
Alter           FORMAT = 8.    LABEL = "Age (years)"
Gender          FORMAT = $6.   LABEL = "Gender"
MaritalStatus   FORMAT = $10.  LABEL = "Marital Status"
Title           FORMAT = $10.  LABEL = "Academic Title"
HasTitle        FORMAT = 8.    LABEL = "Has Title? 0/1"
Branch          FORMAT = $5.   LABEL = "Branch Name"
CustomerSince   FORMAT = DATE9. LABEL = "Customer Start Date"
CustomerMonths  FORMAT = 8.    LABEL ="Customer Duration (months)"
;
ATTRIB /* Accounts */
HasAccounts        FORMAT = 8.  LABEL ="Customer has any accounts"
NrAccounts         FORMAT = 8.  LABEL ="Number of Accounts"
BalanceSum         FORMAT = 8.2 LABEL ="All Accounts Balance Sum"
InterestMean       FORMAT = 8.1 LABEL ="Average Interest"
Loan               FORMAT = 8.2 LABEL ="Loan Balance Sum"
SavingsAccount     FORMAT = 8.2 LABEL ="Savings Account Balance Sum"
Funds              FORMAT = 8.2 LABEL ="Funds Balance Sum"
LoanPct            FORMAT = 8.2 LABEL ="Loan Balance Proportion"
SavingsAccountPctFORMAT = 8.2 LABEL ="Savings Account Balance
Proportion"
FundsPct           FORMAT = 8.2 LABEL ="Funds Balance Proportion"
;
ATTRIB /* Leasing */
HasLeasing         FORMAT = 8.  LABEL ="Customer has any leasing
contract"
NrLeasing          FORMAT  = 8. LABEL ="Number of leasing contracts"
LeasingValue       FORMAT = 8.2 LABEL ="Totals leasing value"
LeasingAnnualRate FORMAT = 8.2 LABEL ="Total annual leasingrate"
;
ATTRIB /* Call Center */
HasCallCenter FORMAT =8.   LABEL ="Customer has any call center
contact"
Calls         FORMAT = 8.  LABEL ="Number of call center contacts"
Complaints    FORMAT = 8.  LABEL ="Number of complaints"
ComplaintPct  FORMAT = 8.2 LABEL ="Percentage of complaints"
;
ATTRIB /* Value Segment */
ValueSegment       FORMAT =$10. LABEL ="Current Value Segment"
LastValueSegment   FORMAT =$10. LABEL ="Last Value Segment"
ChangeValueSegment FORMAT =8.2  LABEL ="Change in Value Segment"
Cancel             FORMAT =8.   LABEL ="Customer canceled"
;
```

Next we merge the data set that we created in the previous steps together and create logical variables for some tables:

```
MERGE Customer (IN = InCustomer)
      AccountSum (IN = InAccounts)
    AccountTypes
    LeasingSum (IN = InLeasing)
    CallCenterContacts (IN = InCallCenter)
    CallCenterComplaints
    ScoreFuture(IN = InFuture)
    ScoreActual
    ScoreLastMonth;
  BY CustID;
  IF InCustomer AND InFuture;
```

Missing values that can be interpreted as zero values are replaced with zeros in the next step. Such missing values occur, e.g. in the case of a transposition, if a subject does not have an entry for a certain category. This can be interpreted as a zero value.

```
ARRAY vars {*} Calls Complaints LeasingValue LeasingAnnualRate
               Loan SavingsAccount Funds NrLeasing;
DO i = 1 TO dim(vars);
 IF vars{i}=. THEN vars{i}=0;
 END;
DROP i;
```

Finally, we calculate derived variables for the data mart:

```
/* Customer Baseline */
HasTitle = (Title ne "");
Alter = (&Snapdate-Birthdate)/ 365.2422;
CustomerMonths = (&Snapdate- CustomerSince)/(365.2422/12);

/* Accounts */
HasAccounts = InAccounts;
IF BalanceSum NOT IN (.,0) THEN DO;
   LoanPct = Loan / BalanceSum * 100;
   SavingsAccountPct = SavingsAccount / BalanceSum * 100;
   FundsPct = Funds / BalanceSum * 100;
END;

/* Leasing */
HasLeasing = InLeasing;

/* Call Center */
HasCallCenter = InCallCenter;
IF Calls NOT IN (0,.) THEN ComplaintPct = Complaints / Calls *100;

/* Value Segment */
Cancel = (FutureValueSegment = '8. LOST');
ChangeValueSegment = (ValueSegment = LastValueSegment);
RUN;
```

Note that we calculated the target variable CANCEL from the fact that a customer has the VALUESEGMENT entry '8. LOST'.

We include only those customers in the final customer data mart that are in the CUSTOMER table per June 30, and that are still active at the beginning of the target window. We infer this from the fact that the CUSTID appears in the SCOREFUTURE table.

In the case of the value segment change, we simply compare the current and the last value segment. It would, however, be more practical to differentiate between an increase and a decrease in value segment. This would be more complicated because we would need to consider all segment codes manually. Here it would be more useful if we had the underlying numeric score available, which would allow a simple numeric comparison.

Referring to Section 16.4, "Time Intervals," we see that we used the "dirty method" to calculate customer age. We divided the time span by the average number of days per year (= 365.2422). For this type of analysis this method is sufficient because we need age only as a rough input variable for analysis.

25.4.6 The Resulting Data Set

The resulting data sets are shown in Table 25.6, Table 25.7, and Table 25.8.

Table 25.6: Customer baseline data in the CUSTOMERMART

	Customer ID	Date of Birth	Age (years)	Gender	Marital Status	Academic Title	Has Title? 0/1	Branch Name	Customer Start Date	Customer Duration (months)
1	1000002	26DEC1958	44	Male	Married		0	Fil1	01JAN2000	41
2	1000005	25JUN1947	56	Male	Single	Ing.	1	Fil4	01APR1999	50
3	1000006	10DEC1945	57	Female	Married		0	Fil4	01SEP1996	81
4	1000007	02JUN1934	69	Male	Married		0	Fil1	01SEP1997	69
5	1000008	15DEC1957	45	Male	Single	Dr.	1	Fil3	01JAN1996	89
6	1000009	11MAR1959	44	Male	Single		0	Fil2	01JUL2001	23
7	1000014	23AUG1952	51	Male	Single		0	Fil4	01MAY1996	85
8	1000015	12MAY1959	44	Male	Single		0	Fil2	01FEB1999	52
9	1000016	11FEB1967	36	Male	Married		0	Fil2	01FEB2001	28

Table 25.7: Account data and leasing data in the CUSTOMERMART

	Customer ID	Customer has any accounts	Number of Accounts	All Accounts Balance Sum	Average Interest	Loan Balance Sum	Savings Account Balance Sum	Funds Balance Sum	Loan Balance Proportion	Savings Account Balance Proportion	Funds Balance Proportion	Customer has any leasing contract	Number of leasing contracts	Total leasing value	Total annual leasingrate
1	1000002	1	2	3100.84	5.0	1550.42	0.00	0.00	50.00	0.00	0.00	1	1	521763.0	254.69
2	1000005	1	1	3775.31	6.0	0.00	0.00	0.00	0.00	0.00	0.00	1	1	855215.0	232.52
3	1000006	1	1	2376.43	2.0	0.00	0.00	2376.43	0.00	0.00	100.00	1	1	560362.0	167.37
4	1000007	1	2	3625.44	5.0	0.00	0.00	1812.72	0.00	0.00	50.00	1	2	1735708	168.75
5	1000008	1	1	3350.65	2.0	0.00	0.00	3350.65	0.00	0.00	100.00	1	1	5276.00	109.15
6	1000009	1	3	3575.46	4.0	1191.82	0.00	1191.82	33.33	0.00	33.33	1	2	591963.0	170.14
7	1000014	1	2	3000.92	4.5	0.00	0.00	0.00	0.00	0.00	0.00	1	1	564728.0	92.51
8	1000015	1	1	2801.09	5.0	0.00	0.00	0.00	0.00	0.00	0.00	1	1	393984.0	189.54
9	1000016	1	2	3325.66	1.0	0.00	0.00	0.00	0.00	0.00	0.00	0	0	0.00	0.00

Table 25.8: Call center data, score data, and the target variable in the CUSTOMERMART

	Customer ID	Customer has any call center contact	Number of call center contacts	Number of complaints	Percentage of complaints	Currenty Value Segment	Last Value Segment	Change in Value Segment	Customer cancelled	SavingsAccount	ScoreID	FutureValueSegment
1	1000002	0	0	0	.	2. SILBER	2. SILBER	1.00	1	1550.42	1000005	8. LOST
2	1000005	0	0	0	.	3. BRONZE	3. BRONZE	1.00	0	3775.31	1000011	3. BRONZE
3	1000006	0	0	0	.	3. BRONZE	3. BRONZE	1.00	0	.	1000017	3. BRONZE
4	1000007	0	0	0	.	2. SILBER	1. GOLD	0.00	0	1812.72	1000023	2. SILBER
5	1000008	0	0	0	.	2. SILBER	2. SILBER	1.00	0	.	1000029	2. SILBER
6	1000009	0	0	0	.	3. BRONZE	4. LEAD	0.00	0	1191.82	1000035	3. BRONZE
7	1000014	1	2	2	100.00	3. BRONZE	1. GOLD	0.00	0	3000.92	1000041	3. BRONZE
8	1000015	0	0	0	.	3. BRONZE	4. LEAD	0.00	0	2801.09	1000047	3. BRONZE
9	1000016	0	0	0	.	2. SILBER	2. SILBER	1.00	0	3325.66	1000053	2. SILBER

25.5 The Results and Their Usage

We created a data mart that can be used to predict which customers have a high probability to leave the company and to understand why customers cancel their service. We can perform univariate and multivariate descriptive analysis on the customer attributes. We can develop a predictive model to determine the likelihood customers will cancel.

Note that we used data that we found in our five tables only. We aggregated and transposed some of the data to a one-row-per-subject structure and joined these tables together. The derived variable that we defined gives an accurate picture of the customer attributes and allows as good an inference as possible of customer behavior.

Chapter 26

Case Study 2—Deriving Customer Segmentation Measures from Transactional Data

26.1 The Business Questions

General

In this case study we will examine point-of-sales data from a "do-it-yourself" retail shop. We have a transaction history of sales with one row for each sale item. Additionally, we assume that each customer has a purchasing card so that we can match each purchase to a customer in order to analyze repeating sales events. These data are usually stored in a star schema (see *Chapter 6 – Data Models*) and are the basis for reporting tasks, either in relational or multidimensional form.

We will show how we can use the transactional point-of-sales data to calculate meaningful derived variables for customer segmentation. We will do this by creating derived variables on the CUSTOMER level. These variables can further be used to categorize and segment customers on the basis of one variable or a set of variables.

The following list shows segmentation variables that turned out to be useful for customer segmentation in retail. We will structure the list of variables by so-called "segmentation axes," each segmentation axis being the logical category of a set of variables.

26.1.1 Axis "Customer Demographics"

In this axis we can include customer information such as age, gender, credit risk status, or region.

26.1.2 Axis "Card Usage"

In this axis we will calculate derived variables on the recency and frequency of purchase card usage such as active period, inactive period, days since last purchase, and average number of purchases per month.

26.1.3 Axis "Sale and Profit"

Here we will calculate sales and profit figures such as the total sale, the sales in the last 12 months, or the average sales per year.

26.1.4 Axis "Promotion and Return"

In this axis we will calculate sums and percentages for special events or actions such as promotion periods or product returns.

26.1.5 Axis "Product Usage"

In this axis we will analyze the usage in different product groups.

26.1.6 Relation to the RFM Segmentation

Note that this segmentation list is similar to an RFM segmentation. RFM stands for recency, frequency, monetary. This type of segmentation describes the customers by their interval since the last purchase, the number of purchases in general, and the monetary value of the purchases.

The difference is that here we propose a set of variables for each category in order to allow for a finer selection of segmentation criteria and that we also include information on product usage. In our example we will classify the product usage only by the product's group proportion on the overall sales amount. Another possibility would be to use market basket analysis to include indicators for the most relevant product combinations in the customer segmentation.

26.2 The Data

Figure 26.1 shows the physical data model for the star schema, which contains the data from sales in a do-it-yourself market. We see a fact table with the name POINTOFSALE and three dimension tables: CUSTOMER, PRODUCT, and PROMOTION.

Note that for simplicity, no separate entity is used for the TIME dimension. Furthermore, the PRODUCT dimension is already de-normalized because we already have the names and IDs of PRODUCT_GROUP and PRODUCT_SUBGROUP in one table.

If a product has been sold in a promotion period, a link to the PROMOTION dimension exists. Otherwise, no PROMO_ID is present for the record. In the case of a return of a product from the customer, the QUANTITY has a negative sign.

Figure 26.1: Data model

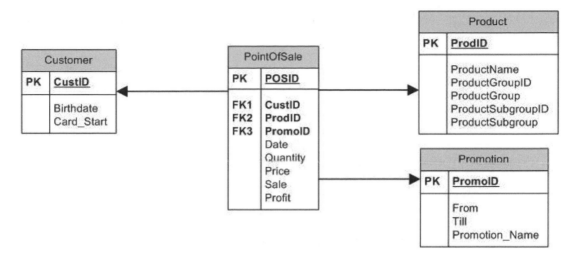

Note that for simplicity the data model is truncated because we do not use the full range of variables that would be available. Tables 26.1 through 26.5 give a short overview of the content of different tables.

Table 26.1: CUSTOMER table with only two attributes per customer

	CustID	Birthdate	Card_Start
1	1	16MAY1970	01JAN2004
2	2	30JUL1948	01FEB2005
3	3	20APR1920	01JUL2002

In Table 26.2, we also have a lookup table for PRODUCTGROUP and PRODUCTSUBGROUP.

Table 26.2: PRODUCT table with a list of products

	ProdID	ProdName	ProductGroupID	ProductSubgroupID
1	10001	Hammer	20	10
2	10012	E-Saw	20	15
3	10207	12V-Battery	40	20
4	13492	Carrot-Seeds	90	25
5	14329	Hosepipe	90	10
6	43003	BBQ-Small	90	20
7	43007	BBQ-Large	90	20

Table 26.3: Lookup tables

	ProductGroupID	ProductGroupName
1	20	Tools
2	40	Car
3	90	Gardening

	ProductGroupID	ProductSubGroupID	ProductSubGroupName
1	20	10	Hand-Tools
2	20	15	E-Tools
3	40	20	Car-Accessories
4	90	25	Garden-Seeds
5	90	10	Garden-Basic
6	90	20	Garden-Leisure

Table 26.4: PROMOTION table that contains a list of promotions

	PromoID	From	Till	PromotionName
1	1	01MAY2003	30JUN2003	Spring2003
2	2	01SEP2003	30NOV2003	Autumn2003

Finally, the fact table POINTOFSALE includes all sale events on a SALE ITEM level.

Table 26.5: PointOfSale data

	PosID	CustID	ProdID	PromoID	Date	Quantity	Price	Sale	Profit
1	26	3	43003	1	01APR03	1	170.00	170.00	23.02
2	27	3	10012	2	15DEC03	1	72.70	72.70	15.66
3	28	3	10001	2	15DEC03	2	5.20	10.40	2.14
4	1	1	43003	.	12APR04	1	170.00	170.00	25.24
5	4	1	14329	.	15APR04	1	23.20	23.20	4.46
6	5	1	14392	.	15APR04	40	23.20	928.00	142.40
7	2	1	43007	.	15APR04	-1	250.00	-250.00	-31.28
8	3	1	43007	.	15APR04	1	250.00	250.00	35.29
9	7	1	10001	1	20APR04	1	5.40	5.40	0.99
10	9	1	10001	1	20APR04	5	5.40	27.00	5.07
11	10	1	10207	.	20APR04	2	60.90	121.80	21.03
12	6	1	14329	1	20APR04	-1	23.20	-23.20	-4.33
13	8	1	14329	.	20APR04	10	23.20	232.00	27.86
14	29	3	14329	1	15MAY04	10	23.20	232.00	32.30
15	14	1	10001	.	16JUN04	5	5.40	27.00	4.18
16	17	1	10001	.	16JUN04	1	5.40	5.40	1.11
17	15	1	10012	.	16JUN04	10	72.70	727.00	104.01
18	12	1	10207	.	16JUN04	-1	60.90	-60.90	-11.55
19	16	1	10207	.	16JUN04	1	60.90	60.90	12.72
20	11	1	14329	1	16JUN04	1	23.20	23.20	4.50
21	13	1	14329	.	16JUN04	1	23.20	23.20	4.16
22	18	2	14329	.	10APR05	1	23.20	23.20	4.52
23	19	2	14329	.	10APR05	1	23.20	23.20	2.83
24	20	2	14329	.	10APR05	1	23.20	23.20	3.60
25	22	2	14329	1	10APR05	3	23.20	69.60	11.55
26	21	2	43003	.	10APR05	1	170.00	170.00	26.96
27	23	2	10001	2	15MAY05	2	5.40	10.80	1.45
28	24	2	13492	.	15MAY05	10	1.20	12.00	1.62
29	25	2	13492	.	15MAY05	10	1.20	12.00	2.38
30	30	3	13492	1	12JUN05	2	1.20	2.40	0.46

26.3 The Programs

General

We will now show how we can start from this data source to create a customer table that contains customer information in a one-row-per-subject structure.

26.3.1 Creating Formats

Starting from the preceding example we first create a format for PRODUCTGROUP and PRODUCTSUBGROUP directly from the data sets. Note that in the case of "real world" retail data the list of products is very long and the option to create a format from a data set is helpful.

```
data FMT_PG(rename =(ProductGroupID=start ProductGroupName=label));
 set ProductGroups end=last;
 retain fmtname 'PG' type 'n';
run;
PROC format library=work cntlin=FMT_PG;
run;

data FMT_PSG(rename =(ProductSubGroupName=label));
 set ProductSubGroups end=last;
 retain fmtname 'PSG' type 'n';
 Start = ProductGroupID*100+ProductSubGroupID;
run;
PROC format library=work cntlin=FMT_PSG; run;
```

In our example for simplicity we will use the format for product group only. If we wanted to use the format for product subgroup we would have to calculate the concatenation of PRODUCTGROUP and PRODUCTSUBGROUP in the data.

26.3.2 Calculating Derived Variables for the Segmentation Axes

First we calculate the sum of sales and profit on the CUSTOMER level and also calculate the first and last usage dates of the purchase card. Note that we do this by using PROC MEANS, the CLASS statement, and the NWAY option, and by specifying explicit statistics and names for each measure. We could also do this by using PROC SQL and a respective SQL statement.

```
PROC MEANS DATA = PointOfSale NOPRINT NWAY;
 CLASS CustID;
 VAR Date Sale Profit;
 OUTPUT OUT = Cust_POS(DROP = _TYPE_ _FREQ_)
                     SUM(Sale)=Total_Sale
                     SUM(Profit)=Total_Profit
                     MAX(Date)=LastVisit
                     MIN(Date)=FirstVisit
 ;
 RUN;
```

In the next step we calculate the number of distinct visit days for each customer:

```
*** Number Visits;
PROC SQL;
 CREATE TABLE Visit_Days AS
 SELECT CustID, COUNT(DISTINCT date) AS Visit_Days
 FROM PointOfSale
 GROUP BY CustID
 ORDER BY CustID;
QUIT;
```

In this step we calculate the amount of sales in a promotion period. Note that the fact that an article was bought within a promotion period is indicated by the presence of a PROMOID value.

```
*** Promotion Proportion;
PROC MEANS DATA = PointOfSale NOPRINT NWAY;
 CLASS CustID;
 VAR Sale;
 WHERE PromoID NE .;
 OUTPUT OUT = Cust_Promo(DROP = _TYPE_ _FREQ_) SUM(Sale)=Promo_Amount;
RUN;
```

In the same fashion we also calculate the number of returned products, which are indicated by a negative quantity value.

```
*** Return ;

PROC MEANS DATA = PointOfSale NOPRINT NWAY;
 CLASS CustID;
 VAR Sale;
 WHERE Quantity < 0;
 OUTPUT OUT = Cust_Return(DROP = _TYPE_ _FREQ_)
SUM(Sale)=Return_Amount;
RUN;
```

Note that in both cases we immediately drop the _TYPE_ and _FREQ_ variables. We could keep the _FREQ_ variable in the number of returned or promotion items that would be needed in segmentation.

26.3.3 Calculating Derived Variables for Product Usage

For product usage we calculate sale sums per product group:

```
PROC SQL;
 CREATE TABLE Cust_PG_tmp AS
 SELECT CustID,
        ProductGroupID FORMAT=PG.,
      SUM(Sale) AS Sale
 FROM PointOfSale AS a,
      Product AS b
 WHERE a.ProdID = b.ProdID
 GROUP BY CustID, ProductGroupID
 ORDER BY CustID, ProductGroupID;
QUIT;
```

And transpose the resulting data set to a one-row-per-subject table:

```
PROC TRANSPOSE DATA = Cust_PG_tmp
                OUT = Cust_PG(DROP = _NAME_);
 BY CustID;
 VAR Sale;
 ID ProductGroupID;
RUN;
```

26.3.4 Creating the Segmentation Data Mart

In the last step we join all resulting tables together and create the segmentation data mart. Note that in this case we use the term data mart for a single table.

We define a macro variable with the snapshot date, which we will use in the calculation of the current age:

```
%LET SnapDate = "01JUL05"d;
```

We also extensively define the variables that are available in the segmentation data mart with the ATTRIB statement. Thus, we have a common place for the definition of variables and a good sorting place for the variables in the data mart:

```
DATA CustMart;

ATTRIB /* Customer Demographics */
CustID        FORMAT=8.
Birthdate     FORMAT=DATE9. LABEL ="Date of Birth"
Age                     FORMAT=8.          LABEL ="Age (years)"
;
ATTRIB /* Card Details */
Card_Start    FORMAT=DATE9. LABEL ="Date of Card Issue"
AgeCardMonths FORMAT=8.1        LABEL ="Age of Card (months)"
AgeCardIssue FORMAT=8.          LABEL ="Age at Card Issue (years)"
;

ATTRIB /* Visit Frequency and Recency */
FirstVisit              FORMAT=DATE9. LABEL = "First Visit"
LastVisit               FORMAT=DATE9. LABEL = "Last Visit"
Visit_Days              FORMAT=8.     LABEL = "Nr of Visit Days"
ActivePeriod            FORMAT=8.     LABEL = "Interval of Card Usage
(months)"
MonthsSinceLastVisit FORMAT = 8.1  LABEL = "Months since last visit
(months)"
VisitDaysPerMonth     FORMAT = 8.2  LABEL = "Average Visit Days per
Month"
;

ATTRIB /* Sale and Profit */
Total_Sale     FORMAT = 8.2 LABEL = "Total Sale Amount"
Total_Profit   FORMAT = 8.2 LABEL = "Total Profit Amount"
ProfitMargin   FORMAT = 8.1 LABEL = "Profit Marin (%)"
SalePerYear    FORMAT = 8.2 LABEL = "Average Sale per Year"
ProfitPerYear  FORMAT = 8.2 LABEL = "Average Profit per Year"
SalePerVisitDay FORMAT = 8.2 LABEL = "Average Sale per VisitDay";

ATTRIB /* Promotion and Return */
BuysInPromo    FORMAT = 8.  LABEL = "Buys in Promotion ever"
```

```
Promo_Amount    FORMAT = 8.2 LABEL = "Sale Amount in Promotion"
PromoPct        FORMAT = 8.2 LABEL ="Sale Proportion (%) in promotion"
ReturnsProducts FORMAT = 8.  LABEL= "Returns Products ever"
Return_Amount   FORMAT = 8.2 LABEL = "Returned Sale Amount"
ReturnPct       FORMAT = 8.2 LABEL = "Return Amount Proportion (%) on
sale"
;

ATTRIB /* Product Groups */
Tools         FORMAT = 8.2 LABEL = "Sale in PG Tools"
Car           FORMAT = 8.2 LABEL = "Sale in PG Car"
Gardening     FORMAT = 8.2 LABEL = "Sale in PG Gardening"
ToolsPct      FORMAT = 8.2 LABEL = "Sale Proportion (%) in PG Tools"
CarPct        FORMAT = 8.2 LABEL = "Sale Proportion (%) in PG Car"
GardeningPct  FORMAT = 8.2 LABEL = "Sale Proportion (%) in PG
Gardening"
;
```

Next we merge all the tables that we created before with the CUSTOMER table and create logical variables for some data sets:

```
MERGE Customer(IN = IN_Customer)
      Cust_pos
      Visit_days
      Cust_promo(IN = IN_Promo)
      Cust_return(IN = IN_Return)
      Cust_pg;
BY CustID;
IF IN_Customer;
```

In the next step we create an ARRAY statement to replace missing values for some variables:

```
ARRAY vars {*} tools car gardening Promo_Amount Return_Amount ;
DO i = 1 TO dim(vars);
 IF vars{i}=. THEN vars{i}=0;
END;
DROP i;
```

Finally, we calculate derived variables from the existing variables:

```
/* Customer Demographics */
Age = (&snapdate - birthdate)/365.25;

/* Card Details */
AgeCardIssue = (Card_Start - birthdate)/365.25;
AgeCardMonths = (&snapdate - Card_Start)/(365.25/12);

/* Visit Frequency and Recency */
VisitDaysPerMonth = Visit_Days / AgeCardMonths;
MonthsSinceLastVisit = (&snapdate - LastVisit)/(365.25/12);
ActivePeriod = (LastVisit - FirstVisit)/(365.25/12);

/* Sale and Profit */
SalePerYear = Total_Sale * 12 / AgeCardMonths;
SalePerVisitDay = Total_Sale / Visit_Days;
ProfitPerYear = Total_Profit * 12 / AgeCardMonths;
ProfitMargin = Total_Profit / Total_Sale * 100;
```

```
/* Promotion and Return */
BuysInPromo = In_Promo;
PromoPct = Promo_Amount / Total_Sale *100;
ReturnsProducts = In_Return;
ReturnPct = -Return_Amount / Total_Sale *100;

/* Product Groups */
ToolsPct = Tools/ Total_Sale *100;
CarPct = Car / Total_Sale *100;
GardeningPct = Gardening/ Total_Sale *100;

RUN;
```

Note that for a long list of product groups using an ARRAY statement is preferable.

Finally, we receive a data mart, which is shown in Tables 26.6 through 26.8.

Table 26.6: The variables of the axis: customer demographics, card details, and visit frequency and recency

	CustID	Date of Birth	Age (years)	Date of Card Issue	Age of Card (months)	Age at Card Issue (years)	First Visit	Last Visit	Nr of Visit Days	Interval of Card Usage (months)	Months since last visit (months)	Average Visit Days per Month
1		16MAY1970	35.13	01JAN2004	17.97	33.63	12APR2004	16JUN2004	4	2.14	12.48	0.22
2	2	30JUL1948	56.92	01FEB2005	4.93	56.51	10APR2005	15MAY2005	2	1.15	1.54	0.41
3	3	20APR1920	85.20	01JUL2002	36.01	82.20	01APR2003	12JUN2005	4	26.38	0.62	0.11

Table 26.7: The variables of the axis: sales and profit, promotion, and return

	CustID	Total Sale Amount	Total Profit Amount	Profit Marin (%)	Average Sale per Year	Average Profit per Year	Average Sale per VisitDay	Buys in Promotion ever	Sale Amount in Promotion	Sale Proportion (%) in promotion	Returns Products ever	Returned Sale Amount	Return Amount Proportion (%) on sale
1	1	2290.00	345.87	15.10	1529.11	230.95	572.50	1	32.40	1.41	1	-334.10	14.59
2	2	344.00	54.89	15.96	837.64	133.66	172.00	1	80.40	23.37	0	0.00	0.00
3	3	487.50	73.58	15.09	162.46	24.52	121.88	1	487.50	100.00	0	0.00	0.00

Table 26.8: The variables of the axis: product groups

	CustID	Sale in PG Tools	Sale in PG Car	Sale in PG Gardening	Sale Proportion (%) in PG Tools	Sale Proportion (%) in PG Car	Sale Proportion (%) in PG Gardening
1	1	791.80	121.80	448.40	34.58	5.32	19.58
2	2	10.80	0.00	333.20	3.14	0.00	96.86
3	3	83.10	0.00	404.40	17.05	0.00	82.95

26.4 The Results and Their Usage

Examples

The preceding results can be used in a variety of ways:

- Reporting and descriptive customer analytics: The data mart can be used to answer questions such as the following:

  □ How is the distribution of customer age of active and inactive customers?

  □ What is the average time interval since a customer last used his purchase card?

  □ How many customers buy more than 20% of their sales in promotions?

The advantage of this data representation is that the preceding questions can be answered without complex joins and subqueries to other tables.

- Data mining and statistical analysis: The data mart can be the input table for data mining analyses to answer questions such as the following:

  □ Which customer attributes are correlated with the fact that a customer buys products in certain product groups or has a high product return rate?

  □ In which clusters can my customers be segmented based on certain input variables?

- Marketing segmentation and selections for direct marketing.

  □ The data mart can be used as the basis for selection of customers for direct marketing campaigns. Because relevant attributes are already pre-calculated the marketer does not need to perform complex joins.

  □ Additional multivariate segmentations such as shopping intensity in the active period compared to the length of the inactive period can help to identify "sleeping potentials" or other relevant marketing groups.

Further variables

Note that the list of variables can be easily increased if we start considering the time axis. For example,

- sales in the last 12 months (last year)
- sales in the second to the last year
- difference between these two figures
- trends in product usage over time

Taking into account the types of derived variables we discussed in *Part 3 – Data Mart Coding and Content*, we can quickly explode our data mart to hundreds of variables. Again we need to mention that it is not the technical ability to create new variables that is important but rather their business relevance to fulfill the underlying business goal.

Chapter 27

Case Study 3—Preparing Data for Time Series Analysis

27.1 Introduction

In this case study we will show how we prepare data for time series analysis. We will use example data from a wholesale trader in the computer hardware industry. In this case study we will show how we can use these data to create longitudinal data marts that allow time series analysis on various hierarchical levels.

Different from the two previous case studies where we created one-row-per-subject data marts with a number of derived variables, the focus in this case study is to prepare the data on the appropriate aggregation level with a few derived variables in order to perform time series analysis.

In this chapter we will show the following:

- An exemplary data model for demand forecast data with hierarchies. This data model is the basis for our analysis data.
- SAS programs that move from transactional data to the most detailed aggregation.
- SAS programs that create aggregations on various hierarchical levels.
- The differences in performance and usability between PROC SQL aggregations and PROC MEANS or PROC TIMESERIES aggregations.
- SAS programs that create *derived variables* for time series modeling.

In this case study we will see the following:

- That using SAS formats instead of full names saves run time and disk space.
- That PROC SQL can aggregate and join tables in one step. It is, however, slower than the corresponding DATA step and SAS procedure steps.

27.2 The Business Context

General

Our business context is the wholesale computer industry. We consider a manufacturer of hardware components for computers. The manufacturer runs shops in various countries that sell products to his customers (re-sellers). These customers then re-sell the hardware components to consumers. Each shop belongs to a sales organization, which is responsible for a number of shops.

The shops pass through the re-seller orders to the central stock. From this central stock, the products are sent directly to the re-sellers.

For each order, we know the shop that has taken the order, the day, the product, the order type the amount, and the price. Note that because a shop can get several orders per day for a product, we can have duplicate lines per SHOP, PRODUCT, DAY, and ORDERTYPE in the data.

27.2.1 The Business Question

The business question is "Should we perform a monthly demand forecast for each product for 12 months ahead?" This demand forecast is needed on the following hierarchical levels:

- overall
- SALESORGANIZATION
- SHOP

27.2.2 Usage of the Results

These forecasts are used to periodically plan the number of produce units for each product. The forecasts are also used once a year in budget planning to plan the sales organization's sales figures for the next year.

27.3 The Data

Tables

We have the following tables as the data basis for our case study:

- Sales organizations
- Shops
- Orders
- OrderType
- ProductGroups
- Products

The physical data models for these six tables are shown in Figure 27.1.

Figure 27.1: The data model

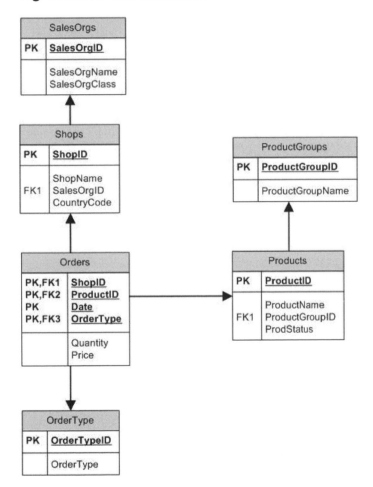

27.3.1 Table Content

Tables 27.1 through 27.5 show the content of the respective tables.

Table 27.1: SALESORGS table

	SalesOrgID	SalesOrgName	SalesOrgClass
1	A	SalesOrganization A	B
2	B	SalesOrganization B	B
3	C	SalesOrganization C	A
4	D	SalesOrganization D	A
5	E	SalesOrganization E	B
6	F	SalesOrganization F	B
7	G	SalesOrganization G	B
8	H	SalesOrganization H	B
9	I	SalesOrganization I	A

Table 27.2: SHOPS table

	ShopID	ShopName	CountryCode	SalesOrgID
1	560	Shop1	MX	M
2	580	Shop2	CH	C
3	590	Shop3	SE	S
4	620	Shop4	HU	H
5	630	Shop5	DK	D
6	650	Shop6	PL	P
7	1060	Shop7	AT	A
8	1100	Shop8	IN	I
9	1150	Shop9	TH	T
10	1200	Shop10	US	U
11	1210	Shop11	TH	T
12	1250	Shop12	CH	C
13	1260	Shop13	CH	C
14	1261	Shop14	CH	C
15	1371	Shop15	PL	P

Table 27.3: ORDERS table

	ShopID	ProductID	Date	Quantity	Price	OrderType
1	580	13101	01SEP2004	8	45.92	2
2	580	13101	02SEP2004	7	45.94	1
3	580	13101	04SEP2004	13	45.98	1
4	580	13101	05SEP2004	4	46.00	1
5	580	13101	07SEP2004	6	46.04	1
6	580	13101	08SEP2004	8	46.06	2
7	580	13101	10SEP2004	7	46.10	2
8	580	13101	11SEP2004	7	46.11	1
9	580	13101	13SEP2004	4	46.15	2
10	580	13101	14SEP2004	5	46.17	1
11	580	13101	15SEP2004	3	46.19	2
12	580	13101	17SEP2004	7	46.22	2
13	580	13101	18SEP2004	5	46.24	1
14	580	13101	18SEP2004	5	46.24	1
15	580	13101	19SEP2004	8	46.25	2
16	580	13101	20SEP2004	6	46.27	2
17	580	13101	21SEP2004	5	46.29	2

Table 27.4: PRODUCTGROUPS table

	ProductGroupID	ProductGroupName
1	16	Printers
2	20	Disk-Drives
3	21	Monitors
4	23	Motherboards
5	26	Keyboards

Table 27.5: PRODUCTS table

	ProdStatus	ProductID	ProductName	ProductGroupID
1	Exist	11350	Product1	26
2	Exist	13101	Product2	26
3	Exist	13105	Product3	16
4	Exist	120664	Product4	26
5	Exist	152723	Product5	16
6	Exist	159652	Product6	16
7	Exist	184507	Product7	16
8	Exist	192605	Product8	16
9	Exist	195995	Product9	16
10	Exist	200357	Product10	16
11	New	206399	Product11	16
12	Exist	213335	Product12	16
13	Exist	216055	Product13	16
14	Exist	222596	Product14	16
15	Exist	237951	Product15	16
16	Exist	244687	Product16	21
17	Exist	251463	Product17	16
18	Exist	256912	Product18	26
19	Exist	268179	Product19	16
20	New	270611	Product20	16

The table ORDERTYPE has only two observations, which are represented by the following SAS format:

```
PROC format lib=sales;
 VALUE OrdType 1 = 'Direct' 2 = 'Indirect';
RUN;
```

27.4 From Transactional Data to the Most Appropriate Aggregation

General

In Table 27.3, lines 13 and 14 have duplicate rows. As mentioned in Section 27.2, "The Business Context," this can come from the fact that one shop receives more than one order for a product per day.

In *Chapter 10 – Data Structures for Longitudinal Analysis*, we saw that transactional data can have observations that can be aggregated to derive the *finest granularity*. In our case the aggregation to the finest granularity level would be DATE, SHOP, PRODUCT, and ORDERTYPE. This means that we would summarize rows 13 and 14 in Table 27.3 by summing QUANTITY to 10 and averaging the two prices to 46.24.

However, the finest granularity level would not solve our business question, because we need the data on the MONTHLY level for monthly forecasts. Therefore, we skip the creation of the finest granularity level and aggregate the data to the *most appropriate aggregation level*, which is SHOP, PRODUCT, ORDERTYPE, and MONTH.

27.4.1 Coding the Aggregation with PROC SQL

With the following code, the data from the preceding tables can be merged and aggregated to a single table:

```
PROC SQL;
 CREATE TABLE sales.ordermart
 AS
 SELECT SalesOrgClass,
      so.SalesOrgID     Format = $so_fmt.,
      CountryCode,
      s.ShopID          Format = shop.,
      p.ProductID       Format = product.,
      p.ProductGroupID Format = pg.,
      OrderType         Format = OrdType.,
      ProdStatus,
      Put(Date, AS MonthYear,
      SUM(quantity) AS Quantity Format = 8.,
      AVG(Price)    AS AvgPrice Format = 8.2
  FROM sales.products     AS p,
       sales.shops        AS S,
       sales.salesorgs    AS SO,
       sales.orders       AS O,
       sales.productgroups AS PG
```

```
WHERE p.productid       = o.productid
  AND s.shopid          = o.shopid
  AND so.salesorgid     = s.salesorgid
  AND p.productgroupid = pg.productgroupid
GROUP BY
      MonthYear,
      so.SalesOrgClass,
      so.SalesOrgID,
      CountryCode,
      s.ShopID,
      p.ProductID,
      p.ProductGroupID,
      OrderType ,
      ProdStatus
  ORDER BY
      s.ShopID,
      p.ProductID,
      OrderType,
      MonthYear;
QUIT;
```

Note the following:

- In order to aggregate from daily data to monthly data we use the following statement: `put(date,yymmp7.) AS MonthYear`. We mentioned in *Chapter 21 – Data Preparation for Multiple-Rows-per-Subject and Longitudinal Data Marts,* that the sole use of FORMAT = yymmp7. is not sufficient in PROC SQL for aggregation levels and that the PUT function can be used. In this case, however, we have a character formatted variable where no DATE functions can be applied. An alternative is to use the following code: `mdy(month(date),1,year(date)) as MonthYear format = yymmp7`.

- In the SELECT clause, we use the variables SHOPID, PRODUCTID, and so on, instead of SHOPNAME and PRODUCTNAME. To display the names instead of the ID, we specify the appropriate format in the FORMAT = expression. The reason is that we decrease both disk space and run time when using IDs and formats. In our example with 732,432 observations in the ORDERS table and 52,858 observations in the aggregated table, the difference in run time and necessary disk space is shown in Table 27.6.

Table 27.6: Disk space and run time

	Disk space of the resulting data	Run time of PROC SQL
Using IDs and SAS formats	4601 kB	32 sec
Using full names	24881 kB	1 min 52 sec

Because we are looking at only a small example and already observe these differences, it is obvious that it makes sense to use the ID and SAS formats method. For this method, however, SAS formats are needed. Their creation is shown in the next section.

27.4.2 Creating the Appropriate SAS Formats

In order to run the preceding code, the appropriate SAS formats need to be created first. This can be done with the following statements. As shown in the case study, it makes sense to create the SAS formats from the data sets.

To create the format for SALESORGS:

```
data salesorgfmt(rename =(salesorgid=start salesorgname=label));
 set sales.salesorgs end=last;
 retain fmtname 'SO' type 'c';
run;
PROC format library= sales cntlin=salesorgfmt;
run;
```

To create the format for SHOPS:

```
data shopfmt(rename =(shopid=start shopname=label));
 set sales.shops end=last;
 retain fmtname 'shop' type 'n';
run;
PROC format library= sales cntlin=shopfmt;
run;
```

To create the format for PRODUCTS:

```
data productfmt(rename =(productid=start productname=label));
 set sales.products end=last;
 retain fmtname 'product' type 'n';
run;
PROC format library=sales cntlin=productfmt;
run;
```

To create the format for PRODUCTGROUPS:

```
data productgroupfmt(rename =(productgroupid=start
productgroupname=label));
 set sales.productgroups end=last;
 retain fmtname 'pg' type 'n';
run;
PROC format library= sales cntlin=productgroupfmt;
run;
```

27.4.3 Using SAS DATA Step and Procedure Code instead of PROC SQL

An alternative to using PROC SQL as we showed earlier is to perform the aggregations with SAS procedures and to perform the table joins with DATA step merges. We will show this option here. We will see that this version saves run time. However, it makes a lot of intermediate steps necessary because we cannot perform a merge on different BY variables in one DATA step.

In our example we assume that we do not have indexes created in our BY variables and we do not use a SAS Scalable Performance Data Engine library; therefore, we need to explicitly sort the data before merging.

First we merge the SHOPS and SALESORGS tables:

```
PROC sort data = sales.shops out= shops (drop=productname);
by salesorgid ;run;
PROC sort data = sales.salesorgs out= salesorgs (drop=salesorgname);
by salesorgid ;run;
data s_so;
 merge shops salesorgs;
 by salesorgid;
run;
```

Then we merge the PRODUCTS and PRODUCTGROUPS tables:

```
PROC sort data = sales.products out=products(drop=productname); by
productgroupid ;run;
PROC sort data = sales.productgroups
out=productgroups(drop=productgroupname); by productgroupid ;run;
data p_pg;
 merge products productgroups;
 by productgroupid;
run;
```

Then we aggregate the data on a SHOP, PRODUCT, ORDERTYPE, and MONTH level using PROC MEANS:

```
PROC means data = sales.orders nway noprint
             ;
class shopid productid ordertype date;
var quantity price;
format date yymmp7. price 8.2 quantity 8.;
output out = orders_month mean(price)=price
                          sum(quantity)=quantity;
run;
```

Next we merge the results of the aggregation with the results of the PRODUCTS and PRODUCTGROUPS merge:

```
PROC sort data = p_pg; by productid; run;
PROC sort data = orders_month ; by productid ;run;
data p_pg_o;
 merge p_pg orders_month(in=in2);
 by productid;
 if in2;
run;
```

And finally we merge these results with the results of the SHOPS and SALESORGS merge:

```
PROC sort data =  s_so ; by shopid ;run;
PROC sort data =  p_pg_o ; by shopid productid ordertype date;run;
data sales.ordermart_DATA step;
 merge s_so p_pg_o(in=in2);
 by shopid;
 if in2;
run;
```

Note that we created a lot of intermediate steps and temporary data sets. This version, however, takes only 5 seconds of real time, whereas the PROC SQL version needed 32 seconds. The reasons for the difference are that PROC MEANS is extremely fast in data aggregation, and after the aggregation the joins are done only on data sets with approximately 50,000 observations. Also note that these performance tests were performed on a laptop with one CPU, so parallelization of PROC MEANS compared to PROC SQL is not the reason for the performance difference.

27.5 Comparing PROC SQL, PROC MEANS, and PROC TIMESERIES

General

In order to compare the performance of PROC SQL, PROC MEANS, and PROC TIMESERIES, a data set ORDER10 was created, where the number of observations was multiplied by 10. This results in 7,324,320 observations. PRODUCTIDs have amended, that way there are 264,290 distinct combinations of PRODUCT, SHOP, ORDERTYPE, and DATE.

In order to compare the performance on sorted and unsorted data, a sorted and a randomly sorted version of ORDER10 was created.

Results

Table 27.7 shows the real run time for the different scenarios.

Table 27.7: Real run time for different aggregation scenarios

Method	Sorted data	Randomly sorted data
PROC SQL	1 min 04 sec	1 min 44 sec
PROC MEANS	24 sec	46 sec
PROC TIMESERIES	29 sec	1 min 59 sec (including 1 min 30 sec for PROC SORT)

27.5.1 Summary

The characteristics of the different methods are as follows. A plus sign (+) indicates an advantage; a minus sign (–) indicates a disadvantage.

PROC SQL

+ Allows implicit sorting; data do not need to be sorted in advance
+ Syntax is understood by programmers who are unfamiliar with SAS
+ Allows to define the aggregation and merge in a VIEW in order to save disk space
+ Allows to join and aggregate tables in one call
– Inferior performance compared to SAS procedures

PROC MEANS

+ Allows implicit sorting; data do not need to be sorted in advance
+ Excellent performance
+ Allows to create several aggregations for combinations (permutations) of CLASS variables
+ SAS formats can be used
– Can operate on a single table only; data need to be joined in advance

PROC TIMESERIES

+ Excellent performance on sorted data
+ Allows to specify aggregation levels for the time variable
+ SAS formats can be used
+ Provides a number of aggregation options such as handling of missing values or alignment of intervals
+ Allows the calculation of time series measures such as trends, seasons, or autocorrelations during data aggregation
– Data need to be sorted
– Can operate on a single table only; data need to be joined in advance
– Runs only with a license of SAS/ETS

27.5.2 Code for PROC TIMESERIES

In order to illustrate the usage of PROC TIMESERIES, we include the syntax for the preceding example here:

```
PROC timeseries data = orders10_sort out = aggr ;
 BY shopid productid ordertype ;
 ID date INTERVAL = MONTH  ACCUMULATE=TOTAL;
 var quantity /ACCUMULATE=TOTAL;
 var price    /ACCUMULATE=average;;
 format date yymmp7. price 8.2 quantity 8. shopid shop.
        productid   product. ordertype ordtype.;
run;
```

Note the following:

- The aggregation levels (except the TIME dimension) have to be specified in a BY statement.
- The ID statement allows for specifying the aggregation level for the time variable, the default accumulation level for the variables, and the treatment of missing values.
- If different aggregation statistics are used for different variables, each variable has to be specified in a separate VAR statement with its ACCUMULATE = option.

27.6 Additional Aggregations

General

With the aggregation ORDERMART that we created earlier, we have the data in an appropriate form to perform time series analysis on the levels PRODUCT, SHOP, ORDERTYPE, and MONTH. This means that in PROC AUTOREG or PROC UCM, for example, we perform a time series analysis on the cross-sectional BY groups PRODUCT, SHOP, and ORDERTYPE.

In our case, however, the aggregation ORDERMART is only the starting point for additional aggregations that we need in order to answer our business question. Earlier we defined that the business question in our case study is to produce monthly demand forecasts on the following levels:

- OVERALL
- SALESORGANIZATION
- SHOP

27.6.1 Aggregations for Overall Demand

The aggregations for overall demand can be performed in the same way as we showed earlier with PROC SQL, PROC MEANS, and PROC TIMESERIES. Here, we will show only a short code example for PROC SQL and PROC MEANS.

To create a table with aggregation with PROC SQL:

```
PROC SQL;
Create table sales.PROD_MONTH
 as
 select
  ProductID AS ProductID format = product.,
  ProductGroupID AS ProductGroupID format = PG.,
  monthyear,
  sum(quantity) as Quantity format=8.,
  avg(avgprice) as Price format = 8.2
 from sales.ordermart
 group by monthyear, ProductID, productgroupid
 order by ProductID, monthyear;
QUIT;
```

Note the following:

- We also use PRODUCTGROUPID in the SELECT clause in order to copy its values down to the PRODUCT level.
- FORMAT = is used to assign the appropriate format to the values.
- The PRICE is not summed but averaged in order to copy its value down to other aggregation levels.

We could simply change the CREATE clause in order to create a view:

```
PROC SQL;
Create VIEW sales.PROD_MONTH_VIEW
 as
 select
  ProductID AS ProductID format = product.,
  ProductGroupID AS ProductGroupID format = PG.,
  monthyear,
  sum(quantity) as Quantity format=8.
 from sales.ordermart
 group by monthyear, ProductID, productgroupid
 order by ProductID, monthyear;
QUIT;
```

This has the advantage that the aggregation is not physically created and therefore disk space is saved. In contrast, however, all analyses on this VIEW take longer because the VIEW has to be resolved. The use of views is of interest if a number of similar aggregations that produce tables with a lot of resulting observations will be created for preliminary analysis, and access to a single view is rarely used.

Alternatively, the aggregation can be performed with PROC MEANS:

```
PROC MEANS DATA = sales.ordermart NOPRINT NWAY;
 CLASS productid productgroupid monthyear;
 VAR quantity avgprice;
 OUTPUT OUT = sales.PROD_MONTH(DROP = _TYPE_ _FREQ_)
               SUM(quantity) = Quantity
               MEAN(avgprice) = Price;
RUN;
```

The resulting table, which is the same for both versions, is shown in Table 27.8.

Table 27.8: PROD_MONTH table

	ProductID	ProductGroupID	MonthYear	Quantity	Price
1	Product1	Keyboards	2003.01	1009	940.18
2	Product1	Keyboards	2003.02	1250	940.18
3	Product1	Keyboards	2003.03	1670	940.18
4	Product1	Keyboards	2003.04	1863	940.18
5	Product1	Keyboards	2003.05	2525	940.18
6	Product1	Keyboards	2003.06	2308	940.18
7	Product1	Keyboards	2003.07	2772	940.18
8	Product1	Keyboards	2003.08	2632	940.18
9	Product1	Keyboards	2003.09	2199	940.18
10	Product1	Keyboards	2003.10	2360	940.18
11	Product1	Keyboards	2003.11	2416	940.18
12	Product1	Keyboards	2003.12	2315	940.18

The remaining aggregations to answer the business question can be created analogously.

27.6.2 Potential Aggregation Levels

Considering the hierarchy *SALES ORGANIZATION* with the dimensions SALESORGCLASS, COUNTRYCODE, SALESORG, SHOP or the hierarchy *TIME* with the values YEAR, QUARTER, and MONTH, we see that a high number of potential combinations of aggregation levels exist. Not all will be relevant to the business question. If, however, the need exists to create a number of different aggregations, it makes sense to plan the names of the aggregations in order to avoid unstructured naming such as PROD1, PROD2, and so forth.

A potential way to do this is to list the relevant aggregations, as in the following example:

```
PROD_MONTH                 ProductID * Month
PROD_MONTH_SORGCLASS       ProductID * SalesOrgClass * Month
PROD_MONTH_CTRY            ProductID * SalesOrgClass * CountryCode *
                           Month
PROD_MONTH_SALORG          ProductID * SalesOrgClass * CountryCode *
                           SalesOrgID * Month
PROD_MONTH_SHOP            ProductID * SalesOrgClass * CountryCode *
                           SalesOrgID * ShopID * Month
PROD_MONTH_SHOP_ORDTYPE    ProductID * SalesOrgClass * CountryCode *
                           SalesOrgID * ShopID * OrderType * Month
PID_MONTH                  ProductGroupID * Month
PROD_QTR                   ProductID * Quarter
```

27.6.3 Calculating New Time Intervals in PROC SQL

The last aggregation (ProductID * Quarter) can be performed easily in PROC MEANS or PROC TIMESERIES:

```
PROC MEANS DATA = sales.ordermart NOPRINT NWAY;
 CLASS productid productgroupid monthyear;
 FORMAT monthyear YYQ5.;
 VAR quantity avgprice;
 OUTPUT OUT = sales.PROD_QTR(DROP = _TYPE_ _FREQ_)
              SUM(quantity) = Quantity
              MEAN(avgprice) = Price;
RUN;
```

If we want to use this in PROC SQL, we need to use the PUT function as we mentioned in Section 21.4, "Aggregating at Various Hierarchical Levels." An alternative would be in the case of quarters to calculate the relevant months with an expression:

```
mdy(ceil(month(monthyear)/3*3),1,year(monthyear))
                              AS QTR_YEAR FORMAT = yyq7.
```

Note that in this case the calculation is redundant because a SAS format exists. It is shown for didactic purposes only, that a clever combination of functions can be an alternative. The result is shown in Table 27.9.

Table 27.9: PROD_QTR table with quarterly aggregations

	ProductID	QTR_YEAR	Quantity
1	Product1	2003Q1	3929
2	Product1	2003Q2	6695
3	Product1	2003Q3	7603
4	Product1	2003Q4	7092
5	Product1	2004Q1	7392
6	Product1	2004Q2	8085
7	Product1	2004Q3	9434
8	Product1	2004Q4	7404
9	Product1	2005Q1	3021

27.7 Derived Variables

General

In the introduction to this chapter we mentioned that this case study focuses more on appropriate aggregation than on the creation of cleverly derived variables. We will, however, show two examples of derived variables in our example.

27.7.1 Indicating a Promotional Period

Very often in time series analysis a certain period in time reflects the existence of certain influential factors. For example, promotions take place that last for a certain time interval. A possibility to make this information available in the aggregation data is to include a condition in order to create an indicator variable.

Consider a promotion that took place from September 1, 2004 through November 30, 2004. The indicator variable PROMOTION can be created with the following DATA step:

```
DATA SALES.ORDERMART;
 SET SALES.ORDERMART;
 IF '01SEP2004'd <= monthyear <= '30NOV2004'd
                                     THEN Promotion =1;
 ELSE Promotion = 0;
RUN;
```

Here is a more elegant alternative:

```
DATA SALES.ORDERMART;
 SET SALES.ORDERMART;
 Promotion=('01SEP2004'd <= monthyear <= '30NOV2004'd);
RUN;
```

This syntax can also be used in PROC SQL directly:

```
PROC SQL;
Create table sales.PROD_MONTH
 as
 select
 ProductID AS ProductID format = product.,
 ProductGroupID AS ProductGroupID format = PG.,
 monthyear,
 sum(quantity) as Quantity format=8.,
  avg(avgprice) as Price format = 8.2,
('01SEP2004'd <= monthyear <= '30NOV2004'd)
                                        AS Promotion

 from sales.ordermart
 group by monthyear, ProductID, productgroupid
 order by ProductID, monthyear;
QUIT;
```

The result is shown in Table 27.10.

Table 27.10: Sales data with a dummy variable for PROMOTION

	ProductID	ProductGroupID	MonthYear	Quantity	Price	PROMOTION
1	Product1	Keyboards	2003.01	1009	940.18	0
2	Product1	Keyboards	2003.02	1250	940.18	0
3	Product1	Keyboards	2003.03	1670	940.18	0
4	Product1	Keyboards	2003.04	1863	940.18	0
5	Product1	Keyboards	2003.05	2525	940.18	0
6	Product1	Keyboards	2003.06	2308	940.18	0
7	Product1	Keyboards	2003.07	2772	940.18	0
8	Product1	Keyboards	2003.08	2632	940.18	0
9	Product1	Keyboards	2003.09	2199	940.18	0
10	Product1	Keyboards	2003.10	2360	940.18	0
11	Product1	Keyboards	2003.11	2416	940.18	0
12	Product1	Keyboards	2003.12	2315	940.18	0
13	Product1	Keyboards	2004.01	2441	940.18	0
14	Product1	Keyboards	2004.02	2240	940.18	0
15	Product1	Keyboards	2004.03	2711	940.18	0
16	Product1	Keyboards	2004.04	2623	940.18	0
17	Product1	Keyboards	2004.05	2745	940.18	0
18	Product1	Keyboards	2004.06	2718	940.18	0
19	Product1	Keyboards	2004.07	2831	940.18	0
20	Product1	Keyboards	2004.08	3882	940.18	0
21	Product1	Keyboards	2004.09	2721	940.18	1
22	Product1	Keyboards	2004.10	2728	940.18	1
23	Product1	Keyboards	2004.11	2288	940.18	1
24	Product1	Keyboards	2004.12	2387	940.18	0
25	Product1	Keyboards	2005.01	1698	940.18	0

27.7.2 Aggregating a Derived Variable from the Data

In some companies a so-called sales strategy exists that regulates which shop is allowed to sell which product. Consider the example of a very high-level product that is sold only in specialty shops or a highly sophisticated technical product that is sold only in shops with specially trained salespeople.

The *number of shops* where a product is sold is obviously a very important predictor for the number of demand forecasts. It is also usually a variable that can be used in prediction because the approximate number of shops that are allowed to sell a product is known for the future.

In some cases, however, the policies that regulate whether a shop is allowed to sell a certain product are not known for the past, especially if the past extends over several years. In this case the number of shops can also be calculated from the data for the past and added to the order data mart. Note that here we are talking only of the shops that effectively sold the article. However, this is usually a good predictor.

In our example we calculate the number of shops (counting the SHOPID) per product and year:

```
PROC sql;
  create table sales.nr_shops as
  select productid,
         mdy(1,1,year(monthyear)) as Year format = year4.,
         count(distinct shopid) as Nr_Shops
  from sales.ordermart
  group by productid,
      calculated Year;
quit;
```

Because we know that the sales policies change only once every year, we calculate the distinct number on a yearly basis, as shown in Table 27.11.

Table 27.11: Number of shops per product

	ProductID	Year	Nr_Shops
1	Product1	2003	14
2	Product1	2004	13
3	Product1	2005	9
4	Product2	2003	15
5	Product2	2004	15
6	Product2	2005	15
7	Product3	2003	13
8	Product3	2004	12
9	Product3	2005	9
10	Product4	2003	17
11	Product4	2004	18
12	Product4	2005	13

This table can now be merged with the ORDERMART with the following statements:

```
PROC sql;
  create table sales.ordermart_enh
  as
  select o.*,
         n.Nr_Shops
  from sales.ordermart as o,
      left join sales.nr_shops as n
    on o.productid = n.productid
     and year(o.monthyear) = year(n.year);
quit;
```

Note that because the variable YEAR in Table 27.11 is formatted only to YEAR but contains a full date value, we need to specify `year(o.monthyear) = year(n.year)` instead of `year(o.monthyear) = n.year`.

If we calculate aggregations on a calendar date such as the number of weekdays per month, we would aggregate them to the ORDERMART in the same fashion. Note that we cannot use SAS formats in this case because the new calculated value depends on both SHOPID and YEAR. If we had aggregates only on SHOPID or YEAR we could use a format to join them to the data.

27.8 Creating Observations for Future Months

General

In time series analysis we want to produce forecasts for future months, also often called lead months. If we want to base our forecasts on exogenous variables such as PRICE and PROMOTION, we need to have these variables populated for historic and future periods. In Section 27.7, "Derived Variables," we saw how to create derived variables for historic periods.

To create this information for future periods we need a list of products that will be forecasted. Additionally, we need information about future promotion periods or price levels.

Coding

In Table 27.12 we see a list of products that will be forecasted, and we also have a new price level for the future.

Table 27.12: List of products with price level

	ProductID	ProductGroupID	Price
1	Product1	Keyboards	968.39
2	Product2	Keyboards	47.57
3	Product3	Printers	15.70
4	Product4	Keyboards	77.21
5	Product5	Printers	19.12
6	Product6	Printers	15.70
7	Product7	Printers	10.99
8	Product8	Printers	47.57
9	Product9	Printers	75.83

Starting from this table we create a table called LEAD_MONTHS that contains the observations for future periods, where the new price level is used and promotional periods from September to November are indicated.

```
%LET FirstMonth = '01Apr2005'd;
DATA sales.Lead_Months;
  FORMAT ProductID product. ProductGroupID pg.
         Monthyear yymmp7. Quantity 8.;
  SET sales.lead_base;
  DO lead = 1 TO 12;
    Quantity = .;
    MonthYear = intnx('MONTH',&FirstMonth,lead);
    IF month(MonthYear) in (9,10,11) THEN Promotion = 1;
                                     ELSE Promotion = 0;
   OUTPUT;
  END;
  DROP lead;
 RUN;
```

Note that our first future month is April 2005, and we are creating 12 lead months per product. The preceding code results in Table 27.13, where we also see values for the input variables PRICE and PROMOTION.

Table 27.13: Twelve future months for the defined products

	ProductID	ProductGroupID	Monthyear	Quantity	Price	Promotion
1	Product1	Keyboards	2005.05	.	968.39	0
2	Product1	Keyboards	2005.06	.	968.39	0
3	Product1	Keyboards	2005.07	.	968.39	0
4	Product1	Keyboards	2005.08	.	968.39	0
5	Product1	Keyboards	2005.09	.	968.39	1
6	Product1	Keyboards	2005.10	.	968.39	1
7	Product1	Keyboards	2005.11	.	968.39	1
8	Product1	Keyboards	2005.12	.	968.39	0
9	Product1	Keyboards	2006.01	.	968.39	0
10	Product1	Keyboards	2006.02	.	968.39	0
11	Product1	Keyboards	2006.03	.	968.39	0
12	Product1	Keyboards	2006.04	.	968.39	0
13	Product2	Keyboards	2005.05	.	47.57	0
14	Product2	Keyboards	2005.06	.	47.57	0
15	Product2	Keyboards	2005.07	.	47.57	0
16	Product2	Keyboards	2005.08	.	47.57	0
17	Product2	Keyboards	2005.09	.	47.57	1
18	Product2	Keyboards	2005.10	.	47.57	1
19	Product2	Keyboards	2005.11	.	47.57	1

This table is then stacked to Table 27.11. From this table SAS Forecast Server and SAS High-Performance Forecasting can start to create ARIMAX and UCM models and produce forecasts for the next 12 months.

27.9 The Results and Their Usage

We mentioned earlier that the resulting data on the respective aggregation level are used in time series forecasting—for example, in PROC AUTOREG or PROC UCM or SAS Forecast Server.

In these procedures, the variable MONTHYEAR is the ID variable; the variable QUANTITY is the analysis variable VAR; variables such as PROMOTION, NR_SHOPS, or PRICE are potential explanatory variables; and the other variables serve as variables to define cross-sectional BY groups in the BY statement.

Chapter 28

Case Study 4—Data Preparation in SAS Enterprise Miner

28.1 Introduction

General

In this case study we will show how SAS Enterprise Miner can be used for data preparation. In earlier chapters we referred to SAS Enterprise Miner for data preparation tasks. Here we will illustrate which functionality of SAS Enterprise Miner is useful for data preparation.

The following examples and descriptions refer to SAS Enterprise Miner 5.2.

28.2 Nodes for Data Preparation

General

When introducing SAS Enterprise Miner for data preparation we will consider the following nodes. We will briefly introduce these nodes and explain their functionality for data preparation. For more details, see the SAS Enterprise Miner Help.

28.2.1 Input Data Source Node

We use the *Input Data Source node* to specify details about the variables in the data source that we are going to use. In this node the specification of the *variable role* and the *variable level* are important. See *Chapter 11 – Considerations for Data Marts*, where we discussed the roles and types of variables. In the examples in the following sections we will explain the most important variable roles.

28.2.2 Sample Node

The *Sample node* allows the sampling of data. In addition to the methods random sampling (=simple sampling), stratified sampling, and clustered sampling, which we explained in *Chapter 22 – Sampling Scoring*, and Automation, the First N and Systematic methods are available. The sampling methods are performed with a restricted sampling method in order to achieve the exact sample count or proportion.

28.2.3 Data Partition Node

The *Data Partition node* splits data into training, validation, and test data. The split criterion can be a random, stratified, or clustered sample. See Section 12.4 for details about validation data, and Chapter 20 – *Coding for Predictive Modeling*, for an example of how to split data in a SAS DATA step.

28.2.4 Time Series Node

The *Time Series node* allows aggregating and analyzing of transactional data. Longitudinal or multiple-rows-per-subject data can be analyzed and aggregated for a specific time interval. Derived variables on a one-row-per-subject level can be created. These derived variables can be measures for season, trend, autocorrelations, and seasonal decomposition. We discussed methods to aggregate multiple-rows-per-subject data and create derived variables from them in *Chapter 18 – Multiple Interval-Scaled Observations per Subject*. Note that the Time Series node used PROC TIMESERIES.

28.2.5 Association Node

The *Association node* allows performing *association and sequence analysis* for multiple-rows-per-subject data. We mentioned these two techniques in *Chapter 19 – Multiple Categorical Observations per Subject*. Because of their complexity, we do not introduce them as macros in SAS code. SAS Enterprise Miner is needed to perform these analyses. The Association node allows for creating an export data set to a one-row-per-subject level that contains indicators or derived variables for selected associations.

28.2.6 Transform Variables Node

The *Transform Variables node* allows performing transformations for variables. For interval variables, simple transformations, binning transformations, and best power transformations are available. For class variables, the grouping of rare levels and the creation of dummy indicators are possible. The definition of interactions can be done in the interaction editor.

We introduced transformation methods in *Chapter 16 – Transformations of Interval-Scaled Variables* and *Chapter 17 – Transformations of Categorical Variables*. The advantage of the Transform Variables node to transformation in SAS code is that no manual intervention such as the writing of the names of groups with low frequency is needed.

The Transform Variables node also allows performing transformations and defining groups that maximize the predictive power if a target variable is specified. These methods can also be done in normal coding, however, they include much more manual work and pre-analysis of the data.

In addition to the Transform Variables node, nodes such as the *Variable Selection node* or the *Tree node* allow you to group data optimally for their relationship to the target variable.

28.2.7 Filter Node

The *Filter node* allows the filtering of observations based on individual limits per variable or by defining general measures such as the number of standard deviations from the mean or the mean absolute deviations.

28.2.8 Impute Node

The *Impute node* allows the imputation of missing values, either by defining a static value or by using imputation methods such as mean imputation, distribution-based imputation, or tree imputation, where a most probable imputation value is calculated based on other variables. In Chapter 16 we showed imputation methods in SAS code. The methods that are offered by the Impute node are much more powerful.

28.2.9 Score Node

The *Score node* allows scoring data within SAS Enterprise Miner and also creates the score code as SAS DATA step code, C code, Java code, or PMML code. Note that PMML is an abbreviation for Predictive Modeling Markup Language. Different from the methods that we discussed in *Chapter 23 – Scoring and Automation*, the Score node creates for SAS scoring the scoring rules in SAS DATA step code only.

All transformations that are performed by standard SAS Enterprise Miner nodes are converted automatically into score code. Transformations that are performed in a SAS Code node have to be prepared explicitly as SAS score code in the appropriate editor in the SAS Code node.

28.2.10 Metadata Node

Similar to the Input Data Source node, the *Metadata node* allows, among other things, the definition of variable roles and variable levels. The Metadata node can be used within a process flow in order to change the metadata.

28.2.11 SAS Code Node

The *SAS Code node* allows the entering of SAS code. The SAS code can be "normal" SAS code that we also use in the Program Editor. The SAS code can make use of the list of macro variables that are available and that are used in SAS Enterprise Miner to define variable roles and input and output data sets. When writing code in the SAS Code node, note that different nodes in SAS Enterprise Miner do not share a common Work library. Consider using a predefined temporary library for temporary data that have to be used across nodes.

28.2.12 Merge Node

The *Merge node* allows match-merging data sets together. The Merge node is important to merge the results with the Time Series node or the Association node per subject ID to a table with one row per subject.

28.3 Data Definition, Sampling, and Data Partition

In the first step we define our input data, draw a sample from the data, and partition the data into training and validation data. In Figure 28.1 we see the process flow with three data mining nodes.

Figure 28.1: Data mining process flow 1

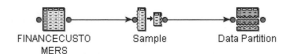

In the first node the data source and the variable metadata are defined. An example is shown in Table 28.1.

Table 28.1: Variable metadata

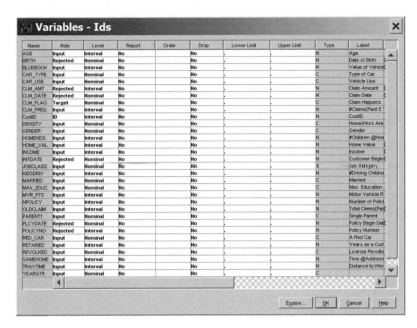

Name	Role	Level	Report	Order	Drop	Lower Limit	Upper Limit	Type	Label
AGE	Input	Interval	No		No			N	Age
BIRTH	Rejected	Nominal	No		No			N	Date of Birth
BLUEBOOK	Input	Interval	No		No			N	Value of Vehicle
CAR_TYPE	Input	Nominal	No		No			C	Type of Car
CAR_USE	Input	Nominal	No		No			C	Vehicle Use
CLM_AMT	Rejected	Interval	No		No			N	Claim Amount
CLM_DATE	Rejected	Nominal	No		No			N	Claim Date
CLM_FLAG	Target	Nominal	No		No			C	Claim Happens
CLM_FREQ	Input	Interval	No		No			N	#Claims(Past 5 Y
CustID	ID	Interval	No		No			N	CustID
DENSITY	Input	Nominal	No		No			C	Home/Work Area
GENDER	Input	Nominal	No		No			C	Gender
HOMEKIDS	Input	Interval	No		No			N	#Children @Hom
HOME_VAL	Input	Interval	No		No			N	Home Value
INCOME	Input	Interval	No		No			N	Income
INITDATE	Rejected	Nominal	No		No			N	Customer Begin[
JOBCLASS	Input	Nominal	No		No			B	Job Category
KIDSDRIV	Input	Interval	No		No			N	#Driving Childre
MARRIED	Input	Nominal	No		No			C	Married
MAX_EDUC	Input	Nominal	No		No			C	Max. Education
MVR_PTS	Input	Interval	No		No			N	Motor Vehicle R
NPOLICY	Input	Interval	No		No			N	Number of Polici
OLDCLAIM	Input	Interval	No		No			N	Total Claims(Pas[
PARENT1	Input	Nominal	No		No			C	Single Parent
PLCYDATE	Rejected	Nominal	No		No			N	Policy Begin Dat[
POLICYNO	Rejected	Interval	No		No			N	Policy Number
RED_CAR	Input	Nominal	No		No			C	A Red Car
RETAINED	Input	Interval	No		No			N	Years as a Cust
REVOLKED	Input	Nominal	No		No			C	License Revolke
SAMEHOME	Input	Interval	No		No			N	Time @Address
TRAVTIME	Input	Interval	No		No			N	Distance to Wor
YEARQTR	Input	Nominal	No		No			C	

Note that here we define the target variable in predictive modeling, reject variables, and define the ID variable. For each input data source it is necessary to define the data set role. This is done in the last step of the input data source or for each data source in the properties. An example is shown in Table 28.2.

Table 28.2: Defining the data source role

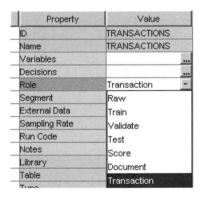

Property	Value
ID	TRANSACTIONS
Name	TRANSACTIONS
Variables	
Decisions	
Role	Transaction
Segment	Raw
External Data	Train
Sampling Rate	Validate
Run Code	Test
Notes	Score
Library	Document
Table	**Transaction**

Note that for one-row-per-subject input data, the role RAW is usually selected. For multiple-rows-per-subject, longitudinal, or transactional data, the role TRANSACTION must be selected. For data sources that will be scored in SAS Enterprise Miner, the role SCORE must be defined.

In the Sample or Data Partition node we chose STRATIFIED sampling. Note that the role of the target variable is automatically set to STRATIFICATION in order to perform a stratified sampling for the target variable. An example is shown in Table 28.3.

Table 28.3: Edit variables in the data partition node

Name	Partition Role	Role
CLM_FLAG	Stratification	Target
CLM_FREQ	Default	Input
CustID	Default	ID
DENSITY	Default	Input
GENDER	Default	Input
HOMEKIDS	Default	Input
HOME_VAL	Default	Input
INCOME	Default	Input
INITDATE	Default	Rejected

28.4 Data Transformation

In the second step we filter outliers, transform variables, and impute missing values. Figure 28.2 shows the data mining process flow with the respective nodes.

Figure 28.2: Data mining process flow 2

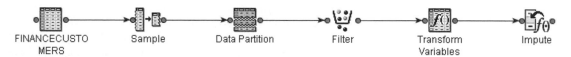

FINANCECUSTO   Sample   Data Partition   Filter   Transform   Impute
MERS   Variables

The Transform Variables node allows the definition of interactions through a graphical user interface, as shown in Figure 28.3.

Figure 28.3: Interactions editor

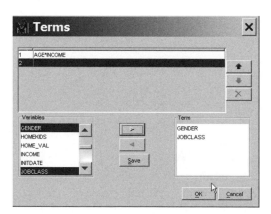

28.5 Scoring in SAS Enterprise Miner

After a model is built, the Score node can be used to retrieve the score code for the model. The Score node can also be used to score fresh data. Note that the training, validation, and test data are scored by default. Figure 28.4 shows how the process flow is continued with two modeling nodes and a Score node.

Figure 28.4: Data mining process flow 3

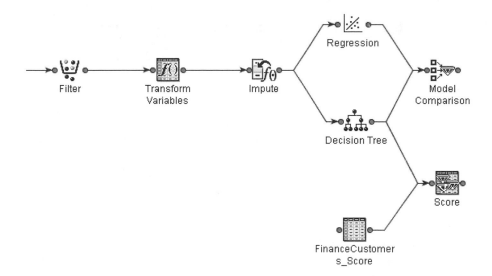

We have to connect the Score node to the Decision Tree node in order to get the score code for this model. Additionally, we defined a new data source FINANCECUSTOMERS_SCORE that contains new data that will be scored. Note that the table role has to be set to SCORE for this data set.

Table 28.4 shows how the data set can be displayed in the results of the Score node. Running the Score node produces an output data set that contains the scores shown in Table 28.5.

Table 28.4: Display the score data set

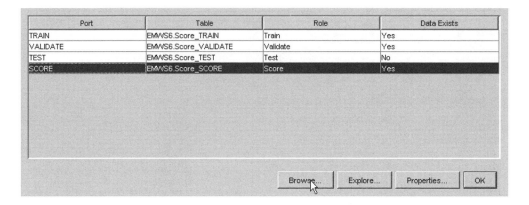

Port	Table	Role	Data Exists
TRAIN	EMWS6.Score_TRAIN	Train	Yes
VALIDATE	EMWS6.Score_VALIDATE	Validate	Yes
TEST	EMWS6.Score_TEST	Test	No
SCORE	EMWS6.Score_SCORE	Score	Yes

Table 28.5: Extraction of the score data set

Transforme...	Imputed: Jo...	Warnings	Into: CLM_F...	Unnormalize...	Node	Leaf	Predicted: CLM_FLAG=Yes	Predicted: C...	Validated: C...	Validated: C...	Segment	Probability f...	Probability o...	Prediction f...
02:0.5-high	Clerical		NO	No	54	5	0.302932	0.697068	0.358025	0.641975	54	0.302932	0.697068	NO
02:0.5-high	Student		YES	Yes	68	27	0.517241	0.482759	0.538462	0.461538	68	0.517241	0.517241	YES
02:0.5-high	Manager		NO	No	67	25	0.11	0.89	0.138211	0.861789	67	0.11	0.89	NO
02:0.5-high	Student		NO	No	53	35	0.029586	0.970414	0.032738	0.967262	53	0.029586	0.970414	NO
02:0.5-high	Blue Collar		NO	No	67	25	0.11	0.89	0.138211	0.861789	67	0.11	0.89	NO
02:0.5-high	Manager		NO	No	73	33	0.070111	0.929889	0.083969	0.916031	73	0.070111	0.929889	NO
02:0.5-high	Professional		YES	Yes	35	11	0.894737	0.105263	0.769231	0.230769	35	0.894737	0.894737	YES
02:0.5-high	Professional		YES	Yes	62	16	0.679245	0.320755	0.721311	0.278689	62	0.679245	0.679245	YES
02:0.5-high	Doctor		NO	No	54	5	0.302932	0.697068	0.358025	0.641975	54	0.302932	0.697068	NO
02:0.5-high	Lawyer		NO	No	54	5	0.302932	0.697068	0.358025	0.641975	54	0.302932	0.697068	NO
02:0.5-high	Home Maker		NO	No	43	21	0.243386	0.756614	0.228261	0.771739	43	0.243386	0.756614	NO
02:0.5-high	Home Maker		NO	No	43	21	0.243386	0.756614	0.228261	0.771739	43	0.243386	0.756614	NO
02:0.5-high	Professional		NO	No	63	17	0.481752	0.518248	0.484848	0.515152	63	0.481752	0.518248	NO
02:0.5-high	Clerical		YES	Yes	40	18	0.517483	0.482517	0.371429	0.628571	40	0.517483	0.517483	YES
02:0.5-high	Clerical		NO	No	53	35	0.029586	0.970414	0.032738	0.967262	53	0.029586	0.970414	NO
02:0.5-high	Manager		NO	No	73	33	0.070111	0.929889	0.083969	0.916031	73	0.070111	0.929889	NO
02:0.5-high	Blue Collar		NO	No	53	35	0.029586	0.970414	0.032738	0.967262	53	0.029586	0.970414	NO
02:0.5-high	Blue Collar		NO	No	53	35	0.029586	0.970414	0.032738	0.967262	53	0.029586	0.970414	NO
02:0.5-high	Blue Collar		NO	No	73	33	0.070111	0.929889	0.083969	0.916031	73	0.070111	0.929889	NO
02:0.5-high	Clerical		YES	Yes	40	18	0.517483	0.482517	0.371429	0.628571	40	0.517483	0.517483	YES
02:0.5-high	Professional		NO	No	73	33	0.070111	0.929889	0.083969	0.916031	73	0.070111	0.929889	NO

Note that the score code creates a number of scoring variables. Depending on the setting "Use Fixed Output Names" in the Score node, the variable that contains the predicted probability in event prediction has the name EM_EVENTPROBABILITY or P_<targetvariablename><targetevent>. For a complete definition of the variables that are created during scoring and of the scoring process as a whole, see the SAS Enterprise Miner Help.

The Score node also generates the score code as SAS DATA step code. Figure 28.5 shows how to retrieve this code from the results of the Score node.

Figure 28.5: Display the score code

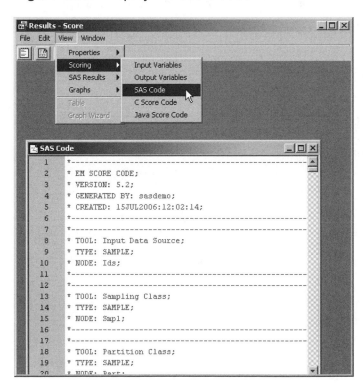

Note that the SCORE CODE contains the score code for all nodes in the path, as shown in Figure 28.6.

Figure 28.6: Score code

```
SAS Code                                                            _ □ ×
  1      *---------------------------------------------------------*;
  2      * EM SCORE CODE;
  3      * VERSION: 5.2;
  4      * GENERATED BY: sasdemo;
  5      * CREATED: 15JUL2006:12:02:14;
  6      *---------------------------------------------------------*;
  7      *---------------------------------------------------------*;
  8      * TOOL: Input Data Source;
  9      * TYPE: SAMPLE;
 10      * NODE: Ids;
 11      *---------------------------------------------------------*;
 12      *---------------------------------------------------------*;
 13      * TOOL: Sampling Class;
 14      * TYPE: SAMPLE;
 15      * NODE: Smpl;
 16      *---------------------------------------------------------*;
 17      *---------------------------------------------------------*;
 18      * TOOL: Partition Class;
 19      * TYPE: SAMPLE;
 20      * NODE: Part;
 21      *---------------------------------------------------------*;
 22      *---------------------------------------------------------*;
 23      * TOOL: Filtering;
 24      * TYPE: MODIFY;
 25      * NODE: Filter;
 26      *---------------------------------------------------------*;
 27      label M_FILTER = 'Filtered Indicator';
 28      if
 29      (AGE eq . or -9<=AGE<= 99)
 30      and (BLUEBOOK eq . or -38270<=BLUEBOOK<= 67570)
 31      and (HOME_VAL eq . or -723055.5556<=HOME_VAL<=1050455.2628)
 32      and (INCOME eq . or -204845.3797<=INCOME<=313741.47056)
 33      and (MVR_PTS eq . or -8<=MVR_PTS<= 10)
 34      and (RETAINED eq . or -23<=RETAINED<= 31)
 35      and (SAMEHOME eq . or -37<=SAMEHOME<= 53)
```

This score code can be applied by saving it to a file and specifying the DATA, SET, and RUN statements to create a SAS DATA step:

```
DATA scored_data;
SET input_data;
%INCLUDE 'scorecode.sas';
RUN;
```

28.6 Merging Multiple-Rows-per-Subject Data

In this section we will extend the process flow from earlier by also considering multiple-rows-per-subject data sources such as ACCOUNTS and TRANSACTIONS, as seen in Table 28.6 and Table 28.7.

Table 28.6: ACCOUNTS data set

CustID	AccountID	Balance	Interest	TYPE
1000002	16853	1550.42	5.00	Saving Account
1000002	16853	1550.42	5.00	Loan
1000005	13296	3775.31	6.00	Saving Account
1000006	21592	2376.43	2.00	Funds
1000007	16789	1812.72	5.00	Funds
1000007	16789	1812.72	5.00	Saving Account
1000008	15567	3350.65	2.00	Funds
1000009	11907	1191.82	4.00	Build assoc saving
1000009	11907	1191.82	4.00	Funds
1000009	11907	1191.82	4.00	Loan
1000014	8800	1500.46	4.00	Saving Account
1000014	13513	1500.46	5.00	Saving Account
1000015	16853	2801.09	5.00	Saving Account
1000016	166	1662.83	1.00	Build assoc saving
1000016	166	1662.83	1.00	Saving Account
1000018	20703	0.00	2.00	Funds
1000019	5065	1487.97	3.00	Funds
1000019	5065	1487.97	3.00	Loan

Table 28.7: TRANSACTIONS data set

CustID	PERIOD	Transactions
1000002	02.02	136
1000002	02.03	170
1000002	02.04	152
1000002	02.05	215
1000002	02.06	231
1000002	02.07	201
1000002	02.08	199
1000002	02.09	225
1000002	02.10	247
1000002	02.11	106
1000002	02.12	194
1000005	02.06	348
1000006	02.11	662
1000006	02.12	630
1000007	02.11	575
1000007	02.12	578
1000008	01.01	702
1000008	01.02	566
1000008	01.03	645
1000008	01.04	913
1000008	01.05	1533

We see that the TRANSACTIONS data set has aggregates on a monthly basis per customer. These data can be further aggregated on a one-row-per-subject basis or they can be used in time series analysis to derive variables that describe the course over time.

The data in the ACCOUNTS data set can be used for association analysis to derive the most frequent account combinations.

We define the data sources in SAS Enterprise Miner and incorporate them into the process flow. Note that both data sets need to have the role TRANSACTION, and the variable roles have to be set as shown in Table 28.8 and Table 28.9.

Table 28.8: Variable roles in the ACCOUNTS data set

Name	Role	Level
AccountID	Rejected	Interval
Balance	Rejected	Interval
CustID	ID	Interval
Interest	Rejected	Interval
Type	Target	Nominal

Table 28.9: Variable roles in the TRANSACTIONS data set

Name	Role	Level
CustID	Cross ID	Interval
Period	Time ID	Interval
Transactions	Target	Interval

Next we create a process flow as shown in Figure 28.6.

Figure 28.6: Data mining process flow 4

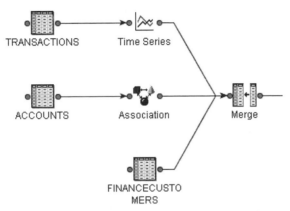

In the Association node we specify to export the rules by ID.

Table 28.10: Node properties for the association node

Property	Value
Node ID	Assoc
Imported Data	...
Exported Data	...
Variables	...
Maximum Number of Item	100000
Rules	...
Association	
Maximum Items	4
Minimum Confidence Level	10
Support Type	Percent
Support Count	2
Support Percentage	5.0
Sequence	
Chain Count	3
Consolidate Time	0.0
Maximum Transaction Dur	
Support Type	Percent
Support Count	2
Support Percentage	2.0
Rules	
Number to Keep	200
Sort Criterion	Default
Number to Transpose	200
Export Rule by ID	Yes

In the Time Series node we specify to create derived variables for the autocorrelations. Note that only one statistic can be exported per node. A separate node has to be used for each statistic per subject. An alternative is to use PROC TIMESERIES in a SAS Code node and specify more output statistics in the code.

Table 28.11: Node using PROC TIMESERIES

Property	Value
Node ID	TIME
Imported Data	
Exported Data	
Variables	
Accumulation	Total
Transformation	None
Box-Cox Parameter	0.0
Apply Differencing	No
Difference Order	1
Time Interval	
Interval Selection	Automatic
Specify an Interval	
Time of Day	No
Seasonal Cycle Selection	Default
Length of Cycle	2
Missing Value	
Set Value	Missing
Constant Value for Missing	0.0
Analysis Method	
Select an Analysis	Correlation
Transpose	Yes
Seasonal Analysis	
Exported Statistics	Sum
Trend Analysis	

In the Merge node we match-merge and select the appropriate BY variable CUSTID. This causes the one-row-per-subject data set FINANCECUSTOMERS to merge with the one-row-per-subject results of the Time Series node and the Association node to form one data set.

In the resulting data set we have variables from the data set FINANCECUSTOMERS, autocorrelations per customer from the Time Series node, and indicators for selected product combinations from the Associations node.

28.7 Conclusion

We see that using SAS Enterprise Miner in data preparation offers a number of advantages. We summarize the most important advantages as follows:

- Documentation of the data preparation process as a whole in the process flow.
- Definition of the variable's metadata that are used in the analysis.
- Powerful data transformations in the Filter node, Transform Variables node, and Impute node. These nodes mostly create "only" SAS DATA step code that could also be produced outside SAS Enterprise Miner. Then, however, a number of intermediate steps and analyses would be needed—for example, for the optimal binning of interval variables.

- Possibility to perform association analysis.
- Creation of score code from all nodes in SAS Enterprise Miner as SAS DATA step code.

Appendix A

Data Structures from a SAS Procedure Point of View

A.1 Introduction

In this appendix we will consider various data mart structures from a SAS language point of view. We will investigate the data structure requirements for selected procedures in Base SAS, SAS/STAT, SAS/ETS, SAS/GRAPH, and SAS/QC. We will also investigate which data mart elements are primarily used by which generic statements in SAS procedures.

A.2 Relationship between Data Mart Elements and SAS Procedure Statements

A.2.1 WHERE Statement

The WHERE statement restricts the analysis to those observations in the data set that satisfy the WHERE condition. Filter conditions are often the following:

- intervals of a numeric variable, such as AGE <= 50
- values of a nominal variable—for example, REGION IN ('NORTH', 'WEST')
- values of a binary indicator such as RESPONSE = YES

See *Chapter 24 – Do's and Don'ts When Building Data Marts* for the optimization of run time with large data sets, if the WHERE condition is passed as a data set option.

A.2.2 BY Statement

A BY statement produces separate analyses on observations in groups defined by the BY variables. When a BY statement appears, the procedure expects the input data set to be sorted in the order of the BY variables. The BY variables are one or more variables in the input data set. The BY variable is usually a categorical or integer variable with discrete values.

Note that the use of the appropriate BY variables in the analysis on multiple-rows-per-subject data marts or longitudinal data marts creates in many cases one-row-per-subject data marts, where the subject entity moves one level lower in the hierarchy.

A.2.3 WEIGHT and FREQ Statements

The FREQ statement names a variable that provides frequencies for each observation in the DATA= data set. Specifically, if n is the value of the FREQ variable for a given observation, then that observation is used n times.

With some procedures such as CORR, FREQ, MEANS/SUMMARY, REPORT, STANDARD, TABULATE, and UNIVARIATE, a WEIGHT statement can be specified that specifies numeric weights for analysis variables in the statistical calculations.

Note that the FREQ statement assumes integer numbers, whereas the WEIGHT statement can be real numbers from 0 to 1.

A.2.4 ID Statement

With some procedures an ID statement can be specified to identify the observations in the analysis results. The ID variable can be a variable that contains an ID variable in the form of a key, as we defined it in *Chapter 6 – Data Models*. However, every other variable in the data set can be defined as a key variable—for example, the variable AGE, which adds the age to each observation in the output.

A.2.5 Missing Values

Missing values in variables that appear in one of the procedure statements have the effect that the observation is omitted from the analysis. The only exceptions are the ID and BY statements. The ID statement causes a missing ID value, and the BY statement treats missing values as a separate group.

Note that an observation is used only in the analysis if none of its variables, as defined in the preceding paragraph, is missing. In multivariate analysis with a list of variables, the number of observations used in the analysis depends on all variables and can be smaller, as in the respective univariate analysis.

A.3 Data Mart Structure Requirements for Selected Base SAS Procedures

A.3.1 General

A few of the Base SAS procedures perform analytical tasks and the differentiation of input data set structures makes sense.

A.3.2 PROC CORR

The CORR procedure can be used with one-row-per-subject data marts to calculate correlations between variables. PROC CORR can also be used with multiple-rows-per-subject data marts or longitudinal data marts to calculate correlations over all observations, or to calculate correlations for each subject or cross ID category if a BY statement is used.

A.3.3 PROC UNIVARIATE

The UNIVARIATE, MEANS, and SUMMARY procedures can be used with any data mart structure. For the sign test, the Wilcoxon test, and the t-test for dependent samples, the input data can be interpreted as a one-row-per-subject data mart.

A.3.4 PROC FREQ

The FREQ procedure can use data in the form of a one-row-per-subject data mart and in the form of pre-aggregated cell statistics of a frequency table. In this case a WEIGHT statement weights the observations according to its cell frequency.

A.4 Data Mart Structure Requirements for Selected SAS/STAT Procedures

A.4.1 One-Row-Per-Subject Data Mart

Most of the procedures in SAS/STAT require data in the form of a one-row-per-subject data mart. The following procedures accept only a one-row-per-subject data mart:

- Regression procedures such as REG, LOGISITC, GENMOD, NLIN, ROBUSTREG, and LIFEREG
- Analysis-of-variance procedures such as GLM, ANOVA, NESTED, and MIXED
- Multivariate procedures such as FACTOR, PRINCOMP, and CORRESP
- Discriminant procedures such as DISCRIM, CANDISC, and STEPDISC
- Cluster procedures such as CLUSTER, FASTCLUS, and VARCLUS
- Survival analysis procedures such as LIFETEST, LIFEREG, and PHREG
- Survey sampling and analysis procedures such as SURVEYSELECT, SURVEYMEANS, SURVEYFREQ, SURVEYREG, and SURVEYLOGISITIC
- Test procedures such as PROC TTEST and NPAR1WAY

A.4.2 One-Row-per-Subject and Multiple-Rows-per-Subject Data Marts

Some analysis-of-variance procedures accept one-row-per-subject data marts as well as multiple-rows-per-subject data marts.

The GLM procedure, for example, accepts input data in the case of repeated measurements as a one-row-per-subject data mart with columns for the repeated observations and a multiple-rows-per-subject data mart with repeats in the rows. If it cannot be assumed that the subjects' measurements are uncorrelated across time, the one-row-per-subject data mart has to be used.

A.4.3 Frequency Data

The CATMOD procedure accepts, in addition to a one-row-per-subject data mart, data in the form of frequency counts. In this case each row of the input data set corresponds to one cell of the input data set. A WEIGHT statement weights the observations according to its cell frequency.

A.4.4 Coordinate or Distance Data

The following procedures require coordinate or distance data:

- PROC MODECLUS requires coordinate or distance data.
- PROC MDS requires coordinate or distance data in a triangular form.
- PROC DISTANCE creates coordinate or distance data in squared or triangular form.

A.4.5 Special Forms of Input Data

The TREE procedure requires data sets in the form of the results of hierarchical clustering as a tree structure. This type of data set is output by PROC CLUSTER or PROC VARCLUS. These data sets must have the following variables: _NAME_ (the name of the cluster), _PARENT_ (the parent of the cluster), _NCL_ (the number of clusters), _VAREXP_ (the amount of variance explained by the cluster), _PROPOR_ (the proportion of variance explained by the clusters at the current level of the tree diagram), _MINPRO_ (the minimum proportion of variance explained by a cluster), _MAXEIGEN_ (the maximum second eigenvalue of a cluster).

A.4.6 Additional Details

The LOGISITC procedure allows the definition of the target variable as compound information of two variables, the EVENT variable and the TRIAL variable.

A.5 Data Mart Structure Requirements for Selected SAS/ETS Procedures

A.5.1 Longitudinal and Multiple-Rows-per-Subject Data Structure

The classic procedures for time series analysis such as FORECAST, ARIMA, and AUTOREG as well as procedures such as PDLREG, SYSLIN, SPECTRA, or UCM require data in the form of a longitudinal data structure. A TIME variable and a VALUE variable need to be present, and additional cross-sectional variables can be provided.

If the data are provided in an *interleaved data structure*, the WHERE statement allows the filtering of the relevant rows of the table.

In the case of a *cross-sectional data mart structure*, a BY statement allows the creation of independent analyses defined by the categories of the BY variables. Also cross-sections can be filtered with a WHERE statement.

A.5.2 PROC TIMESERIES

The TIMESERIES procedure accumulates transactional data into a time series format. The time series data can then be processed with other SAS/ETS procedures or can be processed directly in the TIMESERIES procedure with trend and seasonal analyses.

A.5.3 PROC TCSREG

The TCSREG procedure performs time series cross-sectional analysis. The data set must be in a longitudinal data mart structure, where a variable for TIME, a variable for VALUE, and at least one CATEGORY variable must be available. The data set must be sorted by CATEGORY and TIME.

A.5.4 Switching between Aggregation Levels

The EXPAND procedure allows aggregating and disaggregating time series to different time intervals. Data can be aggregated to lower sampling intervals and interpolated to higher sampling intervals. The EXPAND procedure also allows the replacement of missing observations through interpolation.

A.6 Data Mart Structure Requirements for Selected SAS/QC Procedures

A.6.1 PROC SHEWHART

The SHEWHART procedure requires the data to be in a longitudinal data structure. At least one TIME variable and one VALUE variable need to be available in the data set. The data set can also contain one or more CATEGORY variables to subgroup the analysis or to mark phases of different values of covariables.

P charts can be created from so-called count data, which equals our definition of the longitudinal data set. They can also be created from summary data, where the proportion is already pre-calculated.

In the case of unequal subgroup sizes, the SUBGROUPN option in the p chart, u chart, or np chart requires a variable name in the input data set that contains the subgroup size.

A.6.2 PROC CUSUM

The CUSUM procedure requires data in a longitudinal data structure. At least one TIME variable and one VALUE variable need to be available in the data set. This variable needs to be numeric. The data set can also contain one or more CATEGORY variables to subgroup the analysis or to mark phases of different values of covariables.

A.7 Data Mart Structure Requirements for Selected SAS/GRAPH Procedures

A.7.1 PROC GPLOT

The GPLOT procedure can produce mean plots and single plots (one line for each subject) on the basis of longitudinal and multiple-rows-per-subject data marts. A VALUE variable and a TIME variable have to be provided. CATEGORY variables and ID variables can be used in the BY statement to produce plots per CATEGORY or ID, or they can be used in the PLOT statement to produce a plot with separate lines for each CATEGORY or ID.

A.8 Data Mart Structure Requirements for SAS Enterprise Miner Nodes

A.8.1 One-Row-per-Subject Data Mart

In data mining analysis, in the majority of cases, one-row-per-subject data marts are used. Analyses such as regressions, neural networks, decision trees, or cluster require input data in the form of one-row-per-subject data marts.

All analysis tools in the Model folder of SAS Enterprise Miner, as well as the Variable Selection node, the Cluster node, the SOM/Kohonen node, and the Interactive Grouping node and all SAS Text Miner tools require that structure for the data mart. The Link Analysis node can analyze one-row-per-subject data mart data.

A.8.2 Multiple-Rows-per-Subject Data Mart

The Association node and the Path Analysis node require data in the form of a multiple-rows-per-subject data mart. For a Path Analysis node, the input data set must have a target, ID, and sequence variable. The Association node requires a target variable and an ID variable. The sequence variable is required only for sequence analyses. If multiple-rows-per-subject data are input into the Link Analysis node, an ID variable and a target variable are required. If time will also be analyzed, a time variable has to be provided.

A.8.3 Longitudinal Data Mart Structures

The Time Series node is the only SAS Enterprise Miner node that requires data in the longitudinal data mart structure. A time ID and a target variable have to be defined; a cross ID variable is optional.

A.8.4 General

Note that in SAS Enterprise Miner the type of multiple-rows-per-subject data marts and longitudinal data marts are referred to as transaction data sets. The type of data set has to be defined in the respective Input Data Source node.

Appendix B

The Power of SAS for Analytic Data Preparation

B.1 Motivation

Data management and data preparation are fundamental tasks for successful analytics. We discussed the role and importance of data management in *Part 1 – Data Preparation: Business Point of View*. Here we want to emphasize again the fact that data management and analytics are not sequential phases in the sense that no data management has to be done after analytics has started. The contrary is true. After running the first analyses, we need to determine the following:

- how outliers or skewed distributions will be handled
- how rare classes of categorical values will be treated
- which transformations for target and input variables will be evaluated for different analysis scenarios
- what kinds of derived variables will be created

- how data will be aggregated
- how time series per subject will be condensed into a few meaningful variables per subject

Therefore, it is indispensable for an analytic environment that data management features and analytic tools are seamlessly available.

SAS provides these features to the analyst by integrating its data management capabilities with its analytic procedures. This allows the analyst to perform data management and analysis within one environment without the need to move data between systems.

Data management and analytics are closely linked. During analysis the two phases are iterated over and over until the final results are available. The power of SAS for analytics originates from the fact that powerful data management and a comprehensive set of analytical tools are available in one environment. Therefore, SAS assists the analyst in his natural working process by allowing data management and analysis without barriers.

B.2 Overview

B.2.1 General

In addition to the fact that data management and analytic methods are integrated in one system, SAS offers powerful methods of data management that exceed the functionality of SQL or procedural extensions of SQL. This functionality includes the following:

- shifting a column in a table k rows down
- transposing the rows and columns in a table and transposing the multiple rows per subject to one row per subject
- creating indicators for the first and last observations in a BY group
- a rich set of built data transformation functions such as type and format conversion string matching, and statistical and mathematical transformations
- coding and decoding of text files and complex text files—for example, exports from hierarchical databases

In this appendix we will discuss the power of SAS for data management in more detail. We used these features in earlier chapters of this book when we investigated derived variables and changed data set structures.

The data management capability of SAS is made up of three components:

- the SAS language that includes the SAS DATA step and SAS procedures
- the SAS macro language
- SAS/IML

The SAS language and the SAS macro language are part of Base SAS. SAS/IML is an extra module that allows the application of matrix operations on SAS data sets.

B.3 Extracting Data from Source Systems

Extracting data from other systems is usually the first step in data preparation. SAS assists you by providing data access interfaces to a number of data processing systems.

- The SAS DATA step itself, for example, offers a number of possibilities to import text files, log files, or event data from hierarchical databases.
- SAS/ACCESS interfaces to relational databases such as Oracle, DB2, MS SQL Server, and Teradata allow native access to databases, which speeds up import and export times and the automatic conversion of variable formats and column definitions.
- SAS/ACCESS interfaces to industry standards such as ODBC or OLEDB allow data import and export via these interfaces. Easy data access is therefore possible to any data source that provides an ODBC or OLEDB interface.
- The PC File Server allows access to data format of popular PC file formats such as Excel, Access, Lotus, and dBase.
- SAS provides interfaces, the so-called data surveyors, to enterprise resource planning systems like SAP, SAP BW, PeopleSoft, and Oracle E-Business Suite.

For more details and code examples, see *Chapter 13 – Accessing Data*.

B.4 Changing the Data Mart Structure: Transposing

B.4.1 Overview

Transposing data sets is a significant task in preparing data sets. The structure of data sets might need to be changed due to different data structure requirements of certain analyses, and data might need to be transposed in order to allow a join of different data sets in an appropriate way.

B.4.2 Transposing Data Sets with PROC TRANSPOSE

The TRANSPOSE procedure is part of Base SAS and allows the transposition of data sets in the following ways:

- Complete transposition: The rows and columns of a data set are exchanged. In other words, this type of unrestricted transposition turns a data set at a 45° diagonal.
- Transposing within BY groups: This is very important if data will be transposed per subject ID.

Example: The CUSTOMER table in Table B.1 represents data in a multiple-rows-per-subject structure. With the following statements, the multiple observations per subject are transposed to columns. The results are given in Table B.2.

```
PROC TRANSPOSE DATA  = customer
                OUT  = customer_tp(drop = _name_)
                PREFIX = Month;
  BY  CustID Age;
  VAR Value;
  ID  Month;
RUN;
```

Table B.1: Multiple rows per subject

	CustID	Age	Month	Value
1	1	26	7	45
2	1	26	8	34
3	1	26	9	5
4	2	37	7	34
5	2	37	8	32
6	2	37	9	44
7	3	46	7	56
8	3	46	8	54
9	3	46	9	32

Table B.2: One row per subject

	CustID	Age	Month7	Month8	Month9
1	1	26	45	34	5
2	2	37	34	32	44
3	3	46	56	54	32

The major advantage of PROC TRANSPOSE as part of SAS compared with other SQL languages is that the number and the numbering of repeated measurements are not part of the syntax. Therefore, PROC TRANSPOSE can be used flexibly in cases where the number of repeated measurements or the list of categories is extended.

Chapter 14 deals with different cases of transposing data sets in more detail.

B.5 Data Management for Longitudinal and Multiple-Rows-per-Subject Data Sets

Special data management features for longitudinal and multiple-rows-per-subject data sets are available in SAS. These features are important for the data preparation of multiple-rows-per-subject data sets themselves and for the data preparation of one-row-per-subject data sets, where data have to be aggregated.

B.5.1 Accessing the First and Last Observations per Subject

The logical variables FIRST and LAST allow identifying the first and last observations per subject or BY group. The data set needs to be sorted by subject ID or BY group. The following code selects the last observation per customer (Cust_ID) and outputs it to the CUSTOMER_LAST table:

```
DATA customer_last;
 SET customer;
 BY  CustID;
 IF LAST.CustID THEN OUTPUT;
RUN;
```

Table B.3: CUSTOMER table

	CustID	Month	Value
1	1	7	45
2	1	8	34
3	1	9	5
4	2	7	34
5	2	8	32
6	2	9	44
7	2	10	56
8	3	7	54
9	3	8	32

Table B.4: CUSTOMER_LAST table

	CustID	Month	Value
1	1	9	5
2	2	10	56
3	3	8	32

The FIRST and LAST variables are very powerful for sequential data management operations on a subject level—for example, for the creation of sequence numbers per subject.

B.5.2 Calculating a Sequence Number Subject

The following code adds a sequence number per customer to the CUSTOMER table in the preceding example:

```
DATA customer;
 SET customer;
 BY CustID;
 IF FIRST.CustID THEN sequence = 1;
 ELSE sequence + 1;
RUN;
```

Table B.5: Table with sequences

	CustID	Month	Value	sequence
1	1	7	45	1
2	1	8	34	2
3	1	9	5	3
4	2	7	34	1
5	2	8	32	2
6	2	9	44	3
7	2	10	56	4
8	3	7	54	1
9	3	8	32	2

B.5.3 Shifting Columns in a Table k Rows Down

The LAG function makes the value of a variable available in the rows that follow. The DIF function calculates the difference between the value and the value in the row above. Note that both functions are generic functions—LAG2 provides the value from two rows above and LAGn provides the value of n lines above.

The following code adds the columns LAG_VALUE and DIF_VALUE to the table, where LAG_VALUE contains the LAG value of one row above and DIF_VALUE contains the difference between the actual value and the value one row above.

```
DATA Series;
 SET Series;
 Lag_Value = LAG(Value);
 Dif_Value = DIF(Value);
RUN;
```

Table B.6: Lagged values

	Date	Value	Lag_Value	Dif_Value
1	1999	45	.	.
2	2000	34	45	-11
3	2001	5	34	-29
4	2002	34	5	29
5	2003	32	34	-2
6	2004	44	32	12

B.5.4 Copying Values to Consecutive Rows

In some cases data are available where the static values of an analysis subject are not repeated with each multiple observation but are listed only in the first row per subject. The following code copies the value of CustID and AGE down to consecutive rows if no new subject is started.

Table B.7: Data from a hierarchical structure

	CustID	Age	Month	Value
1	1	26	7	45
2			8	34
3	.	.	9	5
4	2	37	7	34
5	.	.	8	32
6	.	.	9	44
7	3	46	7	56
8	.	.	8	54
9	.	.	9	32

```
DATA customer_filled;
 SET customer_hierarchic;
 RETAIN custid_tmp age_tmp;
 IF CustID NE . THEN DO; custid_tmp = CustID; age_tmp = age;  END;
 ELSE DO; CustID = custid_tmp; age = age_tmp; END;
 DROP custid_tmp age_tmp;
RUN;
```

Table B.8: Table with "filled" blanks of the hierarchical data

	CustID	Age	Month	Value
1	1	26	7	45
2	1	26	8	34
3	1	26	9	5
4	2	37	7	34
5	2	37	8	32
6	2	37	9	44
7	3	46	7	56
8	3	46	8	54
9	3	46	9	32

B.5.5 More Details

For more details, explanations, and examples about data management with longitudinal data in SAS, see the SAS Press title *Longitudinal Data and SAS*.

B.6 Selected Features of the SAS Language for Data Management

B.6.1 SAS Functions

SAS provides a number of important functions for data preparation. These functions allow powerful data management operations—for example, by creating derived variables. The following list gives an overview of the most important function categories:

- character string matching functions that use PERL regular expressions
- general character functions that prepare, process, and extract character strings
- date and time functions that create date and time values and transform date and time variables
- descriptive statistics and quantile functions that calculate, for example, means and/or medians over columns
- mathematical functions that perform mathematical operations
- probability and random number functions that analyze random numbers and probability and density functions for various distributions

B.6.2 Array Processing of Variables

The SAS DATA step allows array processing of variables in the sense that the operations need to be defined only once for a set of variables. In the following example, missing values in the variables MONTH7, MONTH8, and MONTH9 are replaced with 0 by using an array.

```
DATA CUSTOMER_TP;
 SET CUSTOMER_TP;
ARRAY Months {3} Month7 Month8 Month9;
DO I = 1 to 3;
 IF Months{i} = . THEN Months{i}=0;
END;
DROP I;
RUN;
```

B.6.3 Sampling

Sampling can be performed easily by using the random number functions in SAS. The following example shows how to draw a 10% sample:

```
DATA customer_sample_10pct;
SET customer;
IF UNIFORM(1234) < 0.1;
RUN;
```

The following example shows how to draw two 20% samples that do not overlap:

```
DATA customer_sample1
Customer_sample2;
Set customer;
IF UNIFORM(1234) lt 0.2 THEN OUTPUT customer_sample1;
ELSE IF UNIFORM(2345) lt 0.2/(1-0.2) THEN OUTPUT customer_sample2;
RUN;
```

B.6.4 Using Logical IN Variables

Data sets that appear in the SET or MERGE statement can have so-called logical IN variables specified that have a value of 1 if an observation of the resulting DATA step occurs in the respective input data set. These IN variables can also be used to define derived binary variables. The following example shows how a binary variable CALLCENTER can be created that contains information about whether a call center record is available for the customer or not.

```
DATA customer_mart;
 MERGE customer
       Call_center (IN = in2);
 BY Cust_ID;
 CallCenter = in2;
 ---- Other SAS statements ---
RUN;
```

B.6.5 Ranking and Grouping of Observations

Ranking and grouping is important for the preparation of analytic analysis tables, especially in order to calculate derived variables. PROC RANK provides a number of options for ranking and grouping. For example:

- creating a rank variable according to the sort order of a certain variable

- grouping variables into groups—for example, decile groups depending on the value of a certain variable

- calculating the quantile value of an observation depending on the value of a certain variable

The following example shows how a variable can be calculated that contains the deciles of a variable.

```
PROC RANK DATA = customer
OUT = customer
GROUPS = 10;
VAR value;
RANKS value_deciles;
RUN;
```

B.6.6 Aggregating Data for Categories

SAS is extremely powerful in aggregating data for subgroups. PROC MEANS provides the functionality to aggregate data over the whole population for subgroups or for various combinations of subgroups. The following code shows how to aggregate data for COUNTRY and PRODUCT*COUNTRY:

```
PROC MEANS DATA = sashelp.prdsale SUM;
 VAR actual;
 CLASS product country;
 TYPES product*country country;
RUN;
```

The result is shown here.

```
                        The MEANS Procedure
                Analysis Variable : ACTUAL Actual Sales
                              N
                Country      Obs              Sum
                ------------------------------------
                CANADA       480         246990.00
                GERMANY      480         245998.00
                U.S.A.       480         237349.00
                ------------------------------------
              Analysis Variable · ACTUAL Actual Sales
                                 N
            Product     Country      Obs            Sum
            ---------------------------------------------
            BED         CANADA        96        47729.00
                        GERMANY       96        46134.00
                        U.S.A.        96        48174.00
            CHAIR       CANADA        96        50239.00
                        GERMANY       96        47105.00
                        U.S.A.        96        50936.00
            DESK        CANADA        96        52187.00
                        GERMANY       96        48502.00
                        U.S.A.        96        48543.00
            SOFA        CANADA        96        50135.00
                        GERMANY       96        55060.00
                        U.S.A.        96        43393.00
            TABLE       CANADA        96        46700.00
                        GERMANY       96        49197.00
                        U.S.A.        96        46303.00
            ---------------------------------------------
```

This output can also be directed into an output data set and used for further processing.

B.6.7 Re-using the Output of SAS Procedures for Data Preparation

The output data sets of various SAS procedures can be re-used.

Example: For each retail customer, a time series over 12 months with the monthly sales volume is available. PROC REG is used to calculate a linear trend over time of the sales volume for each customer. The following code shows the trend can be calculated and the resulting regression coefficients can be merged with the CUSTOMER table:

```
PROC REG DATA = sales_history
         OUTEST = sales_trend_coeff
               NOPRINT;
 MODEL volume = month;
BY CustID;
RUN;
QUIT;
```

The result is shown in Table B.9. For more details, see Chapter 18.

Table B.9: Regression coefficients per customer

	CustID	Intercept	Month
1	1	66.066666667	-5.971428571
2	2	21.6	1.6857142857
3	3	114.66666667	-2.285714286
4	4	43.744186047	-0.395348837
5	5	20.4	3.4571428571
6	6	16.333333333	1.8571428571
7	7	74	-2.571428571
8	8	89	1
9	9	47	0
10	10	43.6	-2.742857143

```
DATA customer_table;
 MERGE customer
       Sales_trend_coeff (KEEP = CustID time);
 BY CustID;
RUN;
```

PROC REG calculates a linear regression for each customer according to the BY statement and stores the regression coefficient.

B.7 Benefits of the SAS Macro Language

The SAS macro language provides a lot of functionality to make programs more flexible and efficient. For data management the following features are of special interest.

B.7.1 Replacing Text Strings and Variable Names Using Macro Variables

The value of macro variables can be used to replace the value of a text string and also to replace the whole or part of a variable name. With macro parameters, for example, programs can be parameterized to use the actual data set or a special variable. The macro variable assignment

```
%LET Actual_Month = 200606;
```

can be used in a program to select the data set for the correct month. Let's assume that the data set names are in the format CUSTOMER_200605, CUSTOMER_200606, and so on. The DATA step can be written as follows in order to use the appropriate data set:

```
DATA Customer_TMP;
 SET customer_&Actual_Month;
---- Other SAS statements ---
RUN;
```

Macro variables, as in the following example, can be used to define a list of variables that will be processed together:

```
%LET varlist = AGE INCOME NR_CARS NR_CHILDREN;
```

The macro variable VARLIST can be used as a replacement for the variable names.

```
PROC MEANS DATA = customer;
 VAR &varlist;
RUN;

PROC LOGISTIC DATA = customer;
 MODEL claim = &varlist;
RUN;
QUIT;
```

We discussed the importance and application of this feature in *Chapter 23 – Scoring and Automation*.

B.7.2 Generating SAS Code Using Macros

The SAS macro language can also be used to generate SAS code. This can be done simply by replacing a macro invocation with a fixed text in order to make programs shorter and easier to read. Parameterized macros can be used to make SAS programs more flexible and efficient.

The following simple macro shows that a macro can be used to generate SAS code:

```
%MACRO data set_names (name =, number = );
%DO n = 1 %TO &number;
&name&n
%END;
;
%MEND;
```

The macro can be applied as follows:

```
DATA tmp_data;
SET %data set_names(name=customer,number=4);
RUN;
```

This resolves to the following:

```
DATA tmp_data;
SET customer1 customer2 customer3 customer4;
RUN;
```

B.7.3 Using the %SCAN Function to Scan through Value Lists

The %SCAN function can be used to scan through value lists and to use the kth item in an operation. This is very useful if we want to use a list of variables in a macro variable even if the values cannot be specified one after the other in the respective syntax. Consider the following example:

We specify the list of variables in a macro variable and run PROC FREQ:

```
%LET var = product country region;
PROC FREQ DATA = sashelp.prdsale;
 TABLE &var;
RUN;
```

This creates the desired result; we receive a frequency analysis for each variable. However, if we want to save the frequencies in an output data set, amending the code as in the following example does not produce the desired results.

```
PROC FREQ DATA = sashelp.prdsale;
 TABLE &var / OUT = Freqs_OUT;
RUN;
```

We need to specify a TABLE statement for each variable in order to output the results in an output data set. However, in order to use the values in the VAR= variable, we need to use %SCAN to extract them from the syntax:

```
%MACRO loop(data=,var=,n=);
  PROC FREQ DATA = &data;
 %DO i = 1 %TO &n;
   TABLE %SCAN(&var,&i) / OUT = %SCAN(&var,&i)_Table;
 %END;
  RUN;
%MEND;

%LOOP(DATA = sashelp.prdsale,
      VAR = product country region, N   = 3);
```

We see from the SAS log that the macro with the %SCAN function creates the desired code:

```
MPRINT(LOOP):    PROC FREQ DATA = sashelp.prdsale;
MPRINT(LOOP):    TABLE product / OUT = product_Table;
MPRINT(LOOP):    TABLE country / OUT = country_Table;
MPRINT(LOOP):    TABLE region / OUT = region_Table;
MPRINT(LOOP):    RUN;


NOTE: There were 1440 observations read from the data set
SASHELP.PRDSALE
NOTE: The data set WORK.PRODUCT_TABLE has 5 observations and 3
variables.
NOTE: The data set WORK.COUNTRY_TABLE has 3 observations and 3
variables.
NOTE: The data set WORK.REGION_TABLE has 2 observations and 3
variables.
NOTE: PROCEDURE FREQ used (Total process time):
      real time          0.06 seconds
      cpu time           0.06 seconds
```

We have already used this functionality in a number of macros in this book.

B.7.4 More Details

For more details about the SAS macro language, see the *SAS Macro Language: Reference Guide* or SAS Press books on this topic.

B.8 Matrix Operations with SAS/IML

SAS/IML is the abbreviation for interactive matrix language. For data management, the matrix operations features are of interest. SAS data sets are imported to SAS/IML matrices. On those matrices, among others, the following operations can be performed:

- addition, subtractions, and division of matrices
- matrix multiplication and element-wise matrix multiplication
- matrix transposition and matrix inversion
- extractions of sub-matrices and vectors on a subscript basis
- horizontal and vertical concatenations of matrices

For more details, see the SAS/IML Help.

Appendix C

Transposing with DATA Steps

C.1 Transposing and Performance

General

We saw in Chapter 14 that the TRANSPOSE procedure can be used to transpose data in a very flexible way and with a clear syntax. PROC TRANSPOSE and the macros we presented in Chapter 14 are the first choice for data transpositions.

However, in situations where the number of subjects is in the hundreds of thousands, and the number of variables or repetitions is in the thousands, performance issues might arise. In these cases it is advisable to use a SAS DATA step instead of PROC TRANPOSE. In addition to a performance advantage with large data sets, the SAS DATA step also has the advantage that additional data processing can be performed during transposition.

In order to illustrate the differences in performance, a short study was performed that compared the time consumption for a PROC TRANSPOSE and a SAS DATA step transposition. In the following sections, we compare the macros from Chapter 14 with a DATA step transposition. The macro for the DATA step transposition is presented in the following sections.

For simplicity, we will again use the LONG and WIDE terminology for the data set shapes. A one-row-per-subject data set will be called a WIDE data set in this context because the multiple observations are represented by columns. A multiple-rows-per-subject data set is represented as a LONG data set because it has a row per subject and multiple observations.

C.1.1 The Test Environment

The performance tests were performed on a Dell D620 Latitude laptop, with 2 GB RAM and a Genuine Intel CPU T2500.2 GHz and Microsoft Windows XP Professional Edition, Service Pack 2.

For the performance tests, SAS data sets were created, where the transposed VAR variable was filled with a random uniform number. Measurements at n=100,000, 200,000, and so on, were taken with 90 repetitions of the measurement. The duration was measured in seconds by comparing the start time and the end time. Note that the test data sets were already sorted by subject ID.

C.1.2 From a LONG Data Set to a WIDE Data Set

Figure C.1 shows that for large data sets the DATA step transposition (solid line) is faster than the PROC TRANSPOSE version (dotted line). In absolute numbers: for 1,000,000 observations, the difference is 7 minutes (664 seconds for the PROC TRANSPOSE version compared to 244 seconds for the DATA step version).

Figure C.1: Performance comparison for MAKEWIDE transposition

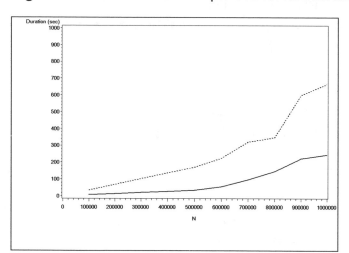

C.1.3 From a WIDE Data Set to a LONG Data Set

We can see from the line plot that the DATA step transposition (solid line) scales almost linearly with the number of observations, and the PROC TRANSPOSE version (dotted line) takes more than twice as long. In absolute numbers: for 1,000,000 observations and 90 repetitions, we see a difference of 6 minutes 20 seconds (172 seconds for the DATA step version and 551 seconds for the PROC TRANSPOSE version).

Figure C.2: Performance comparison for MAKELONG transposition

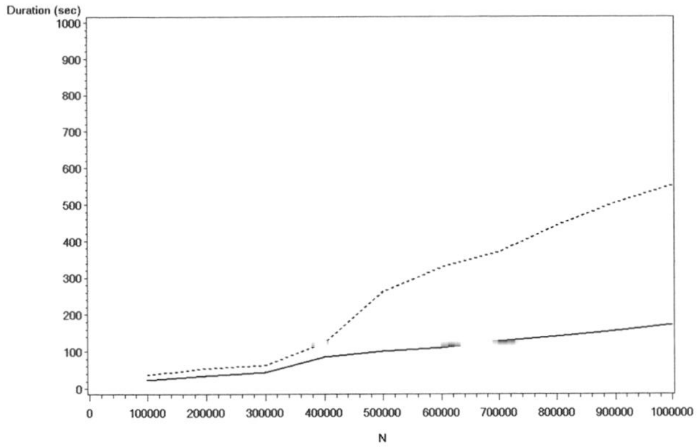

C.1.4 Conclusion

The DATA step version is more time-efficient with increasing data volume. Note, for example, that in data mining analyses not only the repetitions for one variable have to be transposed, but a number of variables. In this case the performance difference quickly multiplies with the number of variables.

The decision whether the DATA step version or the PROC TRANSPOSE version will be selected depends on the number of observations and repetitions that will be transposed and on how important it is to shorten the processing time.

C.1.5 Macros

In the following sections we will introduce the following macros to transpose data using a SAS DATA step:

- %MAKEWIDE_DS to transpose data to a one-row-per-subject structure
- %MAKELONG_DS to transpose data to a multiple-rows-per-subject structure
- %TRANSP_CAT_DS to transpose multiple categories per subject to a one-row-per-subject structure

C.2 From a LONG Data Set to a WIDE Data Set

The following macro %MAKEWIDE_DS transposes a LONG data set into a WIDE data set by using a SAS DATA step:

```
%MACRO makewide_ds(DATA=,OUT=,COPY=,ID=,VAR=,
                   TIME=Measurement);

*** Part 1 - Creating a list of Measurement IDs;
PROC FREQ DATA = &data NOPRINT;
 TABLE &time / OUT = distinct (DROP = count percent);
RUN;
```

```
DATA _null_;
 SET distinct END = eof;
 FORMAT _string_ $32767.;
 RETAIN _string_;
 _string_ = CATX(' ',_string_, &time);
 IF eof THEN DO;
    CALL SYMPUT('list',_string_);
  CALL SYMPUT('max',_n_);
 END;
RUN;

*** Part 2 - Using a SAS DATA step for the transpose;
DATA &out;
 SET &data;
 BY &id;
 RETAIN %DO i= 1 %to &max; &var%SCAN(&list,&i) %END; ;
 IF FIRST.&id THEN DO;
  %DO i= 1 %TO &max; &var%SCAN(&list,&i)=.; %END; ;
 END;
 %DO i = 1 %TO &max;
  IF &time = %SCAN(&list,&i) THEN DO;
    &var%SCAN(&list,&i) = &var;
  END;
 %END;
 IF LAST.&id THEN OUTPUT;
 KEEP &id &copy %DO i= 1 %to &max; &var%SCAN(&list,&i) %END;;
RUN;

%MEND;
```

The main parts of the macro are as follows:

- Using PROC FREQ to create a list of distinct measurement IDs and a _NULL_ DATA step to create a list of measurement IDs into one macro variable. Note that from a performance point of view, PROC FREQ with an OUT= data set is much faster than the equivalent SELECT DISTINCT version in PROC SQL.

- Using a SAS DATA step for the transposition. Here we define new variables with the name from the macro variable &VAR and the numbering from the distinct measurement IDs that are found in the variable &TIME. The variables are retained for each subject and the last observation per subject is output to the resulting data set.

Note that the data must be sorted for the ID variable. The legend for the macro parameters is as follows:

DATA and OUT

The names of the input and output data sets, respectively.

ID

The name of the ID variable that identifies the subject.

COPY

The list of variables that occur repeatedly with each observation for a subject and that will be copied to the resulting data set. We assume here that COPY variables have the same values within one ID.

VAR

> The variable that contains the values that will be transposed. Note that only one variable can be listed here in order to obtain the desired transposition. See below for an example of how to deal with a list of variables.

TIME

> The variable that enumerates the repeated measurements.

Note that the TIME variable does not need to be a consecutive number, but it can also have non-equidistant intervals. See the following data set from an experiment with dogs.

```
%MAKEWIDE_DS(DATA=dogs_long,OUT=dogs_wide_2,
          ID=id, COPY=drug depleted,
          VAR=Histamine,
          TIME=Measurement);
```

C.3 From a WIDE Data Set to a LONG Data Set

The following macro %MAKELONG_DS transposes a WIDE data set into a LONG data set by using a SAS DATA step:

```
%MACRO makelong_ds(DATA=,OUT=,COPY=,ID=,LIST=,MAX=,MIN=,
        ROOT=,TIME=Measurement);

DATA &out(WHERE = (&root NE .));
 SET &data;
 %IF &list NE %THEN %DO;
 *** run the macro in LIST-Mode;

*** Load the number of items in &VARS into macro variable NVARS;
%LET c=1;
%DO %WHILE(%SCAN(&list,&c) NE);
   %LET c=%EVAL(&c+1);
%END;
%LET nvars=%EVAL(&c-1);

   %DO i = 1 %TO &nvars;
       &root=&root.%SCAN(&list,&i);
       &time = %SCAN(&list,&i);
       OUTPUT;
   %END;
 %END;
 %ELSE %DO;
 *** run the macro in FROM/TO mode;
  %DO i = &min %TO &max;
       &root=&root.&i;
       &time= &i;
       OUTPUT;
   %END;
 %END;
 KEEP &id &copy &root &time;
RUN;

%MEND;
```

The macro works as follows:

- The macro uses a DATA step to output one observation for each repetition per subject. The ROOT and TIME value is copied from the respective variable.

Note that the data must be sorted for the ID variable.

Note that the macro can run in two modes: the LIST mode and the FROM/TO mode. The two modes differ in how the time IDs that form the postfix of the variable names that have to be transposed are specified:

- In the LIST mode a list of time IDs has to be explicitly specified in the macro variable LIST = that contains all time IDs where an observation is available. Note that the time IDs do not need to be consecutive numbers.
- In the FROM/TO mode a consecutive list of time IDs is assumed and can be specified with the MIN= and MAX= variables.

If a value is specified for the LIST= variable, the macro runs in LIST mode; otherwise, it runs in FROM/TO mode.

The legend for the macro parameters is as follows:

DATA and OUT

The names of the input and output data sets, respectively.

ID

The name of the ID variable that identifies the subject.

COPY

The list of variables that occurs repeatedly with each observation for a subject and that will be copied to the resulting data set. We assume here that COPY variables have the same values within one ID.

ROOT

The part of the variable name, without the measurement number, of the variable that will be transposed. Note that only one variable can be listed here in order to obtain the desired transposition. See below for an example of how to deal with a list of variables.

TIME

The variable that will enumerate the repeated measurements.

MAX

The maximum enumerations of the variables in the WIDE data set that will be transposed.

MIN

The minimum of the enumerations of the variables in the WIDE data set that will be transposed.

LIST

The list of time IDs in brackets that is used to enumerate the variable names in the WIDE data set (note that the variable names start with the string specified under ROOT=).

See in the following examples two invocations of macros in the LIST mode and in the FROM/TO mode.

```
%MAKELONG_DS(DATA=dogs_wide,OUT=dogs_long2,COPY=drug depleted,ID=id,
             ROOT=Histamine,TIME=Measurement,LIST=(0 1 3 5),MAX=4);

%MAKELONG_DS(data=wide,out=long,id=id,min=1,max=3,
             root=weight,time=time);
```

C.4 Transposing Transactional Categories with a DATA Step

The following macro %TRANSP_CAT_DS transposes transactional data by using a SAS DATA step:

```
%MACRO TRANSP_CAT_DS (DATA = , OUT = out, VAR = , ID =);
*** PART1 - Aggregating multiple categories per subject;
PROC FREQ DATA  = &data NOPRINT;
 TABLE &id * &var / OUT = out(DROP = percent);
 TABLE &var /  OUT = distinct (drop = count percent);
RUN;
*** PART2 - Assigning the list of categories into a macro-variable;
DATA _null_;
 SET distinct END = eof;
 FORMAT _string_ $32767.;
 RETAIN _string_;
 _string_ = CATX(' ',_string_, &var);
 IF eof THEN DO;
    CALL SYMPUT('list',_string_);
   CALL SYMPUT('_nn_',_n_);
 END;
RUN;
*** PART3 - Using a SAS-DATA step for the transpose;
DATA &out;
 SET out;
 BY &id;
 RETAIN &list ;
 IF FIRST.&id THEN DO;
  %DO i= 1 %TO &_nn_; %SCAN((&list),&i)=.; %END; ;
 END;
 %DO i = 1 %TO &_nn_;
  IF &var = "%scan(&list,&i)" THEN  %SCAN((&list),&i) = count;
 %END;
 IF LAST.&id THEN OUTPUT;
 DROP &var count;
RUN;

%MEND;
```

The three main parts of the macro are as follows:

- Aggregating multiple observations per subject. This step is equivalent to the PROC TRANPOSE macro version. Here we additionally create a list of distinct categories in the DISTINCT data set. Note that from a performance point of view, PROC FREQ with an OUT= data set is much faster than the equivalent SELECT DISTINCT version in PROC SQL.

- Assigning the list of categories into a macro variable. Here we use a _NULL_ DATA step to create a macro variable that contains the list of categories. Note the length of all category names separated by columns must not exceed 32,767 characters.

- Using a SAS DATA step for the transposition. Here we define variables from the categories. The variables are retained for each subject and the last observation per subject is output to the resulting data set.

Note that different from the PROC TRANSPOSE macro version, the DATA step version does not allow blanks or leading numbers in the categories!

The legend for the macro parameters is as follows:

DATA and OUT

The names of the input and output data sets, respectively.

ID

The name of the ID variable that identifies the subject.

VAR

The variable that contains the categories, for example, in market basket analysis, the products a customer purchased.

The invocation for our market basket example is as follows:

```
%TRANSP_CAT_DS (DATA = sampsio.assocs, OUT = assoc_tp,
            ID = customer, VAR = Product);
```

Glossary

analysis subject

an entity that is being analyzed. The analysis results are interpreted in the context of the subject. Analysis subjects are the basis for the structure of the analysis table.

analysis table

a single rectangular table that holds the data for the analysis, where the observations are represented by rows, and the attributes are represented by columns.

application layer

a logical layer that enables you to access data from a system using its business logic and application functions.

business question

a type of question that defines the content and rationale for which an analysis is done.

database

an IT system that enables the insertion, storage, and retrieval of data.

data layer

a logical layer where data is directly imported from the underlying database tables.

data mart

a single rectangular table that holds the data for the analysis, where the observations are represented by rows, and the attributes are represented by columns.

de-normalization

a process whereby data is redundantly stored in more than one table in order to avoid joining tables.

derived variable

a type of variable whose values are calculated on the basis of one or more other variables.

dispositive system

a type of IT system that retrieves, stores, and prepares data in order to provide reports, predictions, and forecasts to businesses.

longitudinal analysis

a type of analysis that is based on the time or the sequential ordering of observations.

modeling

the process of creating (learning) a model logic, e.g., in order to predict events and values.

multiple observations

the existence of more than one observation for an analysis subject in the data.

one-to-many relationship

a type of relationship between two tables where the two tables include related rows. For example, in table A and table B, one row in table A might have many related rows in table B (see section 6.2).

operational system

an IT system that is designed to assist business operations by providing the technical infrastructure to process data from their business processes.

relational model

a type of model that structures data on the basis of entities and relationships between them (see section 5.5).

scoring

the process of applying the logic that has been created during modeling to the data.

star schema

a type of relational model. A star schema is composed of one fact table and a number of dimension tables. The dimension tables are linked to the fact table with a one-to-many relationship (see section 6.4).

transposing

a method of rearranging the columns and rows of a table.

Index

S

Books Available from SAS Press

Advanced Log-Linear Models Using SAS®
by **Daniel Zelterman**

Analysis of Clinical Trials Using SAS®: A Practical Guide
by **Alex Dmitrienko, Geert Molenberghs, Walter Offen,** *and*
Christy Chuang-Stein

Annotate: Simply the Basics
by **Art Carpenter**

*Applied Multivariate Statistics with SAS® Software,
Second Edition*
by **Ravindra Khattree**
and **Dayanand N. Naik**

*Applied Statistics and the SAS® Programming Language,
Fifth Edition*
by **Ronald P. Cody**
and **Jeffrey K. Smith**

An Array of Challenges — Test Your SAS® Skills
by **Robert Virgile**

*Carpenter's Complete Guide to the SAS® Macro Language,
Second Edition*
by **Art Carpenter**

The Cartoon Guide to Statistics
by **Larry Gonick**
and **Woollcott Smith**

*Categorical Data Analysis Using the SAS® System,
Second Edition*
by **Maura E. Stokes, Charles S. Davis,**
and **Gary G. Koch**

Cody's Data Cleaning Techniques Using SAS® Software
by **Ron Cody**

*Common Statistical Methods for Clinical Research with
SAS® Examples, Second Edition*
by **Glenn A. Walker**

The Complete Guide to SAS® Indexes
by **Michael A. Raithel**

*Data Management and Reporting Made Easy with
SAS® Learning Edition 2.0*
by **Sunil K. Gupta**

Data Preparation for Analytics Using SAS®
by **Gerhard Svolba**

*Debugging SAS® Programs: A Handbook of Tools and
Techniques*
by **Michele M. Burlew**

*Decision Trees for Business Intelligence and Data Mining: Using
SAS® Enterprise Miner™*
by **Barry de Ville**

*Efficiency: Improving the Performance of Your SAS®
Applications*
by **Robert Virgile**

The Essential Guide to SAS® Dates and Times
by **Derek P. Morgan**

The Essential PROC SQL Handbook for SAS® Users
by **Katherine Prairie**

*Fixed Effects Regression Methods for Longitudinal Data
Using SAS®*
by **Paul D. Allison**

Genetic Analysis of Complex Traits Using SAS®
Edited by **Arnold M. Saxton**

A Handbook of Statistical Analyses Using SAS®, Second Edition
by **B.S. Everitt**
and **G. Der**

Health Care Data and SAS®
by **Marge Scerbo, Craig Dickstein,**
and **Alan Wilson**

The How-To Book for SAS/GRAPH® Software
by **Thomas Miron**

*In the Know ... SAS® Tips and Techniques From
Around the Globe, Second Edition*
by **Phil Mason**

Instant ODS: Style Templates for the Output Delivery System
by **Bernadette Johnson**

*Integrating Results through Meta-Analytic Review Using
SAS® Software*
by **Morgan C. Wang**
and **Brad J. Bushman**

Introduction to Data Mining Using SAS® Enterprise Miner™
by **Patricia B. Cerrito**

Learning SAS® in the Computer Lab, Second Edition
by **Rebecca J. Elliott**

The Little SAS® Book: A Primer
by **Lora D. Delwiche**
and **Susan J. Slaughter**

The Little SAS® Book: A Primer, Second Edition
by **Lora D. Delwiche**
and **Susan J. Slaughter**
(updated to include SAS 7 features)

support.sas.com/pubs

The Little SAS® Book: A Primer, Third Edition
by **Lora D. Delwiche**
and **Susan J. Slaughter**
(updated to include SAS 9.1 features)

The Little SAS® Book for Enterprise Guide® 3.0
by **Susan J. Slaughter**
and **Lora D. Delwiche**

The Little SAS® Book for Enterprise Guide® 4.1
by **Susan J. Slaughter**
and **Lora D. Delwiche**

Logistic Regression Using the SAS® System:
Theory and Application
by **Paul D. Allison**

Longitudinal Data and SAS®: A Programmer's Guide
by **Ron Cody**

Maps Made Easy Using SAS®
by **Mike Zdeb**

Models for Discrete Data
by **Daniel Zelterman**

Multiple Comparisons and Multiple Tests Using SAS®
Text and Workbook Set
(books in this set also sold separately)
by **Peter H. Westfall, Randall D. Tobias,**
Dror Rom, Russell D. Wolfinger,
and **Yosef Hochberg**

Multiple-Plot Displays: Simplified with Macros
by **Perry Watts**

Multivariate Data Reduction and Discrimination with
SAS® Software
by **Ravindra Khattree**
and **Dayanand N. Naik**

Output Delivery System: The Basics
by **Lauren E. Haworth**

Painless Windows: A Handbook for SAS® Users, Third Edition
by **Jodie Gilmore**
(updated to include SAS 8 and SAS 9.1 features)

Pharmaceutical Statistics Using SAS®: A Practical Guide
Edited by **Alex Dmitrienko, Christy Chuang-Stein,**
and **Ralph D'Agostino**

The Power of PROC FORMAT
by **Jonas V. Bilenas**

PROC SQL: Beyond the Basics Using SAS®
by **Kirk Paul Lafler**

PROC TABULATE by Example
by **Lauren E. Haworth**

Professional SAS® Programmer's Pocket Reference,
Fifth Edition
by **Rick Aster**

Professional SAS® Programming Shortcuts, Second Edition
by **Rick Aster**

Quick Results with SAS/GRAPH® Software
by **Arthur L. Carpenter**
and **Charles E. Shipp**

Quick Results with the Output Delivery System
by **Sunil K. Gupta**

Reading External Data Files Using SAS®: Examples Handbook
by **Michele M. Burlew**

Regression and ANOVA: An Integrated Approach Using
SAS® Software
by **Keith E. Muller**
and **Bethel A. Fetterman**

SAS® for Forecasting Time Series, Second Edition
by **John C. Brocklebank**
and **David A. Dickey**

SAS® for Linear Models, Fourth Edition
by **Ramon C. Littell, Walter W. Stroup,**
and **Rudolf J. Freund**

SAS® for Mixed Models, Second Edition
by **Ramon C. Littell, George A. Milliken, Walter W. Stroup,**
Russell D. Wolfinger, *and* **Oliver Schabenberger**

SAS® for Monte Carlo Studies: A Guide for Quantitative
Researchers
by **Xitao Fan, Ákos Felsővályi, Stephen A. Sivo,**
and **Sean C. Keenan**

SAS® Functions by Example
by **Ron Cody**

SAS® Guide to Report Writing, Second Edition
by **Michele M. Burlew**

SAS® Macro Programming Made Easy, Second Edition
by **Michele M. Burlew**

SAS® Programming by Example
by **Ron Cody**
and **Ray Pass**

SAS® Programming for Researchers and Social Scientists,
Second Edition
by **Paul E. Spector**

SAS® Programming in the Pharmaceutical Industry
by **Jack Shostak**

SAS® Survival Analysis Techniques for Medical Research,
Second Edition
by **Alan B. Cantor**

SAS® System for Elementary Statistical Analysis,
Second Edition
by **Sandra D. Schlotzhauer**
and **Ramon C. Littell**

SAS® System for Regression, Third Edition
by **Rudolf J. Freund**
and **Ramon C. Littell**

SAS® System for Statistical Graphics, First Edition
by **Michael Friendly**

support.sas.com/pubs

The SAS® Workbook and *Solutions* Set
(books in this set also sold separately)
by **Ron Cody**

Selecting Statistical Techniques for Social Science Data:
A Guide for SAS® Users
by **Frank M. Andrews, Laura Klem, Patrick M. O'Malley,**
Willard L. Rodgers, Kathleen B. Welch,
and **Terrence N. Davidson**

Statistical Quality Control Using the SAS® System
by **Dennis W. King**

A Step-by-Step Approach to Using the SAS® System
for Factor Analysis and Structural Equation Modeling
by **Larry Hatcher**

A Step-by-Step Approach to Using SAS® for Univariate and
Multivariate Statistics, Second Edition
by **Norm O'Rourke, Larry Hatcher,**
and **Edward J. Stepanski**

Step-by-Step Basic Statistics Using SAS®: Student Guide
and *Exercises*
*(*books in this set also sold separately)
by **Larry Hatcher**

Survival Analysis Using SAS®:
A Practical Guide
by **Paul D. Allison**

Tuning SAS® Applications in the OS/390 and z/OS
Environments, Second Edition
by **Michael A. Raithel**

Univariate and Multivariate General Linear Models:
Theory and Applications Using SAS® Software
by **Neil H. Timm**
and **Tammy A. Mieczkowski**

Using SAS® in Financial Research
by **Ekkehart Boehmer, John Paul Broussard,**
and **Juha-Pekka Kallunki**

Using the SAS® Windowing Environment: A Quick Tutorial
by **Larry Hatcher**

Visualizing Categorical Data
by **Michael Friendly**

Web Development with SAS® by Example, Second Edition
by **Frederick E. Pratter**

Your Guide to Survey Research Using the SAS® System
by **Archer Gravely**

JMP® Books

JMP® for Basic Univariate and Multivariate Statistics: A Step-by-
Step Guide
by **Ann Lehman, Norm O'Rourke, Larry Hatcher,**
and **Edward J. Stepanski**

JMP® Start Statistics, Third Edition
by **John Sall, Ann Lehman,**
and **Lee Creighton**

Regression Using JMP®
by **Rudolf J. Freund, Ramon C. LIttell,**
and **Lee Creighton**